PROTEIN SYNTHESIS

Translational and Post-Translational Events

Experimental Biology and Medicine

Protein Synthesis: *Translational and Post-Translational Events,* edited by **Abraham K. Abraham, Thor S. Eikhom,** and **Ian F. Pryme,** 1983

Ir Genes: *Past, Present, and Future,* edited by **Carl W. Pierce, Susan E. Cullen, Judith A. Kapp, Benjamin D. Schwartz,** and **Donald C. Shreffler,** 1983

Methods in Protein Sequence Analysis, edited by **Marshall Elzinga,** 1982

Inflammatory Diseases and Copper, edited by **John R. J. Sorenson,** 1982

Membrane Fluidity: *Biophysical Techniques and Cellular Regulation,* edited by **Morris Kates** and **Arnis Kuksis,** 1980

PROTEIN SYNTHESIS

Translational and Post-Translational Events

Edited by

ABRAHAM K. ABRAHAM,
THOR S. EIKHOM, AND IAN F. PRYME

University of Bergen, Bergen, Norway

The Humana Press · Clifton, New Jersey

Library of Congress Cataloging in Publication Data
Main entry under title:

Protein synthesis.

(Experimental biology and medicine)
Proceedings of the 13th Linderstrøm-Lang Conference, held at Godøysund, Norway, June 14–18, 1983.
Includes index.
1. Protein biosynthesis—Congresses. 2. Genetic translation—Congresses. I. Abraham, Abraham K. II. Eikhom, T. S. III. Pryme, Ian F. IV. Linderstrøm-Lang Conference (13th : 1983 : Godøysund, Norway) V. Series: Experimental biology and medicine (Humana Press) [DNLM: W3 LI453K 13th 1983p / QU 55 L744 1983p]
QP551.P69774 1984 574.19'296 83-26463
ISBN 0-89603-060-1

Crescent Manor
PO Box 2148
Clifton, NJ 07015

Printed in the United States of America

Proceedings of the 13th Linderstrøm-Lang conference held in Norway, June 14th–18th 1983.

Preface

During the past decade we have witnessed several major discoveries in the area of protein synthesis and post-translational modification of protein molecules. In this volume, many of the latest research developments in these fields are reported by the distinguished international group of scientists who presented their state-of-the-art results at the 13th Linderstrøm-Lang Conference held at Godøysund, Norway, June 14–18, 1983.

We feel that the presentation here of so wide a variety of articles on both the molecular and the cellular aspects of protein synthesis will be of considerable value to many scientists working in the area who were unable to attend, as well as to many who are active in related areas. In addition to the research papers, the contents of the six scientific sessions held during the conference have been summarized by the respective session chairmen. These individual summaries provide insightful syntheses of all the recent progress in each field, identify which current problems remain of special interest, and suggest what the future may hold in the several areas of protein synthesis research covered.

Though this volume obviously cannot provide a complete survey of all important ongoing research on the molecular and cellular biology of translational and post-translational events, we are confident that it will facilitate a much better understanding of many important contemporary problems in research on protein synthesis, including cell differentiation, translational accuracy, protein modification, intracellular transport, and membrane turnover.

A. K. Abraham
T. S. Eikhom
I. F. Pryme

Contents

Translation Fidelity

Intracellular Protein Transport

Protein Glycosylation

Protein Phosphorylation

CONTRIBUTORS

ABRAHAM K. ABRAHAM • *Department of Biochemistry, University of Bergen, Bergen, Norway*

LEIF C. ANDERSSON • *Department of Pathology, University of Helsinki, Helsinki, Finland*

E. ARNOLD • *Institut für Botanik der Universität Regensburg, Regensburg, West Germany*

I. M. ÅSTRAND • *Department of Biochemistry, Arrhenius Laboratory, University of Stockholm and Department of Pathology at Huddinge Hospital, Karolinska Institutet, Stockholm, Sweden*

S. A. AUSTIN • *CRC Group, Department of Biochemistry, St. George's Hospital Medical School, London, England*

R. BASSÜNER • *Zentralinstitut für Genetik und Kulturpflanzenforschung, Berlin-Buch, East Germany*

PER BELFRAGE • *Department of Physiological Chemistry, University of Lund, Lund, Sweden*

R. BENDZKO • *Zentralinstitut für Molekularbiologie der AdW der DDR, Berlin-Buch, East Germany*

T. BERG • *Institute for Nutrition Research, University of Oslo, Blindern, Oslo, Norway*

THOMAS A. BEYER • *Department of Biochemistry, Duke University Medical Center, Durham, North Carolina*

R. BLOMHOFF • *Institute for Nutrition Research, University of Oslo, Blindern, Oslo, Norway*

LEENDERT BOSCH • *Department of Biochemistry, State University of Leiden, Leiden, Holland*

S.-C. CHEN • *Department of Chemistry and Clayton Foudation Biochemical Institute, University of Texas at Austin, Austin, Texas*

MICHAEL J. CLEMENS • *CRC Group, Department of Biochemistry, St. George's Hospital Medical School, London, England*

E. COPPIN-RAYNAL • *Laboratoire de Génétique, Université de Paris, Orsay, France*

G. DALLNER • *Department of Biochemistry, Arrhenius Laboratory, University of Stockholm, and Department of Pathology at Huddinge Hospital, Karolinska Institutet, Stockholm, Sweden*

A. DARVEAU • *Department of Biochemistry, McGill University, Montreal, Canada*

M. DEQUARD-CHABLAT • *Laboratoire de Génétique, Université de Paris, Orsay, France*

B. DOBBERSTEIN • *European Molecular Biology Laboratory, Heidelberg, West Germany*

ALLEN E. ECKHARDT • *Department of Biochemistry, Duke University Medical Center, Durham, North Carolina*

I. EDERY • *Department of Biochemistry, McGill University, Montreal, Canada*

C. EDLUND • *Department of Biochemistry, Arrhenius Laboratory, University of Stockholm and Department of Pathology at Huddinge Hospital, Karolinska Institutet, Stockholm, Sweden*

I. EGGENS • *Department of Biochemistry, Arrhenius Laboratory, University of Stockholm and Department of Pathology at Huddinge Hospital, Karolinska Institutet, Stockholm, Sweden*

MÅNS EHRENBERG • *Department of Molecular Biology, The Biomedical Center, Uppsala, Sweden*

ELIZABETH ANN EIGNER • *University of New Mexico, School of Medicine, Department of Biochemistry, Albuquerque, New Mexico*

T. EKSTRÖM • *Department of Biochemistry, Arrhenius Laboratory, University of Stockholm and Department of Pathology at Huddinge Hospital, Karolinska Institutet, Stockholm, Sweden*

A. FINKELSTEIN • *Institute of Protein Research, Poustchino, Russia*

GUDRUN FREDERIKSON • *Department of Physiological Chemistry, University of Lund, Lund, Sweden*

ERIK FRIES • *Department of Medical and Physiological Chemistry, University of Uppsala, Uppsala, Sweden*

S. FULLILOVE • *Department of Chemistry and Clayton Foundation Biochemical Institute, The University of Texas at Austin, Austin, Texas*

CARL G. GAHMBERG • *Department of Biochemistry and Pathology, University of Helsinki, Helsinki, Finland*

V. GIESELMANN • *Physiologisch-Chemisches Institut der Universität Münster, Münster, West Germany*

JAMIE A. GRIFO • *Department of Biochemistry, School of Medicine, Case Western Reserve University, Cleveland, Ohio*

B. HARDESTY • *Department of Chemistry and Clayton Foundation Biochemical Institute, The University of Texas at Austin, Austin, Texas*

A. HASELBECK • *Institut für Botankik der Universität Regensburg, Regensburg, West Germany*

A. HASILIK • *Physiologisch-Chemisches Institut der Universität Münster, Münster, West Germany*

J. W. B. HERSHEY • *Department of Biological Chemistry, University of California, Davis, California*

ROBERT L. HILL • *Department of Biochemistry, Duke University Medical Center, Durham, North Carolina*

W. HOFFMANN • *Institute of Molecular Biology, Austrian Academy of Sciences, Salzburg, Austria*

T. HULTIN • *Department of Cell Physiology, Wenner-Gren Institute, University of Stockholm, Stockholm, Sweden*

M. HUMBERLIN • *Biocenter, University of Basel, Basel, Switzerland*

A. HUTH • *Zentralinstitut für Molekularbiologie der AdW der DDR, Berlin-Buch, East Germany*

A. HUTTICHER • *Institute of Molecular Biology, Austrian Academy of Sciences, Salzburg, Austria*

SAMSON T. JACOB • *The Milton S. Hershey Medical Center, The Pennsylvania State University, Hershey, Pennsylvania*

MIKKO JOKINEN • *Departments of Biochemistry and Pathology, University of Helsinki, Helsinki, Finland*

L. KÄÄRIÄINEN • *Department of Virology, University of Helsinki, Helsinki, Finland*

RAYMOND KAEMPFER • *Department of Molecular Virology, The Hebrew University-Hadassah Medical School, Jerusalem, Israel*

BAREND KRAAL • *Department of Biochemistry, State University of Leiden, Leiden, Holland*

G. KRAMER • *Department of Chemistry and Clayton Founda-*

tion Biochemical Institute, The University of Texas at Austin, Austin, Texas

G. KREIL • *Institute of Molecular Biology, Austrian Academy of Sciences, Salzburg, Austria*

J. KRUPPA • *Institute of Physiological Chemistry, University of Hamburg, Hamburg, West Germany*

C. G. KURLAND • *Department of Molecular Biology, The Biomedical Center, Uppsala, Sweden*

JAMES A. LAKE • *Molecular Biology Institute and Department of Biology, University of California, Los Angeles, California*

W. LAUER • *European Molecular Biology Laboratory, Heidelberg, West Germany*

K. A. W. LEE • *Department of Biochemistry, McGill University, Montreal, Canada*

J. LIPP • *European Molecular Biology Laboratory, Heidelberg, West Germany*

ROBERT B. LOFTFIELD • *Department of Biochemistry, University of New Mexico School of Medicine, Albuquerque, New Mexico*

I. MALEC • *Institute of Molecular Biology, Austrian Academy of Sciences, Salzburg, Austria*

R. MANTEUFFEL • *Zentralinstitut für Genetik und Kulturpflanzenforschung, Berlin-Buch, East Germany*

KJELD A. MARCKER • *Department of Molecular Biology, University of Aarhus, Århus, Denmark*

MICHAEL MERION • *Department of Pediatrics, Washington University School of Medicine, St. Louis, Missouri*

WILLIAM C. MERRICK • *Department of Biochemistry, Case Western Reserve University School of Medicine, Cleveland, Ohio*

JOAN M. MOEHRING • *Department of Medical Microbiology, University of Vermont, Burlington, Vermont*

THOMAS J. MOEHRING • *Department of Medical Microbiology, University of Vermont, Burlington, Vermont*

C. MOLLAY • *Institute of Molecular Biology, Austrian Academy of Sciences, Salzburg, Austria*

ROBERT J. MULLIN • *Department of Biochemistry, Duke University Medical Center, Durham, North Carolina*

ODD NYGÅRD • *The Wenner-Gren Institute, University of Stockholm, Stockholm, Sweden*

R. OLSEN • *Institute of Medical Biology, University of Tromsö, Tromsö, Norway*

SJUR OLSNES • *Norsk Hydro's Institute for Cancer Research and The Norwegian Cancer Society, Montebello, Oslo, Norway*

HÅKAN OLSSON • *Department of Physiological Chemistry, University of Lund, Lund, Sweden*

P. ORLEAN • *Institut für Botanik der Universität Regensburg, Regensburg, West Germany*

V. M. PAIN • *School of Biological Sciences, University of Sussex, Brighton, Sussex, England*

M. PICARD-BENNOUN • *Laboratoire de Génétique, Université de Paris, Orsay, France*

ALEXANDER PIHL • *Norwegian Cancer Research Institute, Montebello, Oslo, Norway*

R. POHLMANN • *Physiologisch-Chemisches Institut der Universität Münster, Münster, West Germany*

S. PROSCH • *Zentralinstitut für Molekularbiologie der AdW der DDR, Berlin-Buch, East Germany*

C. G. PROUD • *Biological Laboratory, University of Kent, Canterbury, Kent, England*

IAN F. PRYME • *Department of Biochemistry, University of Bergen, Bergen, Norway*

T. A. RAPOPORT • *Zentralinstitut für Molekularbiologie der AdW der DDR, Berlin-Buch, East Germany*

K. RICHTER • *Institute of Molecular Biology, Austrian Academy of Sciences, Salzburg, Austria*

KATHLEEN M. ROSE • *Department of Pharmacology, The University of Texas Medical School at Houston, Houston, Texas*

KIRSTEN SANDVIG • *Norsk Hydro's Institute for Cancer Research and The Norwegian Cancer Society, Montebello, Oslo, Norway*

CAROL A. SATLER • *Department of Biochemistry, Case Western Reserve University, School of Medicine, Cleveland, Ohio*

PAUL SCHLESINGER • *Department of Physiology and Biophysics, Washington University School of Medicine, St. Louis, Missouri*

H. SCHWAIGER • *Institut für Botanik der Universität Regensburg, Regensburg, West Germany*

PETER SIMONS • *Department of Biochemistry, University of New Mexico School of Medicine, Albuquerque, New Mexico*

P. SINGER • *Max-Planck-Institute für Biologie, Tübingen, West Germany*

WILLIAM S. SLY • *Department of Pediatrics, Washington University School of Medicine, St. Louis, Missouri*

N. SONENBERG • *Department of Biochemistry, McGill University, Montreal, Canada*

F. STECKEL • *Physiologisch-Chemisches Institut der Universität Münster, Münster, West Germany*

DEAN A. STETLER • *The Milton S. Hershey Medical Center, The Pennsylvania State University, Hershey, Pennsylvania*

PETER STRÅLFORS • *Department of Physiological Chemistry, University of Lund, Lund, Sweden*

ANDERS SUNDAN • *Norsk Hydro's Institute for Cancer Research and The Norwegian Cancer Society, Montebello, Oslo, Norway*

W. TANNER • *Institut für Botanik der Universität Regensburg, Regensburg, West Germany*

J. TIPPER • *Department of Chemistry and Clayton Foundation Biochemical Institute, The Univerity of Texas at Austin, Austin, Texas*

Ö. TOLLBOM • *Department of Biochemistry, Arrhenius Laboratory, University of Stockholm, Department of Pathology at Huddinge Hospital, Karolinska Institutet, Stockholm, Sweden*

H. TOLLESHAUG • *Institute for Nutrition Research, University of Oslo, Blindern, Oslo, Norway*

H. TRACHSEL • *Biocenter, University of Basel, Basel, Switzerland*

JOHANNES M. VAN NOORT • *Department of Biochemistry, State University of Leiden, Leiden, Holland*

U. VILAS • *Institute of Molecular Biology, Austrian Academy of Sciences, Salzburg, Austria*

R. VLASAK • *Institute of Molecular Biology, Austrian Academy of Sciences, Salzburg, Austria*

K. VON FIGURA • *Physiologisch-Chemisches Institut der Universität Münster, Münster, West Germany*

A. WAHEED • *Department of Chemistry, Purdue University, West Lafayette, Indiana*

KEITH R. WESTCOTT • *Department of Biochemistry, Duke University Medical Center, Durham, North Carolina*

PETER WESTERMANN • *Central Institute of Molecular Biology, Academy of Sciences of GDR, Berlin-Buch, East Germany*

E. WOLLNY • *Department of Chemistry and Clayton Foundation Biochemical Institute, The University of Texas at Austin, Austin, Texas*

G. ZARDENETA • *Clayton Foundation Biochemical Institute and Department of Chemistry, The University of Texas at Austin, Austin, Texas*

1. INITIATION OF PROTEIN SYNTHESIS

STRUCTURAL ORGANIZATION OF INITIATION COMPLEXES INVOLVING THE EUKARYOTIC PROTEIN SYNTHESIS INITIATION FACTORS eIF-2 AND eIF-3

Odd Nygård and Peter Westermann*

The Wenner-Gren Institute, University of Stockholm, S-113 45 Stockholm, Sweden, and
*Central Institute of Molecular Biology, Academy of Sciences of GDR, 1115 Berlin-Buch, GDR

INTRODUCTION

The positioning of a ribosome into the correct reading frame on the messenger RNA is a multiple step process, which in eukaryotes involves at least seven different initiation factors, eIF-1, eIF-2, eIF-3, eIF-4A, eIF-4B, eIF-4C and eIF-5 (Trachsel *et al.*, 1977; Jagus *et al.*, 1981). Consistent with its essential role in the overall translation, the initiation process is an important site for translational regulation. The most extensively studied regulatory systems involve initiation factor eIF-2 but other initiation factors have also been postulated to take part in regulatory mechanisms (for a review see Jackson, 1980). *In vivo* some of the initiation factors, including eIF-2 and eIF-3, are associated with native 40S ribosomal subparticles (Sundkvist & Staehelin, 1975) indicating that the two factors participate in the formation of initiation complexes on the small ribosomal subparticle even under physiological conditions. In order to obtain some further insight into the functional mechanisms of the initiation process we have studied the topographical arrangement of the components involved in the organization of the active site on the 40S particle in which eIF-2, eIF-3, GTP, $Met\text{-}tRNA_f$ and mRNA are specifically joined together.

RESULTS

(a) The Use of Cross-Linking Reagents for Topographic Studies

Cross-linking has successfully been used for determining the structural arrangement of protein and RNA within bacterial ribosomal particles (Sköld, 1981; Sköld, 1982). This approach has also contributed substantially to our present knowledge of the topography of eukaryotic ribosomes and ribosome-associated translational complexes (Uchiumi _et al._, 1980, 1981; Gross _et al._, 1983; Terao _et al._, 1980a, 1980b; Nygård & Nika, 1982). We have used the cross-linking technique to determine the spatial arrangement of eIF-2, eIF-3, rRNA and ribosomal proteins within initiation complexes. By the use of cross-linking reagents with different distances between the reactive groups (Table 1) it was possible to measure the relative distances between the components.

The various reagents also differ in respect to their preferred target molecules. The homobifunctional reagent dimethyl-5,6-dihydroxy-4,7-dioxo-3,8-diaza-decanbisimidate (DBI) reacts only with protein amino groups, while 1,2;3,4-diepoxybutane (DEB) reacts with both proteins and nucleic acids. The heterobifunctional reagents methyl-_p_--azidobenzoylaminoacetimidate (ABAI) and methyl-5-(_p_-azidophenyl)-4,5-dithiapentanimidate (APTPI) are first bound to protein amino groups via the imidate reactive group. Thereafter the azido groups are activated by ultraviolet irradiation, and the resulting nitrene groups react with adjacent proteins or nucleic acids.

(b) Initiation Factors as RNA-Binding Proteins

Some of the initiation factors, including eIF-2 and eIF-3, have been shown to bind to _E. coli_ RNA (Vlasik _et al._, 1980). Although eIF-2 has been claimed to bind specifically to mRNA (Kaempfer _et al._, 1978), in our experiments it interacted to the same extent with 18S rRNA, 28S rRNA and globin mRNA (Nygård _et al._, 1980a). The factor therefore seemed to have general RNA-binding properties and the affinity for RNA could be used in the purification of eIF-2 and eIF-3 by affinity chromatography on rRNA-cellulose (Fig. 1) (Nygård _et al._, 1980a; Nygård & Westermann, 1982a).

TABLE 1. Cross-linking reagents used to identify neighbouring components in initiation complexes involving eIF-2 and eIF-3

Reagents (reactive groups are underlined)		Distance of reactive groups (Å)	Preferred target of reactive groups	
			1st reaction	2nd reaction
DEB	1,2:3,4-Diepoxybutane $H_2C{-}HC{-}CH{-}CH_2$ (two epoxide O bridges, underlined)	4	$-NH_2$ $-SH$	$-NH_2$ $-SH$
ABAI	Methyl-p-azidobenzoylaminoacetimidate $HN{=}C(\underline{H_3CO}){-}CH_2{-}NH{-}CO{-}Ph{-}\underline{N_3}$	10	$-NH_2$	$>NH, -NH_2$ $-SH, >CH,$ $>CH_2, -CH_3$
APTPI	Methyl-5-(p-azidophenyl)-4,5-dithiapentanimidate $HN{=}C(\underline{H_3CO}){-}CH_2{-}CH_2{-}S{-}S{-}Ph{-}\underline{N_3}$	12	$-NH_2$	$>NH, -NH_2$ $-SH, >CH,$ $>CH_2, -CH_3$
DBI	Dimethyl-5,6-dihydroxy-4,7-dioxo-3,8-diazadecanbisimidate $NH{=}C(\underline{H_3CO}){-}CH_2{-}NH{-}CO{-}CHOH{-}CHOH{-}CO{-}NH{-}CH_2{-}C({=}NH)\underline{OCH_3}$	12	$-NH_2$	$-NH_2$

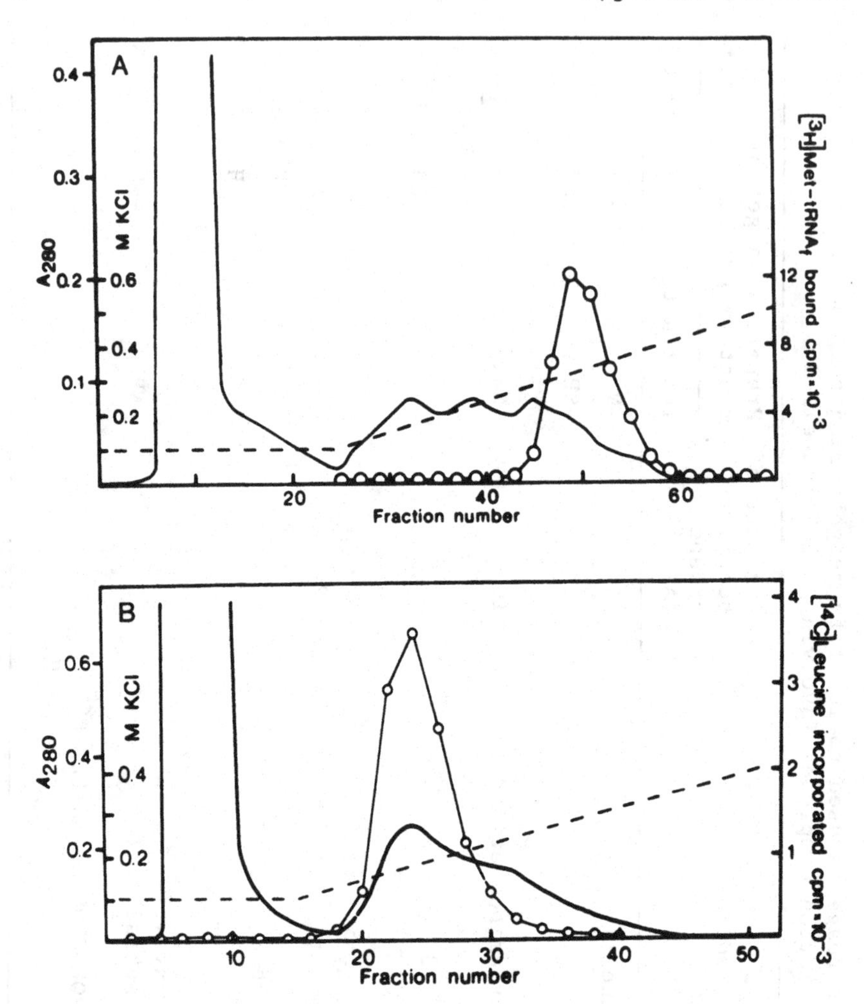

FIG. 1. Affinity chromatography of eIF-2 (A) and eIF-3 (B) on rRNA cellulose. Rat liver microsomal 0.5 M KCl wash was precipitated with ammonium sulfate at 25-40% (B) and 40-50% (A) saturation. The material was dialysed against 100 mM KCl, 20 mM Tris-HCl, pH 7.6, 14 mM 2-mercaptoethanol and 10% (v/v) glycerol and applied to an rRNA-cellulose column equilibrated with the same buffer. Bound proteins were eluted with a linear 0.1-0.5 M KCl gradient in the above buffer. (A) The eIF-2 activity was determined by ternary complex formation. (B) The eIF-3 activity was determined in a globin mRNA-dependent translation system.

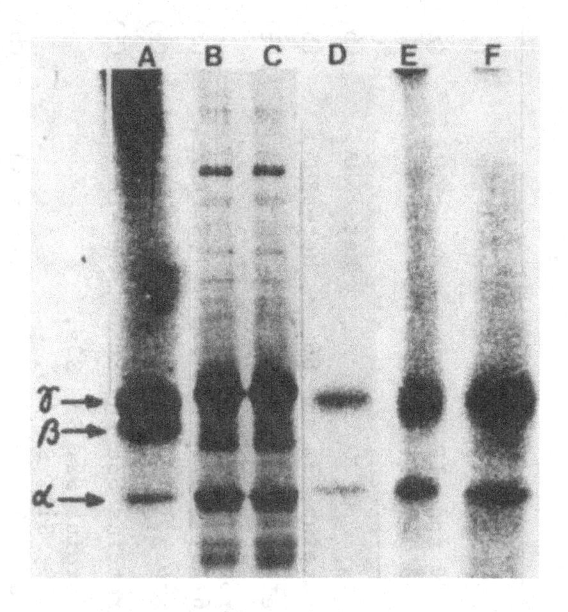

FIG. 2. SDS gel-electrophoresis of eIF-2 in 7-12% (w/v) acrylamide gels (Laemmli, 1970). (A) ^{125}I-labelled eIF-2. Autoradiograph. (B) eIF-2 digested with trypsin for 5 min as described in table 3. (C) eIF-2 in ternary complex digested with trypsin for 5 min as described in table 3. (D-F) Autoradiographs of ^{125}I-labelled eIF-2 cross-linked to 18S rRNA with ABAI (D), APTPI (E) and DEB (F).

(c) Cross-Linking of Initiation Complexes Containing eIF-2

Initiation factor eIF-2 has a total mass of 125 kDa and is composed of three subunits α, β and γ (Fig. 2A) designated in order of increasing isoelectric points (Barrieux & Rosenfeld, 1977). In the presence of Met-tRNA$_f$ and GTP, eIF-2 forms a ternary initiation complex, eIF-2·GTP·Met-tRNA$_f$ which can be bound to derived 40S ribosomal subparticles resulting in a quaternary complex, eIF-2·GTP·Met-tRNA$_f$·40S ribosomal subparticle (Safer _et al._, 1975; Benne _et al._, 1976). In both the ternary and the quaternary initiation complexes GDPCP can be substituted for GTP (Safer _et al._, 1975; Westermann _et al._, 1979).

TABLE 2. Neighbouring components in the ternary and quaternary initiation complexes

Cross-linked complex	Reagent	Neighbouring components A and B		Ref.
		A	B	
eIF-2·GMPPCP·Met-$tRNA_f$	DEB	eIF-2β	Met-$tRNA_f$	(a)
	ABAI or APTPI	eIF-2α, eIF-2β and eIF-2γ	Met-$tRNA_f$	(a)
eIF-2·GMPPCP·Met-$tRNA_f$· 40S ribosomal subunit	DEB	eIF-2β	Met-$tRNA_f$	(b)
		ribosomal proteins S3a and S6	Met-$tRNA_f$	(b)
		eIF-2α and eIF-2γ	18S rRNA	(c)
		ribosomal proteins S3a, S6, S7, S8, S11, S16/18, S23/24 and S25	18S rRNA	(b)
	ABAI	eIF-2α and eIF-2γ	18S rRNA	(c)
		ribosomal proteins S3a and S6	Met-$tRNA_f$	(b)
		ribosomal proteins S2, S3, S3a, S4, S6, S8, S9, S11, S16/18, S23/24, S25 and S26	18S rRNA	(b)
	APTPI	eIF-2α and eIF-2γ	18S rRNA	(c)
	DBI	ribosomal proteins S3, S3a, S6, S13/16 and S15/15a	eIF-2α and/or eIF-2γ	(d)

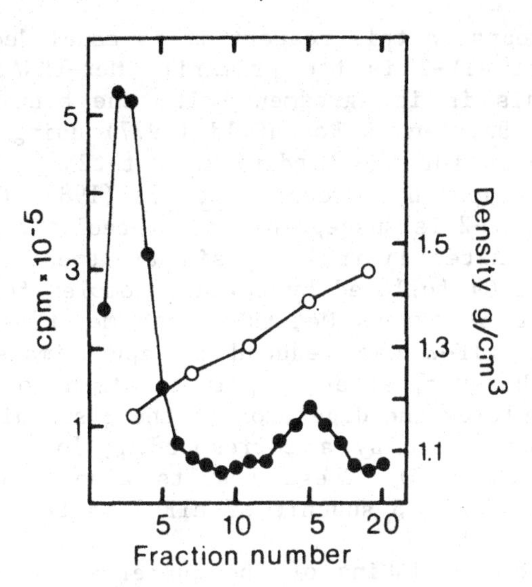

FIG. 3. The use of LiBr density gradient centrifugation for isolating ^{125}I-labelled polypeptides cross-linked to Met-tRNA$_f$ by DEB.

The localization of Met-tRNA$_f$ in the ternary initiation complex was determined by use of cross-linking. Ternary complexes containing ^{125}I-labelled eIF-2 were cross-linked with the heterobifunctional reagents ABAI and APTPI and the homobifunctional reagent DEB. After cross--linking the covalent complexes between Met-tRNA$_f$ and eIF-2 were isolated under denaturating conditions in LiBr density gradients (Fig. 3). After degradation of the initiator tRNA the covalently linked proteins were identified by SDS gel-electrophoresis. With ABAI and APTPI all three subunits of eIF-2 were found to be cross-linked to Met-tRNA$_f$ (Table 2). However, when DEB was used only the β subunit of eIF-2 became covalently linked to Met-tRNA$_f$. Because of the short distance between the

---

Table 2 continued:

(a) Nygård _et_ _al_., 1980b.
(b) Westermann _et_ _al_., 1981.
(c) Westermann _et_ _al_., 1980.
(d) Westermann _et_ _al_., 1979.

reactive groups in this reagent it is concluded that the β subunit of eIF-2 is the primarily Met-$tRNA_f$-binding subunit. This is in agreement with the binding data obtained by Barrieux & Rosenfeld (1977) using separated eIF-2 subunits for the binding of Met-$tRNA_f$. Recently, it has been shown by Zardeneta et al. (1982) that the β subunit of eIF-2 is susceptible to proteolytic degradation by trypsin. After limited trypsin treatment of eIF-2 (Fig. 2B and C) followed by ternary complex formation in the presence of excess Met-$tRNA_f$ the Met-$tRNA_f$-binding capacity of eIF-2 was reduced by approximately 60% (Table 3). However, after trypsin treatment of preformed ternary complexes the digestion of the β subunit (Fig. 2C) was not accompanied by a corresponding loss of bound Met-$tRNA_f$ (Table 3). These results also suggest an involvement of the β subunit of eIF-2 in the binding of Met-$tRNA_f$.

After cross-linking of the quaternary complex with ABAI and DEB, the cross-linked ribonucleoprotein complexes were isolated and the protein components identified by two-dimensional gel-electrophoresis. The β subunit of eIF-2 was found cross-linked to Met-$tRNA_f$ together with the ribosomal proteins S3a and S6 (Fig. 4B). The position of the initiator tRNA in relation to the factor subunits

TABLE 3. Ternary complex formation after trypsin treatment of eIF-2. The trypsin incubation (680 µg eIF-2/µg trypsin) was at 30°C as indicated and the digestion stopped by the addition of trypsin inhibitor (35 µg/µg trypsin) (Zardeneta et al., 1982)

1st incubation	2nd incubation	Trypsin treatment (sec)	pmol Met-$tRNA_f$ / pmol eIF-2	% of control
eIF-2 + Trypsin	Ternary complex formation	0	0.40	100
		100	0.30	75
		200	0.25	62
		300	0.16	40
Ternary complex formation	Trypsin	0	0.37	100
		100	0.34	92
		200	0.32	86
		300	0.32	86

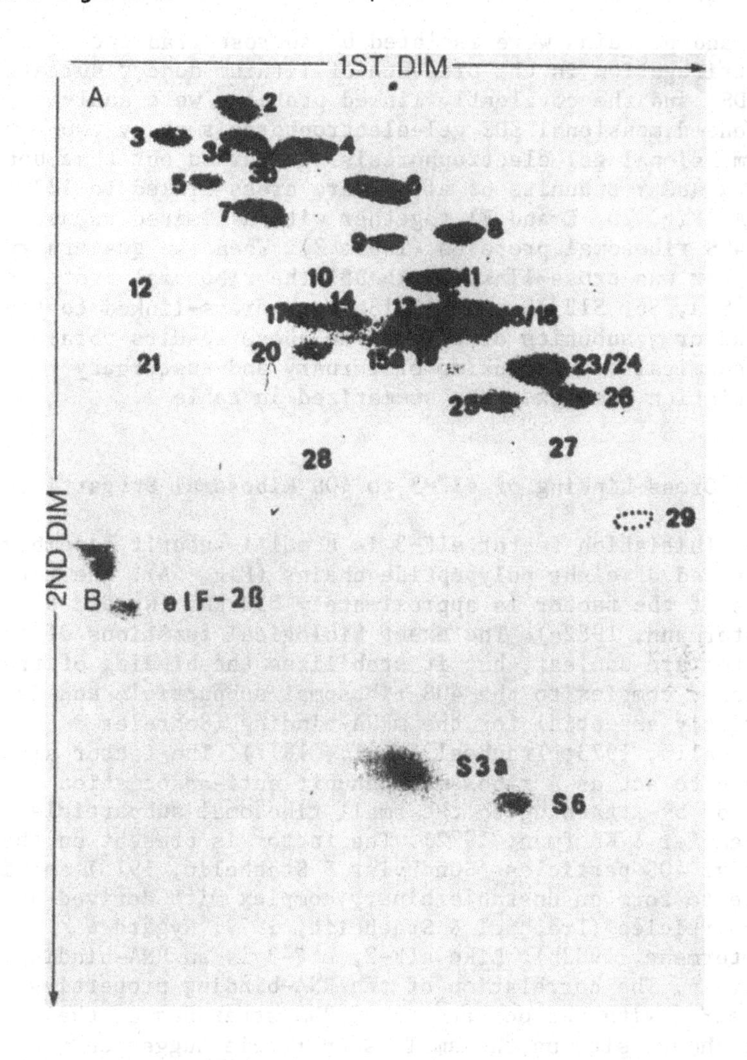

FIG. 4. Two-dimensional gel-electrophoresis of 40S ribosomal proteins (A) and ^{125}I-labelled proteins cross--linked to Met-tRNA$_f$ (B). Electrophoresis in the first dimension was as described by Martini & Gould (1975) and in the second dimension as described by Welfe et al. (1972). (B) Autoradiograph.

therefore seems to be preserved even after the attachment of the ternary complex to the 40S subparticle. In comparable experiments the covalent complexes between 18S

RNA and proteins were isolated by sucrose gradient centrifugation in the presence of lithium dodecylsulfate (LiDS) and the covalently linked proteins were analyzed by one-dimensional SDS gel-electrophoresis or by two--dimensional gel-electrophoresis. It turned out that both the α and γ subunits of eIF-2 were cross-linked to 18S rRNA (Fig. 2D, E and F) together with a limited number of 40S ribosomal proteins (Table 2). When the quaternary complex was cross-linked with DBI the ribosomal proteins S3, S3a, S6, S13/16 and S15/15a were cross-linked to the α and/or γ subunits of eIF-2. The above results obtained by chemical cross-linking of ternary and quaternary initiation complexes are summarized in table 2.

(d) Cross-Linking of eIF-3 to 40S Ribosomal Subparticles

Initiation factor eIF-3 is a multi-subunit assembly composed of eight polypeptide chains (Fig. 5A). The total mass of the factor is approximately 650 kDa (Nygård & Westermann, 1982a). The exact biological functions of the factor are unclear, but it stabilizes the binding of the ternary complex to the 40S ribosomal subparticle and is strictly essential for the mRNA-binding (Schreier & Staehelin, 1973; Trachsel _et al._, 1977). The factor also seems to act as a ribosomal subunit anti-association factor by attaching to the small ribosomal subparticle (Kaempfer & Kaufman, 1972). The factor is present on the native 40S particles (Sundkvist & Staehelin, 1975) and is able to form an unstable binary complex with derived 40S subparticles (Trachsel & Staehelin, 1979; Nygård & Westermann, 1982b). Like eIF-2, eIF-3 is an RNA-binding protein. The correlation of the RNA-binding properties of eIF-2 with the occurrence of RNA stretches at the attachment site on the small subparticle suggested that similar interactions might also occur at the binding site for eIF-3 on the 40S subparticle. In order to verify this hypothesis binary complexes containing ^{125}I-labelled eIF-3 and derived 40S subunits were cross-linked with DEB and ABAI. The covalent 18S rRNA-protein complexes were isolated by sucrose gradient centrifugation in the presence of LiDS (Fig. 6). After degradation of the RNA covalently attached proteins were identified by SDS gel-electrophoresis. With both reagents the 66 kDa subunit of eIF-3 was found covalently linked to 18S rRNA (Fig. 5B and C) indicating a specific orientation of the factor at

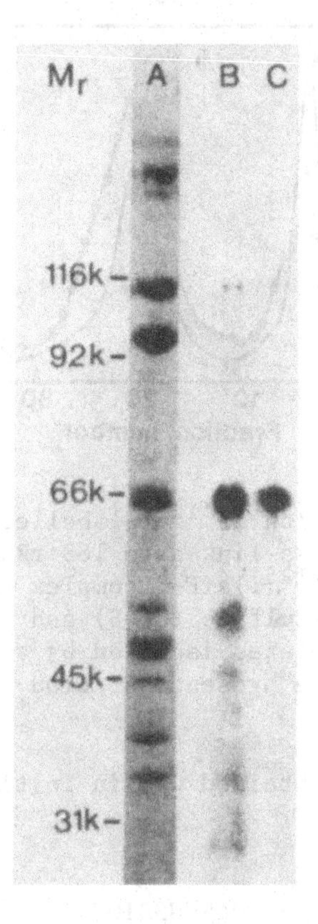

FIG. 5. Identification of ^{125}I-labelled eIF-3 subunits cross-linked to 18S rRNA with DEB (B) and ABAI (C). Electrophoresis was as described in Fig. 2. (A) Factor eIF-3 electrophoresed in parallel.

the attachment site. The binding of eIF-3 to the 40S subparticle would also be expected to involve ribosomal proteins. After cross-linking of binary complexes with DBI followed by isolation of ribosomal proteins cross-linked to eIF-3 by sucrose gradient centrifugation the ribosomal proteins S3a, S4, S6, S7, S8, S9, S10, S23/24 and S27 were found covalently linked to eIF-3 (Fig. 7A). The same proteins were found cross-linked to eIF-3 within the native 40S ribosomal subparticle indicating that the localization of eIF-3 on the 40S subparticle in the binary complex eIF-3·40S subparticle, was identical with that in the native particle (Fig. 7B). Table 4 summarizes the

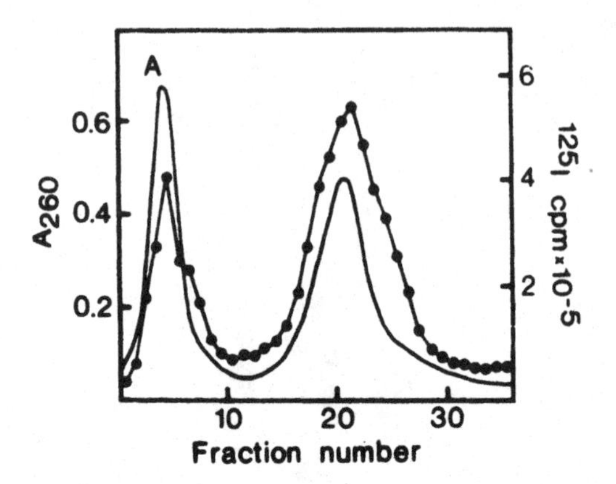

FIG. 6. Isolation of ^{125}I-labelled subunits of eIF-3 covalently cross-linked to 18S rRNA. The cross--linked binary 40S•(^{125}I)eIF-3 complex was dissociated with lithium dodecylsulfate (LiDS) and the covalent 18S RNA-protein complexes isolated by sucrose gradient centrifugation in the presence of LiDS.

cross-linking data obtained within initiation complexes containing eIF-3.

DISCUSSION

The cross-linking experiments described above provide some insight into the structural arrangement of ribosomal proteins, initiation factor subunits and Met-tRNA$_f$ within the initiation complexes investigated. Thus, initiation factor eIF-2 was found to be attached to the 40S ribosomal subparticle through its α and/or γ subunits. The β subunit seems to be primarily responsible for the binding of Met-tRNA$_f$, since it is located in close proximity to the initiator tRNA. The binding sites for both eIF-2 and eIF-3 on the small subparticle contain sequences of 18S rRNA, to which the α and γ subunits of eIF-2 and the 66 kDa subunit of eIF-3 can be cross-linked. In addition, both binding sites partly involve the same set of ribosomal proteins, namely S3, S3a and S6.

Native 40S ribosomal subparticles differ from small subparticles derived from polyribosomes by a lower buoyant density (Hirsch et al., 1973; Sundkvist & Staehelin,

TABLE 4. Neighbouring components in the binary complex, eIF-3·40S subparticle, and in the native small ribosomal subparticle

Cross-linked complex	Reagent	Neighbouring components A and B		Ref.
		A	B	
eIF-3·40S subparticle	DEB	66 kDa subunit of eIF-3	18S rRNA	(e)
	ABAI	66 kDa subunit of eIF-3	18S rRNA	(e)
	DBI	ribosomal proteins S3a, S4, S6, S7, S8, S9, S10, S23/24 and S27	eIF-3	(f)
native 40S subparticle	DBI	ribosomal proteins S3a, S4, S6, S7, S8, S9, S10, S23/24 and S27	eIF-3	(f)

(e) Nygård & Westermann, 1982b.
(f) Westermann & Nygård.

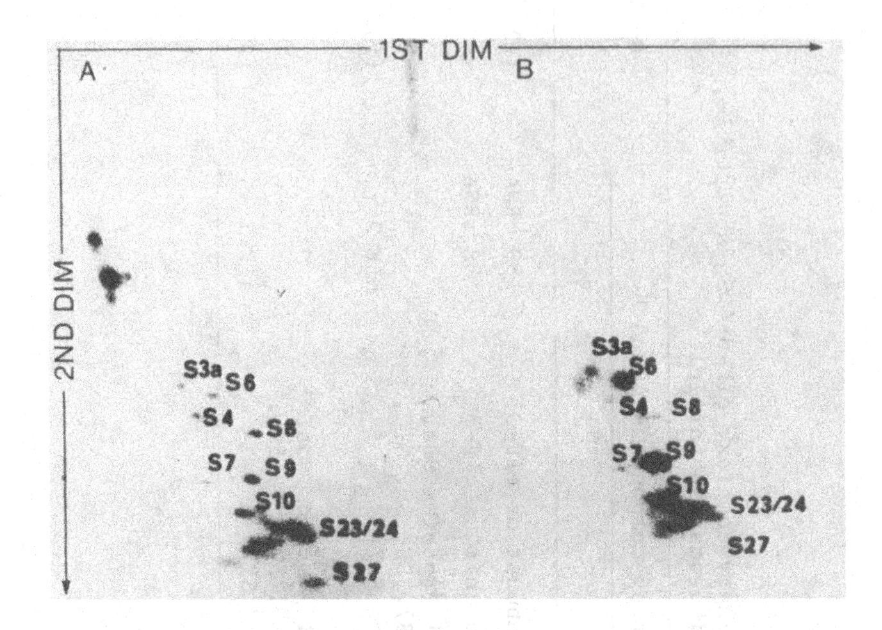

FIG. 7. Two-dimensional gel-electrophoresis of ^{125}I-labelled 40S ribosomal proteins cross-linked to eIF-3. Electrophoresis in the first dimension was as described by Martini & Gould (1975) and in the second dimension as described by Laemmli (1970). (A) 40S proteins cross-linked to eIF-3 with the binary 40S•eIF-3 complex formed in vitro. Autoradiograph. (B) 40S proteins cross-linked to eIF-3 on the native 40S subparticle. Autoradiograph.

1975). This depends on the presence of additional proteins, mainly initiation factors, including eIF-2 and eIF-3 (Sundkvist & Staehelin, 1975). In our experiments the attachment of eIF-3 in native 40S subparticles involved the same ribosomal proteins as in the binary eIF-3•40S subparticle complex obtained in vitro. Preliminary experiments also indicate that the α and γ subunits of eIF-2 and the 66 kDa subunit of eIF-3 can be cross-linked to 18S rRNA within the native 40S subparticle. From the data in tables 2 and 4 in combination with cross-linking studies of the mRNA binding site and the 60S subparticle attachment site on the 40S subparticle (Table 5) it is concluded that the active

TABLE 5. Proteins of the small ribosomal subunit belonging to the attachment sites for natural and synthetic mRNA and the 60S ribosomal subunit as identified by cross-linking

Ligand	Ribosomal proteins involved	Complex investigated	Ref.
mRNA	S3a and S6	polysomes	(g)
poly(U)	S3 and S3a	poly(U)• 40S subparticle	(h)
60S subparticle	S2, S3, S4, S6, S7, S13 and S14	80S ribosome	(i)

(g) Takahishi & Ogata, 1981.
(h) Stahl & Kobets, 1981.
(i) Nygård & Nika, 1982.

domains on the small subparticle responsible for the binding of eIF-2, eIF-3, Met-$tRNA_f$ and mRNA are neighbouring and share some of the ribosomal proteins. The binding site for eIF-3 and the 60S subunit seem to be partly overlapping. The schematic model in Fig. 8 based on cross-linking of ribosomal proteins (Gross et al., 1983), summarizes current knowledge about the structural arrangement of these functional domains.

In electron microscopic studies of native 40S subparticles, eIF-3 was seen located near the protuberance of the ribosomal subparticle (Fig. 9) at or near the ribosomal interface (Emanuilov et al., 1978). By immune electron microscopy (Lutsch et al.,, 1979; Lutsch et al., 1980) it has been shown that the protuberance contains antigenic determinants for proteins S3a and S7. It is therefore suggested that the RNA-rich part of the small ribosomal subparticle (Kühlbrandt & Unwin, 1982) which includes the interface region is involved in the binding of mRNA, Met-$tRNA_f$, eIF-2 and eIF-3. In fig. 9 the structural data are combined with parts of the known protein synthesis initiation sequence (Benne & Hershey, 1978). In the first step eIF-3 is bound to the interface region of the 40S subparticle thereby preventing a

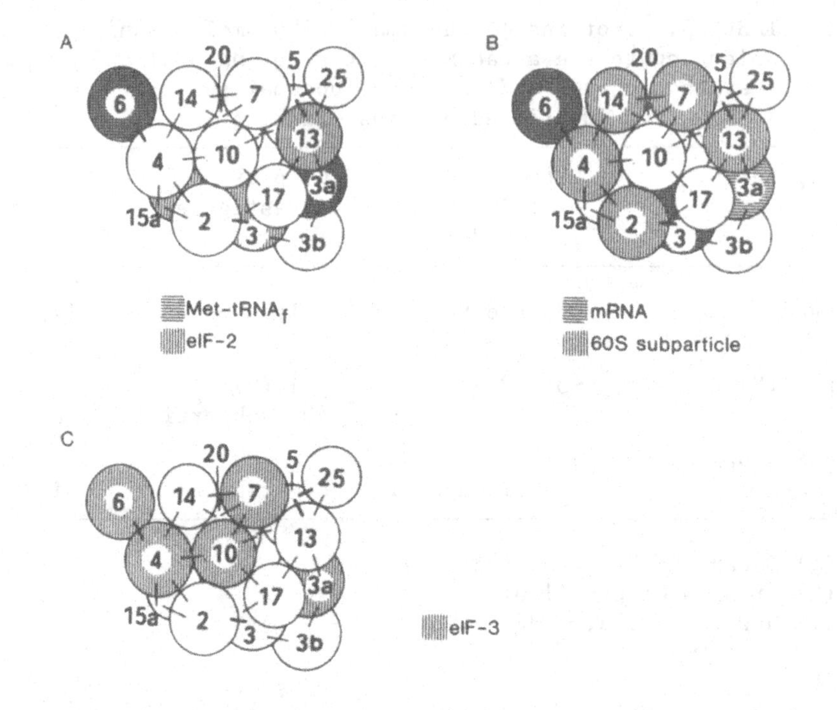

FIG. 8. Schematic representation of the spatial arrangement of 40S ribosomal proteins belonging to the domains active in Met-tRNA$_f$ (A), eIF-2 (A), mRNA (B), 60S subparticle (B) and eIF-3 binding (C).

spontaneous reassociation with 60S subparticles. In the next step eIF-2, GTP and Met-tRNA$_f$ are bound to a ribosomal region close to the eIF-3 attachment site. Thereafter, mRNA becomes bound under the mediation of additional factors. Finally, in the presence of eIF-5 the 40S preinitiation complex is joined with 60S subparticles to form the ultimate 80S initiation complex. In this final step both eIF-2 and eIF-3 are detached from the ribosomal particle (Safer et al., 1977).

ACKNOWLEDGEMENT

Part of this work was supported by a grant (B-Bu 0307-107) from the Swedish Natural Science Research Council.

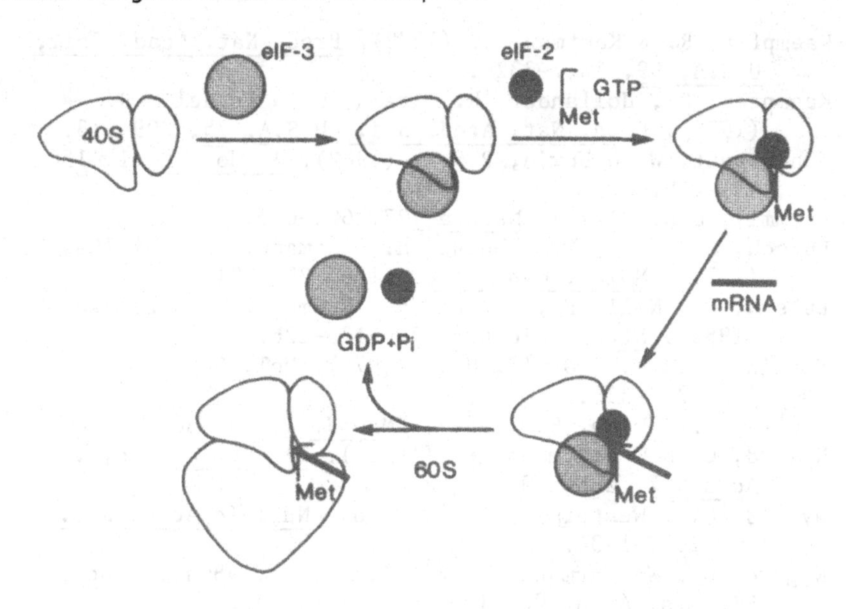

FIG. 9. Apposition of Met-$tRNA_f$, mRNA, eIF-2 and eIF-3 to the 40S ribosomal subparticle within the preinitiation complex.

REFERENCES

Barrieux, A. & Rosenfeld, M.G. (1977). J. Biol. Chem. 252, 3843-3847.
Benne, R. & Hershey, J.W.B. (1978). J. Biol. Chem. 253, 3078-3087.
Benne, R., Wong, C., Luedi, M. & Hershey, J.W.B. (1976). J. Biol. Chem. 251, 7675-7681.
Emanuilov, I., Sabatini, D.D., Lake, J.A. & Freienstein, C. (1978). Proc. Nat. Acad. Sci., U.S.A. 75, 1389-1393.
Gross, B., Westermann, P. & Bielka, H. (1983). EMBO J. 2, 255-260.
Hirsch, C.A., Cox, M.A., van Venrooij, W.C.W. & Henshaw, E.C. (1973). J. Biol. Chem. 248, 4377-4385.
Jackson, R.J. (1980). In Protein Biosynthesis in Eukaryotes (Pérez-Bercoff, R., ed.), pp. 363-418, Plenum Press, New York.
Jagus, R., Anderson, W.F. & Safer, B. (1981). Prog. Nucl. Acid Res. Mol. Biol. 25, 127-185.

Kaempfer, R. & Kaufman, J. (1972). Proc. Nat. Acad. Sci., U.S.A. 69, 3317-3321.
Kaempfer, R., Hollander, R., Arams, W. & Israeli, R. (1978). Proc. Nat. Acad. Sci., U.S.A. 75, 209-213.
Kühlbrandt, W. & Unwin, P.N.T. (1982). J. Mol. Biol. 156, 431-448.
Laemmli, U.K. (1970). Nature 227, 680-685.
Lutsch, G., Noll, F., Theise, H., Enzmann, G. & Bielka, H. (1979). Molec. Gen. Genet. 176, 281-291.
Lutsch, G., Noll, F., Theise, H., Enzmann, G. & Bielka, H. (1980). Studia biophys. 79, 125-126.
Martini, O.H.W. & Gould, H.J. (1975). Mol. Gen. Genet. 142, 299-316.
Nygård, O. & Nika, H. (1982). EMBO J. 1, 357-362.
Nygård, O. & Westermann, P. (1982a). Biochim. Biophys. Acta 697, 263-269.
Nygård, O. & Westermann, P. (1982b). Nucleic Acids Res. 10, 1327-1334.
Nygård, O., Westermann, P. & Hultin, T. (1980a). Biochim. Biophys. Acta 608, 196-200.
Nygård, O., Westermann, P. & Hultin, T. (1980b). FEBS Lett. 113, 125-128.
Safer, B., Peterson, D. & Merrick, W.C. (1977). In Translation of Natural and Synthetic Polynucleotides (Legocki, A.B., ed.), Poznań Agricultural University, Poznań.
Safer, B., Adams, S.L., Anderson, W.F. & Merrick, W.C. (1975). J. Biol. Chem. 250, 9076-9082.
Schreier, M.H. & Staehelin, T. (1973). Nature New Biol. 242, 35-38.
Sköld, S.-E. (1981). Biochimie 63, 53-60.
Sköld, S.-E. (1982). Eur. J. Biochem. 127, 225-229.
Stahl, J. & Kobets, N.D. (1981). FEBS Lett. 123, 269-272.
Sundkvist, I.C. & Staehelin, T. (1975). J. Mol. Biol. 99, 401-418.
Takahishi, Y. & Ogata, K. (1981). J. Biochem. 90, 1549-1552.
Terao, K., Uchiumi, T. & Ogata, K. (1980a). Biochim. Biophys. Acta 609, 306-312.
Terao, K., Uchiumi, T., Kobayashi, Y. & Ogata, K. (1980b). Biochim. Biophys. Acta 621, 72-82.
Trachsel, H. & Staehelin, T. (1979). Biochim. Biophys. Acta 565, 305-314.
Trachsel, H., Schreier, M.H., Erni, B. & Staehelin, T. (1977). J. Mol. Biol. 116, 755-767.

Uchiumi, T., Terao, K. & Ogata, K. (1980). J. Biochem. 88, 1033-1044.
Uchiumi, T., Terao, K. & Ogata, K. (1981). J. Biochem. 90, 185-193.
Welfle, H., Stahl, J. & Bielka, H. (1972). FEBS Lett. 26, 228-232.
Westermann, P. & Nygård, O. Submitted for publication to Biochim. Biophys. Acta.
Westermann, P., Nygård, O. & Bielka, H. (1980). Nucleic Acids Res. 8, 3065-3071.
Westermann, P., Nygård, O. & Bielka, H. (1981). Nucleic Acids Res. 9, 2387-2396.
Westermann, P., Heumann, W., Bommer, U.-A., Bielka, H., Nygård, O. & Hultin, T. (1979). FEBS Lett. 97, 101-104.
Vlasik, T.N., Domogatsky, S.P., Bezlepkina, T.A. & Ouchinnikov, L.P. (1980). FEBS Lett. 116, 8-10.
Zardeneta, G., Kramer, G. & Hardesty, B. (1982). Proc. Nat. Acad. Sci., U.S.A. 79, 3158-3161.

FUNCTIONAL AND STRUCTURAL CHARACTERISTICS OF EUKARYOTIC mRNA CAP BINDING PROTEIN COMPLEX

N. Sonenberg, I. Edery, A. Darveau,

M. Humbelin*, H. Trachsel*, J.W.B. Hershey+

and K.A.W. Lee

Department of Biochemistry, McGill University

3655 Drummond, Montreal, CANADA H3G 1Y6;

*Biocenter, Univ. of Basel, CH-4056, Basel

SWITZERLAND; +Dept. of Biol. Chem., Univ. of

California, Davis, CA. 94616, U.S.A.

We describe a purified mRNA cap binding protein complex comprising three major polypeptides of ∿ 24, 50 and 220 kilodaltons (kDa). The 24 kDa polypeptide corresponds to the 24 kDa cap binding protein previously described (Sonenberg et al., 1978; Tahara et al., 1981), while the 50 kDa polypeptide is eIF-4A. We show some functional characteristics of the CBP complex and present direct evidence that the 220 kDa polypeptide can be proteolyzed in vitro by a component present in extracts from poliovirus-infected cells.

1. INTRODUCTION

It is presently believed that gene expression in eukaryotes is regulated primarily at the transcriptional level. However, it has been clearly demonstrated that qualitative control of protein synthesis is responsible for differential expression of proteins in several instances. Prime examples are: (a) shut-off of host protein synthesis during viral infection (Bablanian, 1975);

(b) Inhibition of translation of most cellular mRNAs following heat-shock with the concomitant expression of heat shock genes (Storti et al., 1980; Hickey and Weber, 1982); and (c) Preferential utilization of maternal mRNPs after fertilization of sea urchin oocytes (Raff, 1980).

These cases indicate regulation of protein synthesis under strictly defined and sometimes extreme conditions. However, there exists ample evidence that differential rates of translation of different mRNAs occur in eukaryotic cells (Lodish, 1976; Walden and Thach, 1982.)The most likely rate limiting step in the initiation of protein synthesis (which is the rate limiting step in translation) is that in which the 40S ribosomal subunit binds the mRNA (Safer et al., 1978). However, it is not clear which structural features of mRNA effect the efficiency of the mRNA binding step and hence determine the intrinsic translational efficiency of a particular mRNA. To study factors which influence translational efficiency one might consider the different elements involved in the mRNA-ribosome interaction. Structural features at the 5' end of the mRNA might play a significant role. These include the 5' cap structure and mRNA primary and secondary structure. Additionally, several initiation factors (eIF) are involved in the binding of mRNA to ribosomes [eIF-4A, 4B and a cap binding protein(s)] and the process requires ATP as an energy source (for a recent review see Jagus et al., 1981).

Numerous studies have established that the cap structure, m^7GpppN, which is present at the 5' terminus of almost all eukaryotic mRNAs, facilitates stable complex formation between mRNA and ribosomes (for reviews see Shatkin, 1976 and Banerjee, 1980). However, the degree of cap dependence for translation initiation is influenced by several parameters including ionic strength, temperature and initiation factor concentration (Weber et al., 1977, 1978; Chu and Rhodes, 1978; Held et al., 1977).

Different approaches have been employed to try and elucidate the nature of mRNA-initiation factor interactions, particularly in relation to cap recognition. Early experiments indicated that eIF-2 (Kaempfer et al., 1978) and eIF-4B (Shafritz et al., 1976) interact with the cap structure, based on their ability to form complexes with radiolabeled mRNA (that were retained on nitrocellulose filters) in a way that was sensitive to inhibition by cap analogs. While expedient, it is difficult to assess cap specificity using this technique, due to non-specific inhibitory effects of cap analogs (Sonenberg and Shatkin, 1978). In addition, one cannot assess, for any particular preparation, the contribution by minor contaminants to the mRNA binding activity.

A direct approach to identify proteins that bind at or near the cap structure was developed by Sonenberg and Shatkin (1977) who used periodate oxidized [^{3}H]methyl-labeled reovirus mRNA to enable chemical cross-linking of initiation factors with mRNA. Any schiff bases formed between the reactive dialdehyde of m^7G and free amino groups of proteins can be stabilized by reduction with $NaBH_3CN$. Cross-linked mRNA-protein complexes are treated with ribonuclease to degrade all but the cap portion of the mRNA and the labeled polypeptides resolved by SDS/polyacrylamide gel electrophoresis and visualized by fluorography. Using this approach Sonenberg _et al._ (1978) initially identified a 24 kDa polypeptide in the high salt wash of rabbit reticulocyte ribosomes which could be specifically cross-linked to the oxidized cap structure of several viral mRNAs. A polypeptide of similar mobility on SDS/polyacrylamide gels and with identical cross-linking characteristics has been detected in initiation factors from mouse Ehrlich Ascites cells (Sonenberg _et al._, 1978) and human HeLa cells (Hansen and Ehrenfeld, 1981; Lee and Sonenberg, 1982). Subsequently, it has been demonstrated that in addition to the 24-CBP, polypeptides of 28, 50 and 80 kDa can be specifically cross-linked to the oxidized cap structure in an ATP-Mg^{++} dependent manner (Sonenberg, 1981). Using purified factors, Grifo _et al._ (1982) suggested that the 50 and 80 kDa polypeptides correspond to eIF-4A and eIF-4B, respectively. This suggestion has been confirmed for eIF-4A in recent experiments (Edery _et al._, submitted) and is most likely true for eIF-4B. Furthermore, it has been suggested by Sonenberg (1981) that the ATP requirement for cross-linking is a consequence of an energy dependent melting of mRNA secondary structure as will be considered elsewhere in this article.

The 24 kDa polypeptide has been purified to near homogeneity by affinity chromatography using m^7GDP coupled to Sepharose 4B (Sonenberg et al., 1979) and was called the 24K cap binding protein (24-CBP). Purified 24-CBP has been shown to stimulate the translation of several capped mRNAs in HeLa cell extracts while having no effect on the translation of naturally uncapped mRNAs (Sonenberg et al., 1980).

Many lines of evidence suggest that the 24-CBP is also able to associate with a high molecular weight protein complex. It was observed that cross-linkable 24-CBP sediments as an ∿ 200 kDa complex in sucrose gradients (Bergmann et al., 1979), and subsequently Tahara et al. (1981, 1982) have isolated a protein complex by m^7GDP affinity chromatography comprising major polypeptides of 48, 55 and 225 kDa in addition to the 24-CBP. This complex is functionally distinct from the 24-CBP in that it has an activity which can restore translation of capped mRNAs in extracts from poliovirus-infected cells (This activity has been termed "restoring activity") as detailed below. Using a similar protocol, we have isolated a CBP complex with similar properties and composition to that of Tahara et al. (1981, 1982) as will be described below (Edery et al., submitted). In addition, Grifo et al. (1983) have isolated a CBP complex using a different purification protocol and found that besides having an activity that can restore capped mRNA translation in extracts from poliovirus infected cells it can stimulate translation in a reconstituted protein synthesis system. The CBP complex has been referred to as CBP II or eIF-4F to distinguish it from the 24-CBP which has been referred to as CBP I or eIF-4E. We will refer to the complex purified in our laboratory as "the CBP complex", since it might not be identical in polypeptide composition to CBP II (eIF-4F) described by Tahara et al. (1981, 1982), Ray et al. (1983) and Grifo et al., (1983).

Poliovirus Infection and the Shut-off of Host Protein Synthesis. Poliovirus infection of HeLa cells results in the shut-off of host-protein synthesis (reviewed by Ehrenfeld, 1982). Extracts prepared from poliovirus-infected HeLa cells mimic the in vivo situation in that they can translate poliovirus RNA but are unable to translate capped RNAs (Rose et al., 1978). Results of Kaufman et al. (1976) indicated that the viral induced lesion in the cellular protein synthesis machinery occurs at the initiation step and consequently, various investigators were led to ask whether one of the known initiation factors could restore the translation of capped mRNAs in extracts from poliovirus-infected cells, or whether the activity of one of these factors was impaired in these extracts. The results from different studies indicated that eIF-4B (Rose et al., 1978) or eIF-3 (Helentjaris et. al., 1979) are implicated in the shut-off of host protein synthesis. Since it was found that poliovirus RNA is naturally uncapped, it has been an attractive hypothesis that inactivation of a CBP explains the shut-off of host protein synthesis (Rose et al., 1978; Helentjaris et al., 1979). Indeed, using a different approach, the restoring activity was purified by conventional techniques and shown to copurify with the 24-CBP, albeit in an unstable form. In contrast, a stable form of restoring activity co-purifies with a CBP complex (CBP II) isolated by Tahara et al. (1981, 1982). The fact that the CBP complex has stable restoring activity implies that it is somehow modified and consequently inactivated in poliovirus-infected cells and that this results in selective inhibition of host protein synthesis. The nature of the defect in CBP complex is not yet clear although recent evidence points to a likely possibility. Using a mono-specific antibody against a ∿ 220 kDa component of eIF-3 (P220), Etchison et al. (1982) have shown that P220 is proteolyzed as a consequence of poliovirus infection. The anti-P220 antibody recognizes the 220 kDa polypeptide of the CBP complex and thus it has been proposed that proteolysis of this polypeptide explains the shut-off of host protein synthesis.

In this article we describe the characterization of a CBP complex with respect to its polypeptide composition and biological activity in HeLa cell extracts. We also present direct evidence that the 220 kDa polypeptide of the CBP complex can be proteolyzed *in vitro* by a component present in extracts from poliovirus-infected cells.

2. RESULTS

(a) Purification of the CBP Complex

The protocol for purification of the 24-CBP developed by Sonenberg et al. (1979) used as starting material a 0.5 M KCl wash of rabbit reticulocyte ribosomes (RSW). This preparation was fractionated in low salt (0.1 M KCl) sucrose gradients and the upper fractions (not containing the multi-subunit initiation factor eIF-3) were applied on a m^7GDP-Sepharose 4B column to purify the 24-CBP. However, a substantial amount of the 24-CBP polypeptide in the RSW preparations appears to cosediment with eIF-3 (Trachsel et al., 1980) under low salt conditions. The 24-CBP as detected by cross-linking can be released from eIF-3 by sucrose density gradient centrifugation in high salt (0.5 M KCl, Trachsel et al., 1980). Subsequently, Tahara et al. (1981, 1982) have modified the protocol for the isolation of polypeptides with affinity for the cap structure, by fractionation of RSW preparations on sucrose gradients containing 0.5 M KCl and applying different gradient fractions on a m^7GDP-Sepharose 4B affinity column. This modification enabled the investigators to isolate a CBP complex comprising the 24-CBP and major polypeptides of 48, 55 and 225 kDa. This complex is functionally distinct from the 24-CBP in that it has activity which can restore the translation of capped mRNAs in extracts from poliovirus-infected cells.

We followed the protocol of Tahara et al. (1981, 1982) to isolate the CBP complex except for the omission of protease inhibitors (other than PMSF) during the purification procedures and the use of a m^7GTP-agarose column for the last step. The CBP complex preparation obtained comprised three major polypeptides of 24, 50 and 220 kDa (Fig. 1A). The different polypeptides eluted from the m^7GTP-agarose column are probably in the form of a complex, since they cosediment in sucrose gradients in 0.1 M KCl and 0.5 M KCl (I.E., unpublished observations). The 24 kDa polypeptide of the CBP complex comigrates on SDS gels with purified 24-CBP isolated from rabbit reticulocyte S-100 fraction, and the 50 kDa component comigrates with eIF-4A (data not shown), as previously observed by Tahara et al., (1982) and Grifo et al. (1983). To test the possibility that the 50 kDa polypeptide is eIF-4A, we used an anti-eIF-4A monoclonal antibody (Edery et al., submitted) as a probe in an immunoblot assay. Cap binding complex proteins were resolved by SDS/polyacrylamide gel electrophoresis and transferred to nitrocellulose paper, followed by reaction with anti-eIF-4A monoclonal antibody and visualization of immunoreactive species using peroxidase conjugated second antibody. Only the 50 kDa polypeptide of the CBP complex reacts with the anti-eIF-4A monoclonal antibody (Fig. 1B).

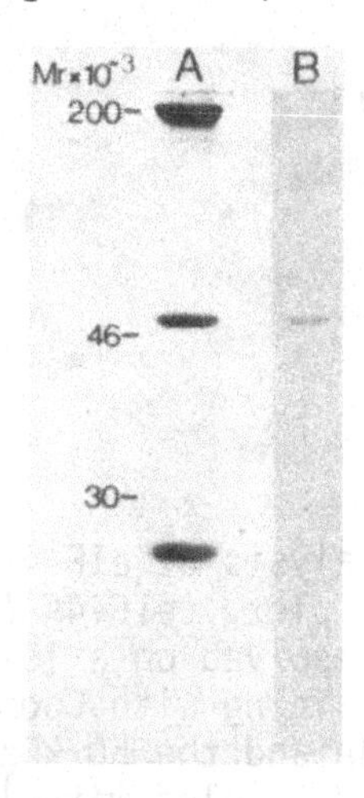

Fig. 1. (A) Analysis of the CBP complex by SDS/polyacrylamide gel electrophoresis. The CBP complex was purified from rabbit reticulocyte RSW as previously described (Etchison et al., 1982; Edery et al., submitted). A fraction (∿ 2 μg) of the CBP complex was resolved on a 12.5% SDS/polyacrylamide gel followed by Coomassie blue staining. (B) Immunoblot analysis of the CBP complex using anti-eIF-4A antibody. CBP complex (∿ 4 μg) was resolved on a 10-18% gradient SDS/polyacrylamide gel and polypeptides were transferred onto nitrocellulose paper essentially as described by Towbin et al. (1979). The nitrocellulose paper was incubated with 2.5% BSA and 5% horse serum in Tris (pH 7.5) buffered saline (TBS) and washed with TBS, followed by incubation with anti-eIF-4A monoclonal antibody (prepared as described by Edery et al., submitted). Bound antibody was detected by incubating the blot with peroxidase conjugated IgG (1:500 dilution in TBS), followed by a color reaction with diaminobenzidine (Towbin et al., 1979).

To support the immunological data indicating structural similarity between the 50 kDa polypeptide present in the CBP complex and eIF-4A, peptide analysis of the two polypeptides was performed. Fig. 2 shows the tryptic maps of eIF-4A (panel A) and the 50 kDa polypeptide (panel C). It is clear that the majority of peptides are common to eIF-4A and the 50 kDa polypeptide (the peptides are indicated by small arrowheads). However, one consistent and possibly significant difference in the peptide maps of the two polypeptides is noted by the heavy and thin arrows.

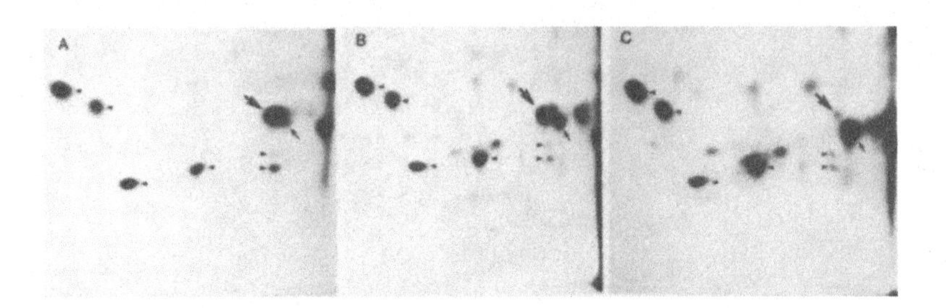

Fig. 2. Peptide map analysis of eIF-4A and the 50 kDa component of the CBP complex. eIF-4A (1 μg) and CBP complex (∿ 3 μg) were resolved on a 10-18% SDS/polyacrylamide gel followed by staining with Coomassie blue. The gel pieces containing eIF-4A and the 50 kDa component of CBP complex were excised and labeled with ^{125}I (0.4 mCi/slice) by the chloramine-T method. The gel pieces were washed to remove free ^{125}I and the proteins digested in the gel with 25 μg of trypsin (Worthington). The resulting peptides were eluted from the gel and lyophilized. Peptides ($1.5x10^5$ to $2x10^5$ cpm) were analyzed by electrophoresis in the first dimension and chromatography on cellulose coated thin layer plates in the second dimension. Electrophoresis was in pyridine/acetic acid/acetone/water (1:2:8:40 v/v) at pH 4.4 for 75 min at 800 volts. Chromatography was in n-butanol/acetic acid/water/pyridine (15:3:12:10 v/v) for 5-6 hrs. Plates were exposed to Cronex-4 X-ray film with Cronex Hi-plus intensifying screens for 16-24 hrs. A. eIF-4A; B. eIF-4A + 50 kDa polypeptide of CBP complex; C. 50 kDa polypeptide of the CBP complex.

Whereas the peptide indicated with the heavy arrow appears to be prominent in the eIF-4A preparation (panel A), the peptide indicated with the thin arrow is prominent in the 50 kDa polypeptide of the CBP complex (panel C). This difference may reflect a modification of eIF-4A that could contribute to the observed distribution between the free form and the CBP complex.

An important question raised by these findings concerns the functional significance of the partition of eIF-4A between the CBP complex and the free polypeptide. As described earlier, a 50 kDa polypeptide can be specifically cross-linked to the oxidized cap structure of reovirus mRNA if ATP is available (Sonenberg, 1981), and recent results have suggested that this polypeptide is eIF-4A

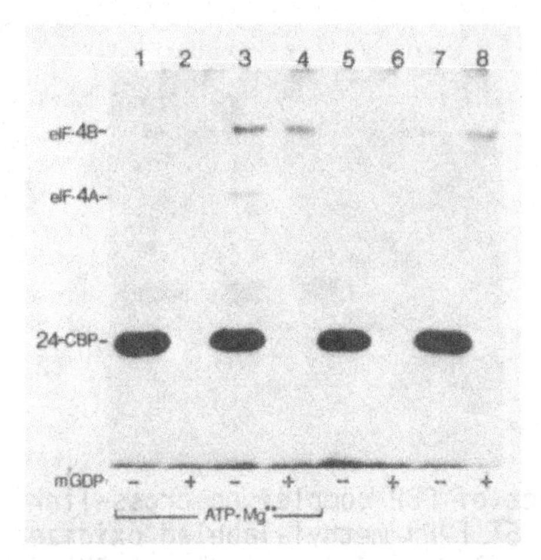

Fig. 3. Cross-linking of the CBP complex to [^{3}H] methyl-labeled oxidized reovirus mRNA. [^{3}H] methyl-labeled oxidized reovirus mRNA was incubated with ∿ 2 μg of CBP complex and 6 μg of BSA (as carrier) for 10 min at 30°C in a final volume of 40 μl under conditions described before (Lee and Sonenberg, 1982). Following addition of $NaBH_3CN$ and RNase A to degrade the mRNA (Sonenberg and Shatkin, 1977), samples were resolved on SDS/polyacrylamide gels, and labeled bands detected by fluorography. Cross-linking was performed in presence or absence of 0.7 mM m^7GDP and in the presence or absence of 1 mM ATP and 0.5 mM Mg^{++} as indicated in the figure. Lanes 3,4,7 and 8 contained 1 μg of eIF-4B in addition to CBP complex.

(Grifo et al., 1982). Consequently, it was of interest to determine the cross-linking characteristics of eIF-4A in relation to its presence in the CBP complex, and since eIF-4B has been implicated in the cap recognition process (Grifo et al., 1982), we examined its involvement here. The results (Fig. 3) demonstrate that the eIF-4A component of the CBP complex can be specifically cross-linked to a significant extent only in the presence of eIF-4B and ATP-Mg^{++} (lane 3). Cross-linking of eIF-4A under these conditions is cap specific, since it is completely inhibited by m^7GDP as is the cross-linking of the 24-CBP, while that of eIF-4B is not wholly specific (Fig. 3, compare lane 4 to 3). Control experiments show that in the absence of ATP-Mg^{++} (lanes 5 to 8) or in the absence of eIF-4B (lanes 1,2) there is very little or no cross-linking of eIF-4A. These results indicate that the eIF-4A in the CBP complex can

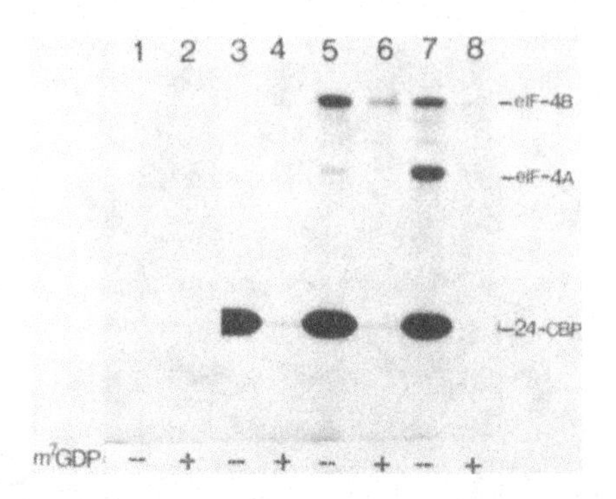

Fig. 4. Effect of CBP complex on cross-linking of eIF-4A and eIF-4B to 5' [^{3}H]-methyl-labeled oxidized reovirus mRNA. [^{3}H]-methyl-labeled oxidized reovirus mRNA was incubated with initiation factors and samples processed for SDS/polyacrylamide gel electrophoresis and fluorography as in Fig. 3. Cross-linking was performed in the presence of 1 mM ATP and 0.5 mM Mg^{++} and in the presence or absence of 0.7 mM m^7GDP as indicated in the figure. The following amounts of factors were used: eIF-4A, 0.6 µg; eIF-4B, 0.5 µg; CBP complex, 0.8 µg. Lanes 1 and 2, eIF-4A + eIF-4B. Lanes 3 and 4, CBP complex. Lanes 5 and 6, eIF-4B + CBP complex. Lanes 7 and 8, eIF-4A + eIF-4B + CBP complex.

interact with the cap structure. On the other hand, Grifo et al. (1982) have shown that eIF-4A in its free form can be cross-linked to oxidized cap structures in the presence of eIF-4B and ATP-Mg^{++}, implying that the free form of eIF-4A can interact with the cap structure of the mRNA. We have repeated these experiments in our laboratory and found that eIF-4A can not be cross-linked to mRNA in the presence of eIF-4B and ATP-Mg^{++} (Fig. 4, lane 1). However, the addition of the CBP complex (lane 7) does promote the cap specific cross-linking of eIF-4A and eIF-4B. These results indicate that the CBP complex contains an activity that is required for the cap specific cross-linking of both eIF-4A and eIF-4B, and that eIF-4B mediates cap recognition by eIF-4A in the CBP complex.

It is also of interest to consider these results with respect to the hypothesis that a cap binding protein(s) facilitates ribosome binding by melting mRNA secondary structure (Sonenberg, 1981). Lee et al. (1983) have

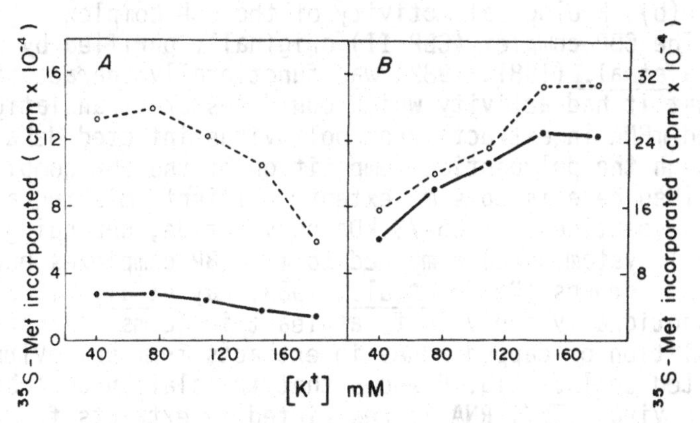

Fig. 5. Effect of CBP complex on translation of TMV RNA in extracts from poliovirus-infected and uninfected cells. Conditions of in vitro protein synthesis were as previously described (Lee and Sonenberg, 1982). Reaction mixtures containing 30 μCi of [35S]-methionine (> 1000 Ci/mmol) and 2 μg of TMV RNA in a final volume of 25 μl were incubated for 60 min at 30°C and aliquots of 5 μl were withdrawn to determine incorporation of radioactivity into TCA precipitable polypeptides. Final concentration of K^+ as indicated in the figure was adjusted by addition of KOAc to the 30 mM KOAc contributed by the extract. Panel A, translation in extracts from poliovirus-infected HeLa cells. Panel B, translation in extracts from uninfected cells. Translation was performed in absence (●——●) or presence of 1 μl (∿0.15 μg) of CBP complex (o-----o). Background incorporation of radioactivity without addition of exogenous mRNA (7670 cpm for A and 5580 cpm for B) was subtracted.

recently shown that cap specific cross-linking of eIF-4A in crude preparations of rabbit reticulocyte RSW has a reduced dependence on ATP if the mRNA is devoid of stable secondary structure. This suggested that eIF-4A can be cross-linked to the cap structure only after the energy dependent melting of 5' proximal sequences of mRNA. The observation that the eIF-4A in the CBP complex cannot be cross-linked to the cap structure unless eIF-4B is present (Figs. 3 and 4) implies, therefore, that any putative melting activity is not solely present in the CBP complex but is dependent on eIF-4B or alternatively that eIF-4B directly mediates cap recognition by eIF-4A.

(b) Biological Activity of the CBP Complex

The CBP complex (CBP II) originally purified by Tahara et al. (1981, 1982) was functionally characterized in that it had activity which could restore translation of capped mRNA in extracts from poliovirus-infected HeLa cells. Although the polypeptide composition of the CBP complex described here is to some extent deficient (missing a 55 kDa polypeptide or a 65-73 kDa polypeptide, depending on the gel system used) compared to the CBP complexes purified by other groups (Grifo et al., 1983; Ray et al., 1983), it is functionally equivalent, at least in terms of restoring translation of capped mRNAs in extracts from poliovirus infected cells. Fig. 5 shows that translation of tobacco mosaic virus (TMV) RNA is restricted in extracts from poliovirus-infected cells as compared to mock-infected (compare panel A to B) and the addition of CBP complex significantly stimulates translation of TMV RNA (3 to 5-fold) over a wide range of K^+ concentrations (Fig. 5A). It is of interest that translation of TMV RNA in extracts from mock-infected cells is not stimulated significantly by the CBP complex (Fig. 5B) in contrast to previous reports showing stimulation of Sindbis and globin mRNAs by CBP II (Tahara et al., 1981, 1982). Further experiments considering this point are described below. Translation of other capped viral mRNAs [Vesicular stomatitis virus (VSV) and reovirus] and globin mRNA was also significantly stimulated in the infected lysate by addition of the CBP complex (data not shown).

In light of our finding (Sonenberg et al., 1982) that the capped AMV-4 RNA can be translated with similar efficiency in either extracts from mock-infected or infected cells, it was of interest to determine the dependence of this mRNA on the CBP complex for translation. We have so far attributed efficient translation of AMV-4 RNA in infected lysates (Sonenberg et al., 1982) to the fact that it is devoid of any stable secondary structure at its 5' proximal region (Gehrke et al., submitted). We found that the CBP complex had no effect on AMV-4 RNA translation at 125 mM K^+ and stimulated by two-fold at 180 mM K^+ (data not shown). These data indicate that AMV-4 RNA has a lower requirement for the CBP complex than other capped mRNAs. As previously demonstrated in several laboratories (Rose et al., 1978; Bonatti et al., 1980), translation of the naturally uncapped RNAs of poliovirus and mengovirus was efficient in extracts from poliovirus-infected cells and

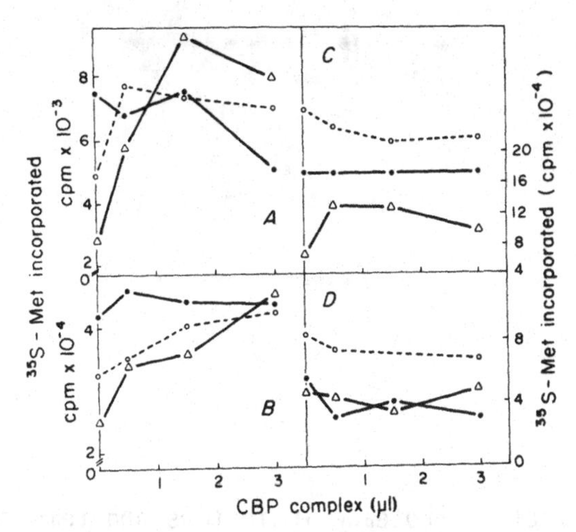

Fig. 6. Effect of CBP complex on the translation of capped and naturally uncapped mRNAs in extracts from mock-infected cells. Translation was performed as in Fig. 5 with 0.5 µg each of the following mRNAs: VSV (panel A), reovirus (B), mengovirus (C) and poliovirus (D) in a total incubation mixture of 12.5 µl containing the following final K^+ concentrations: 75 mM K^+ (●——●); 145 mM K^+ (o----o); 215 mM K^+ (Δ——Δ). Incorporation of radioactivity was determined in 5 µl aliquots and background radioactivity incorporated in the absence of exogenous mRNA (A, 4730 cpm; B, 4730 cpm; C, 5613 cpm; D, 21051 cpm) was subtracted. Concentration of added CBP complex was 0.15 µg/µl.

addition of CBP complex had no stimulatory effect (poliovirus RNA) or only a slight stimulatory effect (mengovirus RNA) on their translation (data not shown).

The 24-CBP and CBP II have been reported to stimulate translation of capped mRNAs in extracts from uninfected HeLa cells (Sonenberg et al., 1980; Tahara et al., 1982) but translation of TMV RNA, in our system, is only marginally stimulated by the CBP complex (Fig. 5B). It was pertinent, therefore, to examine the CBP complex dependence of other capped and naturally uncapped mRNAs in extracts from mock-infected cells under different salt concentrations. As expected, the CBP complex has no effect on the translation of the naturally uncapped RNAs of poliovirus (Fig. 6, panel D) and mengovirus (Fig. 6, panel C). As seen with TMV RNA, CBP complex does not stimulate trans-

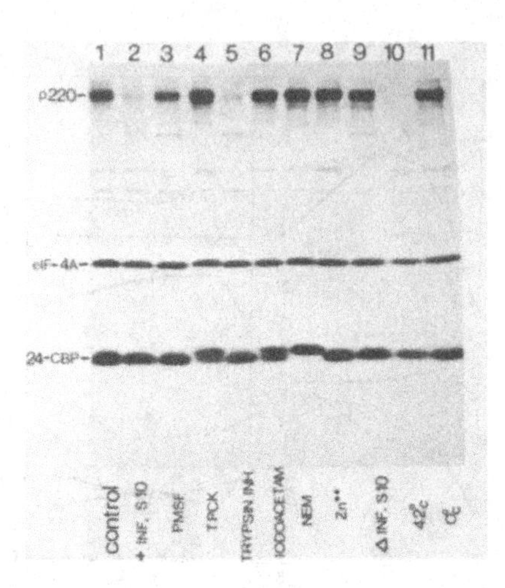

Fig. 7. Effect of protease inhibitors and temperature on proteolysis of the 220 kDa polypeptide of the CBP complex. Cell extracts from poliovirus-infected cells were preincubated for 15 min at 37°C with various protease inhibitors, as indicated and 10 µl were added to 1 µg [^{14}C] labelled CBP complex. Reaction mixtures were incubated for 30 min at 37°C in buffer containing 90 mM KOAc, 1.5 mM $Mg(OAc)_2$, 10 mM Hepes (pH 7.5) and 1 mM DTT in a total volume of 13.5 µl. Another 10 µl of preincubated poliovirus infected extracted was added and incubation continued for a further 30 min. Reactions were terminated by the addition of gel electrophoresis sample buffer (Sonenberg and Shatkin, 1977), and samples resolved on a 15% SDS/polyacrylamide gel (acrylamide/bisacrylamide=166) followed by fluorography. Lane 1, control with cell extract buffer instead of cell extract; Lane 2, preincubated cell extract only; Lane 3, + 2 mM phenylmethanesulfonyl fluoride (PMSF); Lane 4, + 2 mM tosyl phenylchloroketone (TPCK); Lane 5, + 1 mg/ml soybean trypsin inhibitor; Lane 6, + 10 mM iodoacetamide; Lane 7, + 10 mM N-ethylmaleimide (NEM); Lane 8, + 0.5 mM $Zn(OAc)_2$; Lane 9, + heat inactivated extract from poliovirus infected cells (65°C for 5 min). Lane 10 and 11, CBP complex was incubated with extracts from poliovirus-infected cells at 42°C and 0°C, respectively.

lation of VSV and reovirus mRNAs at low salt concentrations (75 mM K^+, Fig. 6 Panels A and B, respectively). The lack of stimulatory activity of the CBP complex indicates that the endogenous amount of this factor saturates the system used here in contrast to the deficient systems previously described (Sonenberg et al., 1980; Tahara et al., 1982). At higher salt concentrations (145 mM and 215 mM K^+) translation of VSV and reovirus mRNAs is reduced, and the CBP complex contains an activity which can reverse this inhibition (Fig. 6A and B), consistent with our working model for a function of the CBP complex, as will be considered in the Discussion.

When looking at the sum total of indirect evidence to hand, there seems little doubt that the CBP complex is somehow inactivated by poliovirus infection. The intriguing question that now arises concerns the structural nature of the defect in the CBP complex. As outlined before, the most recent evidence indicates that a 220 kDa polypeptide, which is antigenically related to the 220 kDa polypeptide of the CBP complex, is proteolyzed as a consequence of poliovirus-infection. We wanted to obtain direct evidence that the 220 kDa polypeptide of the CBP complex is a substrate for the viral (or induced) protease, and to this end we labeled the CBP complex by reductive methylation and incubated it with extracts from poliovirus-infected and mock-infected cells. Incubation of [^{14}C]-labeled CBP complex with extracts from mock-infected cells did not have any effect on the integrity of the different labeled polypeptides, whereas incubation with extracts from poliovirus-infected cells revealed that some extracts have an activity which can degrade the 220 kDa polypeptide while having no effect on the 24-CBP or eIF-4A (data not shown). This result indicates that degradation of the 220 kDa polypeptide is specific, consistent with the earlier results of Etchison et al. (1982).

To characterize the proteolytic cleavage reaction and to exclude some trivial explanations for the disappearance of the 220 kDa (e.g. loss due to sticking on vessel walls) we performed several experiments. (a) Kinetic analysis shows that the disappearance of the 220 kDa follows first order kinetics consistent with proteolytic cleavage and not non-specific loss of the protein (data not shown). (b) The loss of label did not take place at 0°C (Fig. 7, lane 11) or following heat inactivation (65°C for 5 min) of the extract from infected cells (Fig. 7, lane 9). (c) Several protease inhibitors abolished the proteolytic activity e.g. TPCK (a serine protease inhibitor) (lane 4), iodoacetamide and N-ethylmaleimide (inhibitors of thiol-dependent proteases) (lanes 6-7). The inhibitors which are effective at blocking the proteolysis of the 220 kDa polypeptide are known to inhibit the protease activity which processes poliovirus precursor polypeptides (Korant, 1979). It seems, therefore, a likely possibility that the protease involved in poliovirus protein processing is also responsible for the cleavage of the 220 kDa polypeptide. In this respect, it is of interest that poliovirus protein processing (Korant, 1979) and the cleavage of the 220 kDa polypeptide (Fig. 7, lane 8) are inhibited by Zn^{++} ions.

3. DISCUSSION

It has been previously proposed that a CBP(s) mediates the energy dependent melting of mRNA 5' sequences to facilitate ribosome binding (Sonenberg, 1981; Sonenberg et al., 1981). The results described in this report and elsewhere (to be detailed below) provide indirect evidence that the CBP complex (CBP II) has activity which is required for such a function. Our current working hypothesis is that the CBP complex binds initially to the 5' end of the mRNA (probably via interaction of the 24K CBP) followed by ATP hydrolysis and concomitant melting of mRNA secondary structure. The melting step may also be dependent on eIF-4B. After opening up the base-paired regions, the cap becomes accessible to eIF-4A (either as a component of the CBP complex or as the free polypeptide) and eIF-4B and this interaction might then facilitate ribosome binding.

Many observations suggest that ATP and the cap structure are mediators of a step involving melting of mRNA secondary structure to facilitate ribosome binding. The requirements for ATP (Marcus, 1970; Trachsel et al., 1977; Benne and Hershey, 1979) and the cap structure (Shatkin, 1976) for ribosome binding are partially obviated for mRNAs with reduced secondary structure (Kozak, 1980; Morgan and Shatkin, 1980). Moreover, naturally uncapped mRNAs also exhibit a reduced dependence on ATP for ribosome binding (Jackson, 1982) again suggesting that the requirement for the cap structure and for ATP are related aspects of translation initiation. In addition, a monoclonal antibody with anti-CBP activity can inhibit initiation complex formation with native reovirus mRNA but not with inosine-substituted reovirus mRNA (Sonenberg et al., 1981).

One parameter which affects the stability of mRNA secondary structure is ionic strength (Holder and Lingrel,

1975). The observation that inhibition of translation by cap analogs is augmented at higher salt concentrations (Weber et al., 1977; Chu and Rhodes, 1978) and that ribosome binding of reovirus mRNA is inhibited at higher salt concentrations (Lee et al., 1983) might reflect the fact that these conditions favour stable mRNA secondary structure and consequently increase the dependence on the cap structure. This rationale receives support from the finding that ribosome binding to the irreversibly denatured, inosine-substituted reovirus mRNA is not sensitive to inhibition by high salt concentrations (Sonenberg et al. 1982). Similarly, binding of AMV-4 RNA, which is devoid of stable secondary structure at its 5' end (Gehrke et al., submitted) is insensitive to high salt concentrations (Gehrke et al. , submitted). If the activity which becomes more limiting for ribosome binding at elevated salt concentrations resides in the CBP complex it is predicted that the complex might be able to alleviate the inhibition of translation seen under these conditions. The data shown in Fig. 6 show that this is the case for the translation of reovirus and VSV mRNAs.

Extracts from poliovirus-infected HeLa cells are deficient in an activity required for translation of capped mRNAs (Ehrenfeld, 1982). However, capped mRNAs with reduced secondary structure can function in extracts from poliovirus-infected cells (Sonenberg et al., 1982). Since the lesion in infected cells is in the CBP complex, one is led to believe that this complex has an activity that is required to a lesser extent, if at all, for the translation of capped mRNAs with reduced secondary structure. In accord with this idea, the translation of AMV-4 RNA appears to have a reduced dependence on the CBP complex (unpublished observations).

The molecular mechanism by which cap recognition factors act is an open question. The observation that there is a reduced dependence on ATP for cross-linking of the 50 and 80 kDa polypeptides (most probably eIF-4A and eIF-4B, respectively) when the mRNA has reduced secondary structure (Lee et al., 1983), suggested that ATP hydrolysis results in a conformational change in the mRNA such that the cap structure is more accessible to eIF-4A and eIF-4B. Subsequently, in light of the observation that it is possible to reconstitute the specific cross-linking profile obtained when using crude initiation factors, by using the CBP complex and eIF-4B (Edery et al., submitted) it appears that the activity required to melt mRNA secondary structure resides in the CBP complex, but may require eIF-4B. One possibility is that the CBP complex interacts with the cap structure but cannot melt mRNA secondary structure in the absence of eIF-4B. Alternatively, the CBP complex could have melting activity by itself but the interaction of eIF-4A with the cap structure is directly mediated by eIF-4B. If the melting activity resides in the CBP complex, then the loss of this activity in poliovirus-infected cells would prevent interaction of eIF-4A (either as a subunit of CBP complex or as a free polypeptide) and eIF-4B with the cap structure, consistent with the finding that these latter factors cannot be cross-linked to oxidized cap structures when using crude initiation factors from poliovirus-infected cells (Lee and Sonenberg, 1982). There is good evidence that neither eIF-4A or eIF-4B are structurally modified (Duncan et al., 1983) or functionally impaired (Helentjaris et al. 1970) as a result of poliovirus infection, again indicating that these factors are dependent on an activity of the CBP complex for interaction with the cap structure.

A pertinent question arising from these studies concerns the mechanism of translation initiation for naturally uncapped mRNAs particularly in relation to any involvement of CBP. One possible model is that their 5' non-coding region is devoid of extensive secondary structure. This is true for STNV RNA in which the 5' end can only form an energetically unstable hairpin loop [$\Delta G°(25°C)=-5.8$Kcal/mol; Leung et al., 1979). However, in contrast, a relatively stable hairpin loop can be formed close to the 5' end of poliovirus RNA (Larsen et al., 1981). Another possible model is that ribosomes do not scan picornavirus RNA in order to reach the initiation codon, but bind directly (internal binding) to the initiator AUG. This provocative hypothesis has been suggested previously (Perez-Bercoff, 1982), the most compelling evidence being that ribosomes do not accumulate on the long leader sequence of poliovirus RNA when elongation of protein synthesis is blocked. Interestingly in this respect, the 5' non-coding sequence of poliovirus RNA contains an adenine-uridine-rich region (about 80 nucleotides) just preceding the initiator codon (Dorner et al., 1982).

There is currently little evidence to indicate whether the CBP complex is dispensable for translation of naturally uncapped mRNAs or whether they are able to utilize a modified form of CBP (possibly containing a proteolytic cleavage fragment of the 220 kDa polypeptide). Recent work (Ray et al., 1983) has indicated that the cap structure is not the sole recognition determinant for the CBP complex, since the differential stimulatory activity exhibited by the CBP complex on translation of capped reovirus mRNAs is not altered when uncapped reovirus mRNAs are used. However, the cap structure greatly enhances the interaction between the CBP complex and the mRNA,which thus explains the lower binding potential of decapped mRNAs. A low affinity of decapped mRNAs for the CBP complex could also explain the preferential stimulation of decapped VSV mRNA translation by a 24-CBP containing protein complex in reticulocyte lysate (Bergmann et al., 1979).

We thank Richard Gallo for help in preparation of the CBP complex. Part of this work was supported by the MRC and NCI of CANADA and a Terry Fox Cancer Research award to N.S.. I.E. and K.A.W.L. are predoctoral research fellows of the Cancer Research Society (Montreal) and A.D. is a recipient of a F.R.S.Q. fellowship.

REFERENCES

Bablanian, R. (1975) In "Progress in Medical Virology". Melnick, J.L. ed. Karger, Basel, Vol. 29, pp. 101-176.

Banerjee, A.K. (1980) Bacteriol. Rev. 44, 175-205.

Benne, R. and Hershey, J.W.B. (1978) J. Biol. Chem. 253, 3078-3087.

Bergmann, J.E., Trachsel, H., Sonenberg, N., Shatkin, A.J. and Lodish, H.F. (1979) J. Biol. Chem. 254, 1440-1443.

Bonatti, S., Sonenberg, N., Shatkin, A.J. and Cancedda, R. (1980) J. Biol. Chem. 255, 11473-11477.

Chu, L.-Y. and Rhodes, R.E. (1978) Biochem. 17, 2450-2454.

Dorner, A.J., Dorner, L.F., Larsen, G.R., Wimmer, E. and Anderson, C.W. (1982) J. Virol. 42, 1017-1028.

Duncan, R., Etchison, D. and Hershey, J.W.B. (1983) J. Biol. Chem., in press.

Edery, I., Humbelin, M., Darveau, A., Lee, K.A.W., Milburn, S., Hershey, J.W.B., Trachsel, H. and Sonenberg, N. (1983) Submitted.

Ehrenfeld, E. (1982) Cell 28, 435-436.

Etchison, D., Milburn, S.C., Edery, I., Sonenberg, N. and Hershey, J.W.B. (1982) J. Biol. Chem. 257, 14806-14810.

Gehrke, L., Auron, P.E., Quigley, G.J., Rich, A. and Sonenberg, N. (1983) Submitted.
Grifo, J.A., Tahara, S.M., Morgan, M.A., Shatkin, A.J. and Merrick, W.C. (1983) J. Biol. Chem., in press.
Grifo, J.A., Tahara, S., Leis, J., Morgan, M.A., Shatkin, A.J. and Merrick, W.C. (1982) J. Biol. Chem. 257, 5246-5252.
Hansen, J.L. and Ehrenfeld, E. (1981) J. Virol. 38, 438-445.
Held, W.A., West, K. and Gallagher, J.F. (1977) J. Biol. Chem. 252, 8489-8497.
Helentjaris, T., Ehrenfeld, E., Brown-Leudi, M.L. and Hershey, J.W.B. (1979) J. Biol. Chem. 254, 10973-10978.
Hickey, E.D. and Weber, L.A. (1982) Biochem. 21, 1513-1521.
Holder, J.W. and Lingrel, J.B. (1975) Biochem. 14, 4209-4215.
Jackson, R.J. (1982) In "Protein Synthesis in Eukaryotes" Perez-Bercoff, R. ed. Plenum Press, New York, pp.363-418.
Jagus, R., Anderson, W.F. and Safer, B. (1981) Nucl. Acid Res. and Mol. Biol. 25, 127-185.
Kaempfer, R., Rosen, H. and Israeli, R. (1978) Proc. Natl. Acad. Sci. USA 75, 650-654.
Kaufman, Y., Goldstein, E. and Penman, S. (1976) Proc. Natl. Acad. Sci. USA 73, 1834-1838.
Korant, B.D. (1979) In "The Molecular Biology of Picornaviruses" Perez-Bercoff, ed. Plenum Publishing Corp., New York, pp. 113-125.
Kozak, M. (1980) Cell 22, 459-467.
Larsen, G.R., Semler, B.L. and Wimmer, E. (1981) J. Virol. 37, 328-335.
Lee, K.A.W., Guertin, D. and Sonenberg, N. (1983) J. Biol. Chem. 258, 707-710.
Lee, K.A.W. and Sonenberg, N. (1982) Proc. Natl. Acad. Sci. USA 79, 3447-3451.
Leung, D.W., Browning, K.S., Heckman, J.E., RajBhandary, U.L. and Clark, J.M. Jr. (1979) Biochem. 18, 1361-1366.
Lodish, H.F. (1976) Ann. Rev. Biochem. 45, 39-72.
Marcus, A. (1970) J. Biol. Chem. 245, 955-961.
Morgan, M.A. and Shatkin, A.J. (1980) Biochem. 19, 5960-5966.
Perez-Bercoff, R. (1982) In "Protein Biosynthesis in Eukaryotes" Perez-Bercoff, R. ed. Plenum Press, New York pp. 245-252.
Raff, R. (1980) In "Cell Biology" Prescott, D.M. and Goldstein, L. eds. Academic Press, New York, Vol. 4, pp. 107-129.
Ray, B.K., Brendler, T.G., Adya, S., Daniels-McQueen, S., Miller, J.K., Hershey, J.W.B., Grifo, J.A., Merrick, W.C. and Thach, R.E. (1983) Proc. Natl. Acad. Sci. USA 80 663-667.

Rose, J.K., Trachsel, H., Leong, K. and Baltimore, D. (1978) Proc. Natl. Acad. Sci. USA 75, 2732-2736.
Safer, B., Kemper, W. and Jagus, R. (1978) J. Biol. Chem. 253, 3384-3386.
Shatkin, A.J. (1976) Cell 9, 645-653.
Sonenberg, N., Guertin, D. and Lee, K.A.W. (1982) Mol. Cell Biol. 2, 1633-1638.
Sonenberg, N., Guertin, D., Cleveland, D. and Trachsel, H. (1981) Cell 27, 563-572.
Sonenberg, N. (1981) Nucleic Acids Res. 9, 1643-1656.
Sonenberg, N., Trachsel, H., Hecht, S.M. and Shatkin, A.J. (1980) Nature 285, 331-333.
Sonenberg, N., Rupprecht, K.M., Hecht, S.M. and Shatkin, A.J. (1979) Proc. Natl. Acad. Sci. USA 76, 4345-4358.
Sonenberg, N., Morgan, M.A., Merrick, W.C. and Shatkin, A.J. (1978) Proc. Natl. Acad. Sci. USA 75, 4843-4847.
Sonenberg, N. and Shatkin, A.J. (1978) J. Biol. Chem. 253, 6630-6632.
Sonenberg, N. and Shatkin, A.J. (1977) Proc. Natl. Acad. Sci. USA 74, 4288-4292.
Shafritz, D.A., Weinstein, J.A., Safer, B., Merrick, W.C., Weber, L.A., Hickey, E.D. and Baglioni, C. (1976) Nature 261, 291-294.
Storti, R.V., Scott, M.O., Rich, A. and Pardue, M.L. (1980) Cell 22, 825-834.
Tahara, S.M., Morgan, M.A. and Shatkin, A.J. (1981) J. Biol. Chem. 256, 7691-7694.
Tahara, S.M., Morgan, M.A., Grifo, J.A., Merrick, W.C. and Shatkin, A.J. (1982) in "Interaction of Translational and Transcriptional Controls in the Regulation of Gene Expression," (Grunberg-Manago, M. and Safer, B. eds.) Elsevier, New York, pp. 359-372.
Trachsel, H., Sonenberg, N., Shatkin, A.J., Rose, J.K., Leong, K., Bergmann, J.E., Gordon, J. and Baltimore, D. (1980) Proc. Natl. Acad. Sci. USA 77, 770-774.
Trachsel, H., Erni, B., Schreier, M.H. and Staehelin, T. (1977) J. Mol. Biol. 116, 755-767.
Towbin, H., Staehelin, T. and Gordon, J. (1979) Proc. Natl. Acad. Sci. USA 76, 4350-4354.
Walden, W.E. and Thach, R.E. (1982). In: "Interaction of Translational &Transcriptional Controls in the Regulation of Gene Expression." Eds. M. Grunberg-Manago and B. Safer, Elsevier, New York, Vol. 24, pp. 399-413.
Weber, L.A., Hickey, E.D., Nuss, D.L. and Baglioni, C. (1977) Proc. Natl. Acad. Sci. USA 74, 3254-3258.

INITIATION FACTORS WHICH RECOGNIZE mRNA

William C. Merrick, Jamie A. Grifo,
and Carol A. Satler
Dept. of Biochemistry, School of Medicine
Case Western Reserve University, Cleveland,
Ohio 44106 U.S.A.

A number of discoveries in the last few years have added new and clarifying information to the interpretation of events in eukaryotic protein synthesis initiation (Jagus et al., 1981). Of those discoveries, perhaps the most notable are those that relate to the binding of mRNA to 40S ribosomal subunits as this area had generated only limited information on the mRNA recognition process and often there appeared to be conflicting reports on the specific properties of given initiation factors, especially eIF-3 and eIF-4B. Several independent lines of research have subsequently generated substantial clarification. The first line was the development of an assay by Shatkin and co-workers to identify those proteins which appeared to specifically recognize the m^7GDP cap structure of eukaryotic mRNAs (Sonenberg & Shatkin, 1977; Sonenberg et al., 1978). A separate methodology developed by Marilyn Kozak has probed the ATP requirement for the binding of mRNA to 40S subunits. Her conclusion from these studies was that there are two ATP-dependent steps, one in which an area of the mRNA near the cap structure is bound to the 40S ribosomal subunit and a second step where the mRNA is "scanned" to locate the initiating AUG codon (Kozak, 1978; Kozak, 1980a; Kozak, 1980b). A third line of research was the development by Thach and co-workers of a mathematical model to study mRNA competition and an assay system to test various initiation factors for their role in mRNA recognition (Walden et al., 1981; Brendler et al., 1981a; Brendler et al., 1981b; Godefroy-Colbrun & Thach, 1981). The final line of research was the identification and characterization of a new initiation factor, eIF-4F, which was a common contaminant of most preparations of eIF-3 or eIF-4B (Grifo et al., 1983).

Presented in Table I is part of the data which indicated eIF-4F was required for the optimal translation of natural mRNA. Additional experiments indicated that adding increased levels of eIF-4A, eIF-4B or eIF-3 could not substitute for the requirement for eIF-4F. Another interesting point is that there appears to be similar factor requirements for the translation of a naturally capped (globin) and uncapped (STNV) mRNA. While this small sampling is not enough to justify an extension of these results to all capped and uncapped mRNAs, data of this type would seem to be consistent with the observation that initiation factors which can discriminate between competing mRNAs recognized more than just the m^7GDP cap structure (Ray et al., 1983; and see below).

The data in Table I indicated a requirement for eIF-4A, eIF-4B, and eIF-4F for the optimal translation of mRNA, but did not give any hint at the function of these factors. As an initial attempt to examine function, each factor was tested for its ability to bind radiolabelled globin mRNA under circumstances compatible with hemoglobin synthesis (pH, Mg^{+2} and K^{+1} concentrations) and characteristic of the binding process (dependent on ATP and sensitive to inhibition by an analog of the cap structure, m^7GDP). Table II presents some of the data obtained from such an analysis. The most striking result is that both eIF-4A and eIF-4F are capable of effecting an ATP-dependent mRNA binding which is stimulated by eIF-4B and is sensitive to inhibition by m^7GDP. This is most noticeable with eIF-4A which binds little or no mRNA by itself while the binding by eIF-4F is more sensitive to inhibition by m^7GDP (note: m^7GDP inhibits eIF-4F-directed binding either in the presence or absence of ATP and eIF-4B). The observation of similar mRNA binding activities is consistent with other comparisons of eIF-4A and eIF-4F which include: a similar, but not identical ability to relieve mRNA competition (Ray et al., 1983); the presence of a similar if not identical eIF-4A-like peptide in eIF-4F as judged by two dimensional gel electrophoresis (Grifo et

TABLE I

Factor Requirements for Natural mRNA Translation

	mRNA	eIF-4A	eIF-4B	eIF-4F	% control
A.	globin	+	+	+	100
	globin	–	+	+	27
	globin	+	–	+	20
	globin	+	+	–	37
B.	globin	+	+	+	108
	STNV	+	+	+	100
	STNV	–	+	+	45
	STNV	+	–	+	33
	STNV	+	+	–	38

Translation of natural mRNAs was determined using [^{14}C]leucine and valine (A) or [^{35}S]-methionine (B) and determination of hot trichloroacetic acid precipitable radioactivity as described (Grifo *et al.*, 1983). Fifty microliter reaction mixtures contained a crude preparation of aminoacyl-tRNA synthetases, partially purified initiation factors (eIF-1, eIF-2, eIF-3, eIF-4C, eIF-4D, eIF-5), purified EF-1 and EF-2, reticulocyte tRNA (0.3 A_{260} units), ribosomes (0.2 A_{260} units), 20 mM Tris·HCl, pH 7.5, 5 mM magnesium acetate, 100 mM KCl, 1 mM ATP, 0.2 mM GTP, 4 mM phosphoenolpyruvate, 0.3 IU pyruvate kinase, 1 mM dithiothreitol, 40 µm unlabelled amino acids, 20 µM [^{14}C]leucine and valine or 0.2 µM [^{35}S]methionine, and 5 µg of globin mRNA or 8 µg of STNV RNA (generously provided by Drs. John Clark and Scott Butler). In experiment A, 100% = 34 pmol of [^{14}C] leucine/valine incorporated into product and in experiment B, 100% = 0.28 pmol of [^{35}S]methionine incorporated into product. The levels of initiation factors used were 3 µg eIF-4A, 2.3 µg eIF-4B and 3.6 µg eIF-4F.

TABLE II

Initiation Factor-Dependent Retention of Globin mRNA on Nitrocellulose Filters

	eIF-4A	eIF-4B	eIF-4F	m^7GDP	Mg^{++}/ATP	mRNA Bound (cpm)
1.	+	–	–	–	–	525
2.	–	+	–	–	–	743
3.	+	+	–	–	–	2,157
4.	+	+	–	–	+	7,184
5.	+	+	–	+	+	3,362
6.	–	–	+	–	–	3,315
7.	–	–	+	–	+	3,100
8.	–	–	+	+	–	871
9.	–	+	+	–	–	6,445
10.	–	+	+	–	+	12,306
11.	–	+	+	+	+	1,558

Assays for the retention of [^{32}P]labelled globin mRNA on nitrocellulose filters were performed in 50 µl reaction mixtures which contained 20 mM Tris·HCl, pH 7.5, 90 mM KCl, 3 mM magnesium acetate, 1 mM dithiothreitol, 3 mM phosphoenolpyruvate, 0.2 IU pyruvate kinase, 33,000 cpm [^{32}P] globin mRNA and 3 µg eIF-4A, 1.4 µg eIF-4B, and 0.55 µg eIF-4F as indicated. When present the concentration of m^7GDP and Mg^{++}/ATP were 2 mM. After 2 min at 37° the reactions were stopped by the addition of 2 ml of cold buffer (20 mM Tris HCl, pH 7.5, 1 mM dithiothreitol, 0.1 mM EDTA, 10% glycerol, 100 mM KCl, 2.5 mM magnesium acetate) and immediately applied to nitrocellulose filters. The filters were then dried and radioactivity determined using 10 ml Formula 963 (New England Nuclear) and scintillation spectrometry.

al., 1983); an ATP- and eIF-4B-dependent crosslinking to oxidized reovirus mRNA (Grifo *et al.*, 1983; Tahara *et al.*, 1982).

Another feature of the ATP-dependent mRNA binding was the observation that a non-hydrolyzable ATP analog (AMP-P(NH)P) would not substitute for ATP, consistent with the findings of others on the binding of mRNA to 40S subunits (Benne & Hershey, 1978; Trachsel *et al.*, 1977). This finding plus a curiousity as to the protein responsible for the ATP requirement prompted an attempt to establish what should be the reciprocal assay to ATP-dependent mRNA binding, namely RNA-dependent ATP hydrolysis. Some of our results using this assay are indicated in Table III. From this table several observations can be made. First, although the value for eIF-4B is not zero, it would appear that eIF-4A and eIF-4F are responsible for most of the ATP hydrolysis activity. This is apparent in experiment A using globin mRNA to stimulate ATP hydrolysis from lines 1, 3, and 6. When poly(U) is used to stimulate ATPase activity, eIF-4A is clearly the most active initiation factor; eIF-4F is not inactive, but rather only active at the same level observed with globin mRNA. Finally, with globin mRNA, the combination of all three initiation factors catalyzes more ATP hydrolysis (3 to 4-fold) than might be expected by additivity implying a certain uniqueness of having all of the initiation factors present.

From the above results it might appear that eIF-4A was the predominant initiation factor responsible for ATP hydrolysis, especially when its activity with poly(U) was so much greater than that of eIF-4F. However, on a molar basis eIF-4F (M_r estimated to be 300,000) is as active or more active than eIF-4A (M_r = 46,000, [Grifo *et al.*, 1982]). With either RNA, eIF-4F catalyzed the hydrolysis of 2 moles of ATP per mole of eIF-4F whereas eIF-4A achieved this high level only with poly(U) and was 10-fold less active with globin mRNA. While the ratio of eIF-4A to eIF-4F activity should not be taken as absolute as it is not possible to know the percentage of active molecules in each preparation, it should be apparent that eIF-4A and eIF-4F have similar levels of activity with poly(U), but that globin mRNA is rather ineffective with eIF-4A in causing ATP hydrolysis.

TABLE III

RNA-Dependent ATPase Activity

A.	eIF-4A	eIF-4B	eIF-4F	globin mRNA-dependent P_i released (pmol)
1.	+	–	–	15
2.	–	+	–	5
3.	–	–	+	12
4.	+	+	–	35
5.	–	+	+	21
6.	+	–	+	63
7.	+	+	+	221
B.	**eIF-4A**	**eIF-4B**	**eIF-4F**	**poly(U)-dependent P_i released (pmol)**
1.	+	–	–	107
2.	–	+	–	6
3.	–	–	+	11
4.	+	+	–	119
5.	–	+	+	30
6.	+	–	+	118
7.	+	+	+	187

Assays were performed in 20 µl incubated at 37° for 15 min and contained the following: 15 mM HEPES-KOH, pH 7.5; 80 mM KCl, 2.5 mM magnesium acetate, 1 mM dithiothreitol, 100 µM [γ-^{32}P]ATP, specific activity 3,000 to 5,000 cpm per pmol. Where indicated, reaction mixtures also contained 3.3 µg eIF-4A, 1.8 µg eIF-4B, 1.65 µg eIF-4F, and 0.12 A_{260} units of globin mRNA (A) or 0.29 A_{260} units of poly(U) (B). P_i release in the absence of added protein (3.4 to 5.9 pmol) was subtracted from each value. P_i release was quantitated by extraction into isobutanol/benzene solution as a molybdate complex (Merrick, 1979).

As indicated earlier, other investigators had practically set the stage for the arrival of eIF-4F and it has been much of their effort that has encouraged us with the validity of the above work on the new initiation factor and the role of ATP in initiation. At this point it is appropriate to fit our results with theirs. The most direct comparison is with that of Thach and co-workers. Using their model translation system (Walden et al., 1981; Brendler, et al., 1981a; Brendler et al., 1981b; Godefroy-Golburn & Thach, 1981), they were able to show that both eIF-4A and eIF-4F have the ability to relieve mRNA competition and therefore represent the proteins capable of mRNA discrimination (Ray et al., 1983). From the work of Shatkin and co-workers, eIF-4F has been identified as the same protein as CBP II, a protein which restores translation of host mRNAs in a polio virus infected HeLa cell extract and specifically crosslinks to oxidized reovirus mRNA (Grifo et al., 1983; Tahara et al., 1981). The above two observations are independent support of the need for eIF-4A and eIF-4F in natural mRNA translation and also indicate some differences in the possible function of eIF-4A and eIF-4F.

Based upon the work of Marilyn Kozak (Kozak, 1978; Kozak, 1980a; Kozak, 1980b) it would seem that eIF-4A and/or eIF-4F might well be responsible for the ATP-dependent step which leads to the binding of mRNA to 40S subunits. At present we can not tell whether one or both of the factors catalyze ATP hydrolysis during this step nor can we estimate the stoichiometry of the reaction (i.e. ATPs hydrolyzed per mRNA bound). Preliminary studies with radiolabelled eIF-4A, eIF-4B or eIF-4F have not indicated any stable complex with the 40S subunit as analyzed by sucrose density gradients (data not shown). If indeed these factors do not continue to be complexed with the mRNA after binding to the 40S subunit, then some other factor must be responsible for Kozak's proposed second ATP-dependent step, mRNA "scanning " (Kozak, 1978; Kozak, 1980a; Kozak, 1980b). The identification of a "scanning" factor has not yet been accomplished.

The importance of eIF-4A, eIF-4B, and eIF-4F in mRNA recognition has also been explored by chemical cross-linking of cap oxidized mRNA to proteins in a manner that is sensitive to inhibition by the cap structure analog,

m^7GDP. Shatkin and co-workers developed this technique for ATP-independent crosslinking (Sonenberg & Shatkin, 1977; Sonenberg et al., 1978) and subsequently ATP-dependent crosslinking was reported by both Sonenberg (Sonenberg, 1981) and Shatkin and co-workers (Grifo et al., 1983; Tahara et al., 1982; Grifo et al., 1982). These studies indicated an ATP-independent crosslinking to the 24,000 dalton peptide of eIF-4F and an ATP-dependent crosslinking to the eIF-4A peptide (46,000) and eIF-4B, although in mixtures it was not possible to determine whether the crosslinked 46,000 dalton peptide was from eIF-4A or eIF-4F (Grifo et al., 1983; Sonenberg, 1981).

Our current interpretation of the mRNA binding process is presented in Figure 1. In this scheme eIF-4A, eIF-4B and eIF-4F interact with mRNA in an ATP-dependent manner forming an intermediate mRNA·initiation factor complex. This complex is subsequently able to bind to a 43S preinitiation complex which contains a 40S subunit, eIF-3, eIF-4C and a ternary complex of eIF-2, GTP and Met-$tRNA_f$. In this scheme eIF-4A, eIF-4B and eIF-4F recognize, discriminate, and bind to the mRNA in solution which is consistent with the assay data in Table II and the translational competition studies of Thach and co-workers (Ray et al., 1983). The following ATP-dependent movement of the mRNA to the initiating AUG codon, as proposed by Kozak (Kozak, 1978; Kozak, 1980a; Kozak, 1980b), represents completion of the positioning of the mRNA on the 40S subunit and presumably then allows the subunit joining steps to be initiated.

A possible model which would extend this scheme is presented below with the caution that too little information is available to support it, but several observations have generated this "working model". Based upon the ability of eIF-4F to crosslink to oxidized mRNA in the absence of ATP, it is likely that eIF-4F would be the first initiation factor to bind to the mRNA and do so at the 5' cap structure. Based upon the ability of eIF-4B to interact with eIF-4A and its ability to bind nucleic acids, it would seem reasonable to suppose that eIF-4B binds next due partially to its interaction with the mRNA and partially to its interaction with eIF-4F. Finally, eIF-4A which has the least ability to bind to nucleic acids joins this complex as a function of its ability to

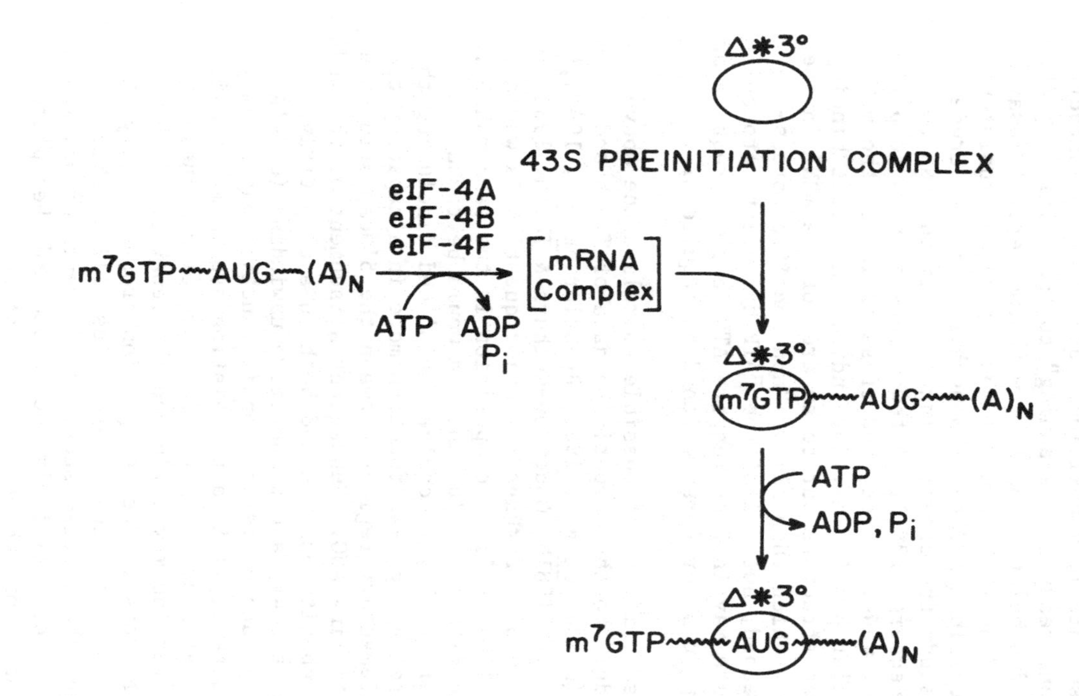

Figure I - Possible Pathway for the Binding of mRNA to 43S Preinitiation Complexes. The symbols used are as follows: ⬭, 40S ribosomal subunit; Δ, eIF-3; *, eIF-4C 3°, a ternary complex of eIF-2, GTP, and Met-tRNA$_f$; m^{7}GTP, the 5' cap structure; AUG, the initiating codon in the mRNA; $(A)_n$, the poly(A) tail.

interact with eIF-4B. Clearly it should be noted that any of the protein-protein interactions (especially eIF-4F - eIF-4B) may occur prior to binding to the mRNA. Subsequent transfer of this complex might then place the mRNA on the 40S subunit in a position close to or in contact with eIF-3 and ready for "scanning" to the initiating AUG codon. Besides being compatible with the partial assay data presented here, this model would also be consistent with the crosslinking data of Sonenberg who used crude systems to study the crosslinking of oxidized mRNA to protein (Sonenberg, 1981). A few points on this comparison should be made. First, the crosslinking of the mRNA to a particular peptide may not indicate a direct binding to the 5' cap structure but to an area of the mRNA in the near vicinity which then allows "non-specific" crosslinking. Second, not all of the "ATP-dependent" crosslinking events may in fact require ATP, but may require an earlier step to occur first which does require ATP.

At present it is not possible to relate the above "working model" to the competition data of Thach and co-workers (Walden *et al.*, 1981; Brendler *et al.*, 1981a; Brendler *et al.*, 1981b; Godefroy-Colburn & Thach, 1981; Ray *et al.*, 1983). Perhaps the major question is whether eIF-4A and eIF-4F relieve competition in the same manner. One possible difference in function would be to have eIF-4F differentially recognize the 5' end (including the cap structure) and eIF-4A function more in a capacity to unwind complementary regions between the 5' cap structure and the initiating AUG. Subsequent attachment of the mRNA to the 40S subunit would depend both on a competitive recognition (eIF-4F) and competitive unwinding (eIF-4A). Unfortunately such a distinction in function may be very difficult to test, as in a translation system mRNAs exist as mRNPs and at present little is known of how these additional proteins may influence the behavior of the initiation factors in the recognition process. Only the continued efforts of many laboratories will be able to better define the precise sequence of events of how mRNAs (as mRNPs) are competitively recognized and then placed on the 40S subunit for subsequent translation.

Supported in part by grant T32-GM-07250 (J.A.G. and C.A.S.) and GM-26795 (W.C.M.).

REFERENCES

Benne, R. & Hershey, J.W.B. (1978) J. Biol. Chem. 253, 3078-3087
Brendler, T.G., Godefroy-Colburn, T., Carlill, R.D. & Thach, R.E. (1981a) J. Biol. Chem. 256, 11747-11754.
Brendler, T.G., Godefroy-Colburn, T., Yu,S. & Thach, R.E. (1981b) J. Biol. Chem. 256, 11755-11761.
Godefroy-Colburn, T. & Thach, R.E. (1981) J. Biol. Chem. 256, 11762-11773.
Grifo, J.A., Tahara, S.M., Leis, J.P., Morgan, M.A., Shatkin, A.J. & Merrick, W.C. (1982) J. Biol. Chem. 257, 5246-5252.
Grifo, J.A., Tahara, S.M., Morgan, M.A., Shatkin, A.J. & Merrick, W.C. (1983) J. Biol. Chem. 258, 5804-5810.
Jagus, R.W., Anderson, W.F. & Safer, B. (1981) Prog. Nucl. Acid Res. Mol. Biol. 25, 127-185.
Kozak, M. (1978) Cell 15, 1109-1123.
Kozak, M. (1980a) Cell 22, 459-467.
Kozak, M. (1980b) J. Mol. Biol. 144, 291-304.
Merrick, W.C. (1979) Methods Enzymol. 60, 108-123.
Ray, B.K., Brendler, T.G., Adya, S., Daniels-McQueen, S., Miller, J.K., Hershey, J.W.B., Grifo, J.A., Merrick, W.C. & Thach, R.E. (1983) Proc. Natl. Acad. Sci. U.S.A. 80, 663-667.
Sonenberg, N. & Shatkin, A.J. (1977) Proc. Natl. Acad. Sci. U.S.A. 74, 4288-4292.
Sonenberg, N., Morgan, M.A., Merrick, W.C. & Shatkin, A.J. (1978) Proc. Natl. Acad. Sci. U.S.A. 75, 4843-4847.
Sonenberg, N. (1981) Nucleic Acids Res. 9, 1643-1656.
Tahara, S.M., Morgan, M.A. & Shatkin, A.J. (1981) J. Biol. Chem. 256, 7691-7694.
Tahara, S.M., Morgan, M.A., Grifo, J.A., Merrick, W.C. & Shatkin, A.J. (1982) Interaction of Translational and Transcriptional Controls in the Regulation of Gene Expression (Grunberg-Manago, M. & Safer, B., eds) Elsevier North Holland, Inc., New York, pp 359-372.
Trachsel, H., Erni, B., Schreier, M.H. & Staehelin, T. (1977) J. Mol. Biol. 116, 755-767.
Walden, W.E., Godefroy-Colburn, T. & Thach, R.E. (1981) J. Biol. Chem. 256, 11739-11746.

DIFFERENTIAL GENE EXPRESSION BY MESSENGER RNA COMPETITION FOR EUKARYOTIC INITIATION FACTOR eIF-2

RAYMOND KAEMPFER

Department of Molecular Virology
The Hebrew University-Hadassah Medical School
91 010 Jerusalem, Israel

ABSTRACT

Eukaryotic initiation factor 2 (eIF-2) has a dual function in initiation of protein synthesis: it binds Met-$tRNA_f$ to the small ribosomal subunit, and it binds directly to mRNA. The binding of eIF-2 to mRNA is highly specific and occurs in satellite tobacco necrosis virus RNA and Mengovirus RNA at the nucleotide sequences that constitute the ribosome binding sites. These findings support the concept that, during translation, eIF-2 may guide the ribosome to this site. Cl^- or OAc^- ions inhibit the direct binding of globin mRNA to eIF-2 in a manner that closely resembles their inhibitory effect on the translation of globin mRNA, an inhibition that is relieved by excess eIF-2. Hence, these anions may act to inhibit the interaction between mRNA and eIF-2 during protein synthesis. In mRNA-dependent reticulocyte lysates, a molecule of Mengovirus RNA competes in translation 35-fold more strongly than (on average) a molecule of globin mRNA. This competition is relieved by excess eIF-2. Mengovirus RNA binds directly to eIF-2 with 30-fold higher affinity than does globin mRNA. These results reveal a direct correlation between the affinity of a given mRNA species for eIF-2 and its ability to compete in translation. Indeed, the translational competition between α- and β-globin mRNA is also relieved by excess eIF-2, and in direct binding analysis, β-globin mRNA exhibits greater affinity for eIF-2 than does α-globin mRNA.

We have asked if translational competition occurs in a less extensively differentiated tissue than reticulocytes, the liver, that synthesizes a more complex spectrum of proteins, in order to test the prediction that the more evenly different species of mRNA are expressed, the less extreme must be the differences in competing ability between them. Using quantitative immunoprecipitation of products of cell-free translation, the existence of translational competition between individual mRNAs of the liver was demonstrated: the mRNA species encoding haemopexin, ferritin and albumin possess a progressively greater ability to compete in translation. Competition between the liver mRNA species likewise appears to involve initiation factor eIF-2 as a target.

INTRODUCTION

Eukaryotic gene expression is often regulated by the selective translation of specific mRNA templates over other ones. Examples of this type of control, involving competition between mRNA species, are encountered in cellular differentiation and virus infection. Messenger RNA competition is thought to occur mainly at initiation of translation, during the recognition of mRNA and its binding to ribosomes.

Among the initiation factors for eukaryotic translation, eIF-2 stands out by its importance to translational control. The activity of eIF-2 is regulated in response to a large variety of biological stress conditions, such as heme deprivation (Kaempfer, 1974; Clemens et al., 1975), the presence of double-stranded RNA (Kaempfer, 1974; Clemens et al., 1975), or after treatment with interferon (Kaempfer et al., 1979b). Here, evidence will be presented in support of the concept that eIF-2 interacts directly with mRNA during protein synthesis, recognizing in mRNA molecules the sequence and conformation that constitute the ribosome binding sites, and that the ability of a given mRNA species to compete in translation is correlated directly with its affinity for eIF-2. These properties impart on eIF-2 an essential function in differential gene expression at the level of translation.

RESULTS

The mRNA-Binding Activity of eIF-2

eIF-2 binds with absolute specificity to methionyl-$tRNA_f$ (Met-$tRNA_f$), the initiator species. This binding depends on

GTP and leads to formation of a ternary complex, eIF-2/Met-tRNA$_f$/GTP, that subsequently binds to the 40 S ribosomal subunit (e.g., Levin et al., 1973). Only when Met-tRNA$_f$ is bound to this subunit can binding of mRNA take place (Schreier & Staehelin, 1973; Darnbrough et al., 1973). Thus, the unique property of providing Met-tRNA$_f$ already imparts on eIF-2 a crucial role in the binding of mRNA. It is important to realize that while additional initiation factors participate in the stable binding of mRNA (e.g., Trachsel et al., 1977; Grifo et al., 1982), none can act in the absence of eIF-2.

In addition to binding Met-tRNA$_f$, eIF-2 itself can bind to mRNA (Kaempfer, 1974; Barrieux & Rosenfeld, 1977, 1978; Kaempfer et al., 1978a; Kaempfer et al., 1979a). This binding is specific in that all mRNA species tested possess an effective binding site for eIF-2, including mRNA species lacking the 5'-terminal cap or 3'-terminal poly (A) moieties (Kaempfer et al., 1978a), while RNA species not serving as mRNA, such as tRNA (Barrieux & Rosenfeld, 1977, 1978; Kaempfer et al., 1978a, 1979a; Rosen & Kaempfer, 1979), ribosomal RNA (Barrieux & Rosenfeld, 1977; Kaempfer et al., 1981), or negative-strand viral RNA (Kaempfer et al., 1978a) do not possess such a site.

The mRNA-binding property is a feature of eIF-2 itself. This can be demonstrated in a number of ways, but perhaps most convincingly by the finding that binding of mRNA to purified eIF-2 preparations can be inhibited completely by competing amounts of Met-tRNA$_f$, provided GTP is present (Rosen & Kaempfer, 1979). Conversely, mRNA competitively inhibits binding of Met-tRNA$_f$ to eIF-2 (Kaempfer et al., 1978b; Rosen et al., 1981). Thus, mRNA and Met-tRNA$_f$ are mutually exclusive in their binding to eIF-2, suggesting that during initiation of translation, the interaction of a molecule of mRNA with eIF-2 on the 40 S ribosomal subunit displaces the previously bound Met-tRNA$_f$ from this factor. We have proposed that during initiation, three processes may occur in one step: binding of mRNA to eIF-2, displacement of Met-tRNA$_f$ from eIF-2, and base-pairing between mRNA and Met-tRNA$_f$ (Rosen & Kaempfer, 1979).

The Sequence in mRNA Recognized by eIF-2

Globin mRNA molecules lacking the 3'-terminal poly (A) tail or an additional 90 nucleotides from the 3'-untranslated region bind to eIF-2 as tightly as native globin mRNA

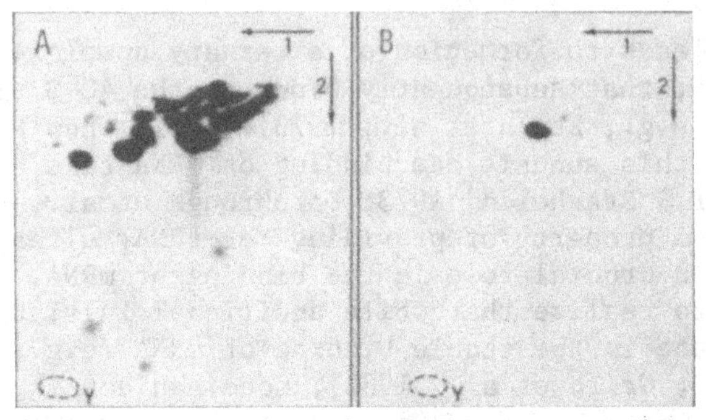

FIG. 1. Analysis of T1 RNase digests of total (A) and eIF-2-selected (B) 5' end-labeled STNV RNA by electrophoresis at pH 3.5 in the first dimension and by homochromatography in the second (arrows). Intact RNA gave one spot co-migrating with the major one in B (not shown). Y, methyl orange. From Kaempfer et al. (1981).

(Kaempfer et al., 1979a). On the other hand, cap analogs inhibit binding of both mRNA and Met-tRNA$_f$ to eIF-2 (Kaempfer et al., 1978b). While this could suggest that the cap interacts with eIF-2, studies with cap analogs can give rise to artifacts. Indeed, the genomic RNA species from Mengovirus or satellite tobacco necrosis virus (STNV) bind extremely well to eIF-2, in fact even better than globin mRNA, yet they do not carry a cap structure (Kaempfer et al., 1978b, 1981). This and other observations led to the suggestion that binding of eIF-2 to mRNA occurs primarily at an internal sequence (Kaempfer et al., 1978b).

STNV RNA is particularly suitable for analyzing this sequence because the RNA, 1,239 nucleotides long, has an unmodified 5' end that can be labeled with polynucleotide kinase. RNA isolated from STNV virions migrates, after 5' end labeling, as a heterogeneous collection of fragments in gels, with only a minor amount of label in fully intact viral RNA. Such preparations contain some 35 different 5' end sequences, as judged by fingerprinting after digestion with T1 (Fig. 1A) or pancreatic ribonuclease (Kaempfer et al., 1981), attesting to the presence of many fragments originating from internal regions of the viral RNA molecule. When this RNA is offered to eIF-2 and RNA bound by eIF-2 is isolated and fingerprinted with either RNase, one major spot is observed, migrating precisely as the 5' end of intact viral RNA, i.e., pApG after T1 (Fig. 1B) or pApGpU after

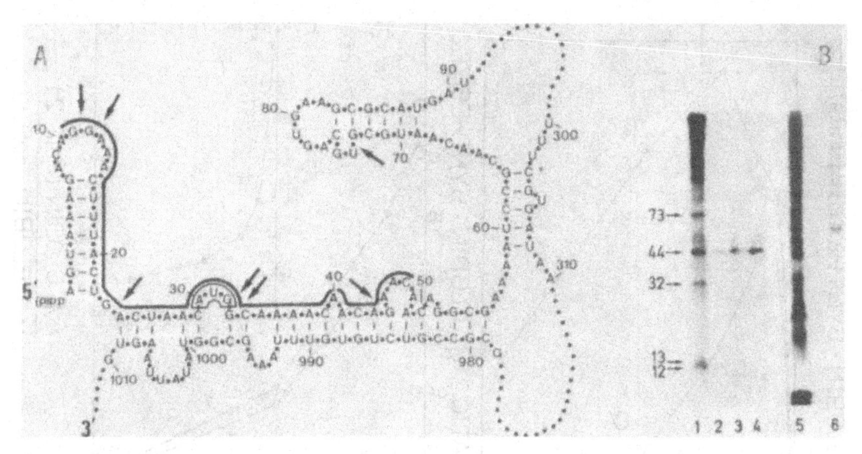

FIG.2(A) Secondary structure model for the 5' end of STNV RNA. The model depicts stable secondary interactions. Line, nucleotides protected by 40 S ribosomal subunits against nucleases; arrows, prominent sites of RNase T1 cleavage. See text. (B) Effect of eIF-2 on nuclease digestion of STNV RNA. Intact ^{32}P-labeled STNV RNA was digested with T1 RNase (lanes 1-4) or P1 nuclease (lanes 5 and 6). An aliquot of each digest was analyzed on a 12% polyacrylamide gel in 7 M urea (lanes 1 and 5). To other aliquots, increasing amounts of eIF-2 were added during digestion, and RNA bound to eIF-2 was isolated before electrophoresis. From Kaempfer et al., (1981).

pancreatic RNase treatment (Kaempfer et al., 1981). Sequence verification confirmed that eIF-2 selectively binds to STNV fragments starting with the 5' end of intact STNV RNA (Kaempfer et al., 1981).

To map the eIF-2 binding site more exactly, intact 5' end-labeled STNV RNA was isolated and digested partially with RNase T1, to generate a nested set of labeled RNA fragments, all containing the 5' end of intact RNA and extending to various points within the molecule. Fragments of discrete size were isolated by gel electrophoresis and their ability to bind to eIF-2 was studied. Fig. 2A depicts the 5'-terminal sequence of STNV RNA, including the unique AUG translation initiation codon located at positions 30-32. Arrows denote G residues sensitive to RNase T1 attack. We found that eIF-2 does not bind the 32-nucleotide 5'-terminal fragment ending with the AUG codon, or shorter ones, but it does

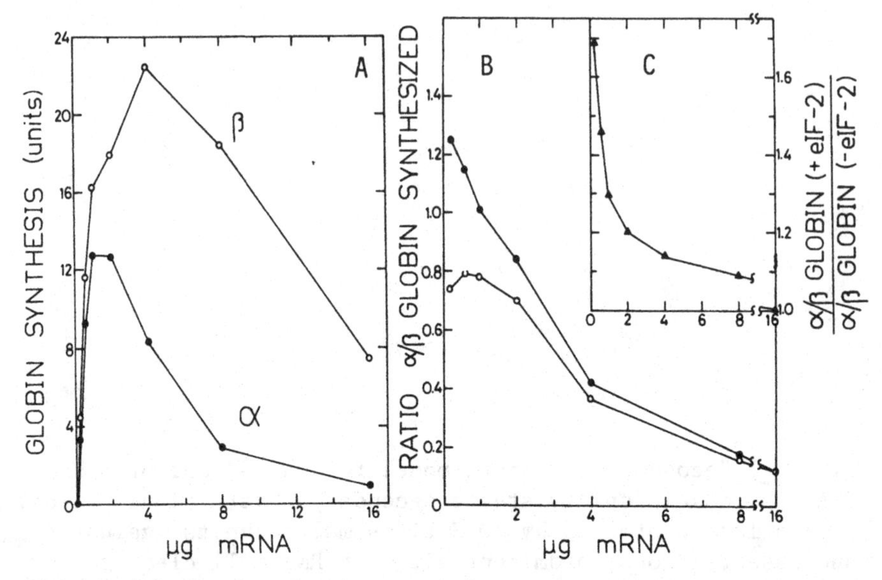

FIG. 3 (A) Synthesis of α- and β-globin as a function of globin mRNA concentration. (B) Effect of eIF-2 on the α/β synthetic ratio, computed for the samples incubated without added eIF-2 (A) (o) or for parallel samples that received a constant amount of eIF-2 (•). (C) The effect of eIF-2 on the α/β ratio was determined by dividing the ratio observed in the presence of added eIF-2 by that observed in its absence, and is plotted as a function of the amount of mRNA present. From Di Segni et al. (1979), with permission. Copyright 1979 American Chemical Society.

bind to the 44-nucleotide fragment or larger ones, and with the same specificity as to intact viral RNA. This places the 3'-proximal boundary of the eIF-2 binding site at or near nucleotide 44. Indeed, binding of eIF-2 to intact STNV RNA greatly increases the sensitivity of the RNA to cleavage by RNase T1 at nucleotide 44, or by nuclease P1 near position 60 (Fig. 2B), attesting to a conformational change induced at these points by the binding of the initiation factor molecule. On the 5'-terminal side, eIF-2 shields positions 11, 12, 23, 32 and 33 against digestion (Fig. 2A), placing the boundary at or before position 10. Since the G residues at position 2 and 7 are hydrogen-bonded and thus resistant to nuclease attack (Leung et al., 1979), it is not certain if the eIF-2 binding site extends to the physical

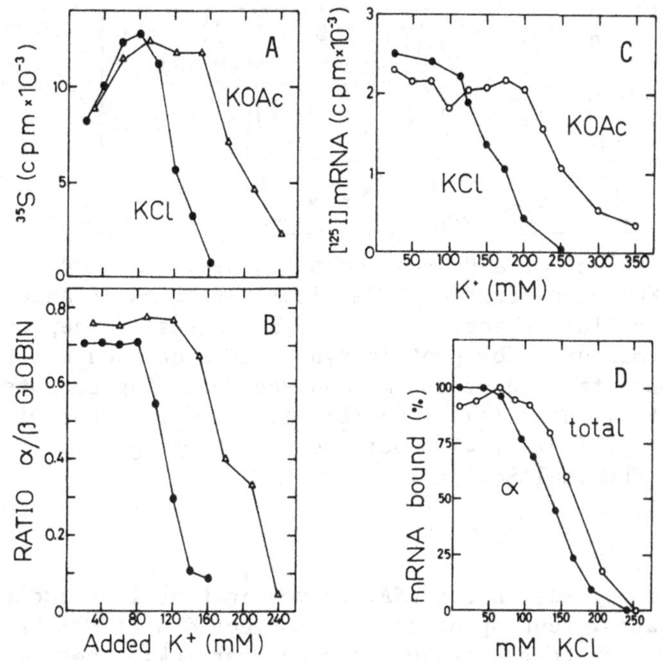

FIG. 4. Effect of KCl and KOAc on translation of globin mRNA (A), on the α/β globin synthetic ratio (B), and on complex formation between ^{125}I-labeled globin mRNA and eIF-2 (C); effect of KCl on complex formation between eIF-2 and labeled α-globin mRNA or total globin mRNA (D). From Di Segni et al. (1979), with permission. Copyright 1979 American Chemical Society.

5' end. The striking aspect of the eIF-2 binding site is, however, that it overlaps virtually completely with the binding site for 40 S ribosomal subunits (Browning et al., 1980), depicted by the line in Fig. 1. Thus, eIF-2 by itself recognizes virtually the same nucleotide sequence that is bound by 40 S ribosomal subunits carrying eIF-2, Met-tRNA$_f$ and all other components needed for initiation of translation. This finding suggests that the binding of ribosomes to STNV RNA may be guided directly by eIF-2.

We then mapped the eIF-2 binding site in Mengovirus RNA. This RNA has a length of about 7,500 nucleotides and contains a poly (C) tract located several hundred nucleotides from the 5' end. A map was constructed of the major T1 oligonucleotides in Mengovirus RNA, ordered relative to the 3' end (Perez-Bercoff & Kaempfer, 1982). In 40 S or 80 S ini-

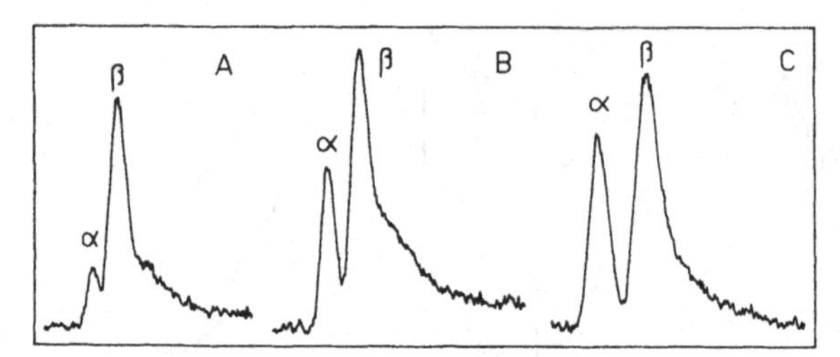

FIG. 5. Effect of eIF-2 on the synthesis of α- and β-globin at high KCl concentration. Densitometer scans of autoradiograms of cellulose acetate electropherograms are depicted. Reaction mixtures for protein synthesis contained KCl to an added concentration of 130 mM and the following amounts of added eIF-2: none (A), 1 μg (B), 1.5 μg (C). From Di Segni et al. (1979), with permission. Copyright 1979 American Chemical Society.

tiation complexes on the RNA, formed in L cell or ascites cell lysates, four specific oligonucleotides (15-28 bases long) were protected against nuclease attack. These map into at least two widely separated domains at internal sites downstream from the poly (C) tract. When intact Mengovirus RNA was offered to eIF-2 and the sequences protected by the initiation factor were isolated, three specific oligonucleotides were recovered, out of this very large sequence, and they were identical with 3 of the 4 protected in initiation complexes (Perez-Bercoff & Kaempfer, 1982). The virtual identity of the binding sites in Mengovirus RNA for ribosomes on one hand, and for eIF-2 on the other, reinforces the results with STNV RNA and points to a critical role for eIF-2 in recognition of mRNA by ribosomes.

Translational Competition for eIF-2

To study the functional implications of the interaction between mRNA and eIF-2, we analyzed translational competition, choosing to work with the mRNA-dependent reticulocyte lysate because it allows the precise quantitation of each mRNA species during translation and is capable of efficient and repeated initiation (Pelham & Jackson, 1976). The addition of increasing amounts of rabbit globin mRNA to this system generates conditions of increasing mRNA competition pressure that considerably magnifies even small differences

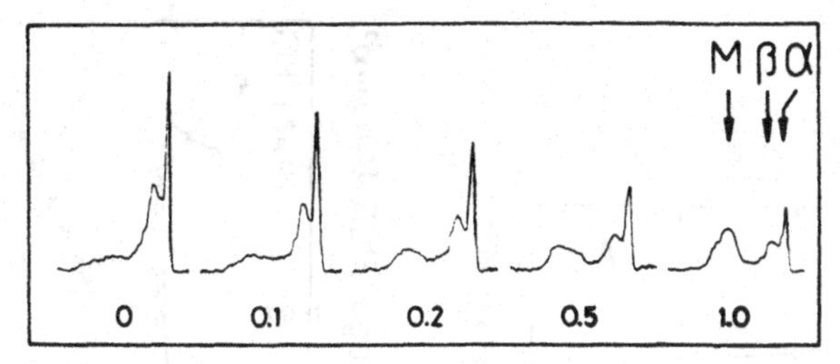

FIG. 6. Cellulose acetate electrophoresis analysis of products synthesized during simultaneous translation of globin mRNA and Mengovirus RNA. Translation mixtures contained 1.1 μg of globin mRNA and the indicated μg amounts of Mengovirus RNA. Densitometer scans of the autoradiogram of ^{35}S-labeled products are shown. α, α-globin; β, β-globin; M, Mengovirus RNA-directed products of translation. From Rosen et al. (1982).

in competing ability between individual mRNA species. As seen in Fig. 3A, α- and β-globin are synthesized in about equimolar yield when mRNA is sub-saturating (<1 μg), but beyond that point, β-globin synthesis occurs at the progressive expense of α-globin mRNA translation. The addition of a constant amount of purified eIF-2 does not change overall translation at any mRNA concentration (Di Segni et al., 1979), but, as seen in Fig. 3B (upper curve), increases the α/β synthetic ratio as compared to the control (lower curve). At low mRNA concentrations, the addition of eIF-2 causes a shift from about 0.8 to about 1.3, approaching the molar ratio of α- to β-globin mRNA (about 1.4-1.5) (Lodish, 1971). When the relative translation yield equals the molar ratio of the mRNAs under study, there is no competition in translation; hence eIF-2 acts to relieve competition. We note here that at very low concentrations of mRNA, the α/β synthetic ratio exceeds 1, and eIF-2 has no effect, but as the counts are very low in such conditions, the data are not shown (note that the curves do not start at zero mRNA). The relieving effect of eIF-2 is abolished by increasing the mRNA concentration (Fig. 3C). Translation of globin mRNA is inhibited by increasing concentrations of Cl^- or OAc^-, the former being more inhibitory (Fig. 4A). These anions primarily inhibit the translation of α-globin mRNA, resulting in a decrease in the α/β globin synthetic ratio (Figs. 4B and 5A). The inhibition of α-globin mRNA translation by Cl^- is readily relieved by the addition of eIF-2 (Figs. 5B, C).

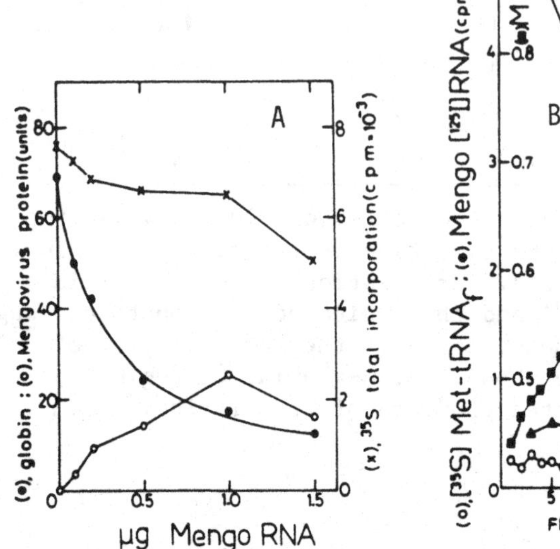

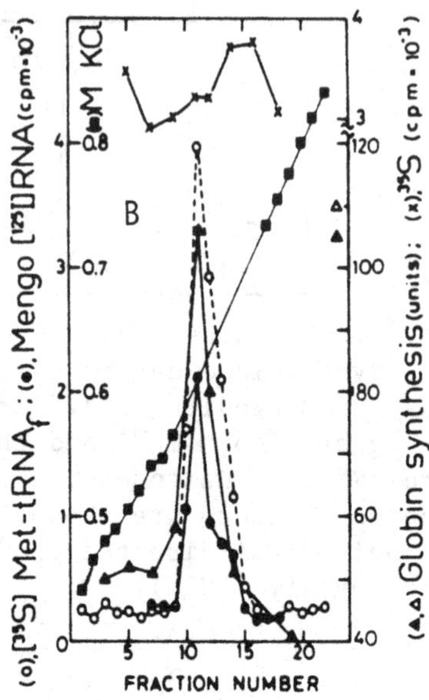

FIG. 7. (A) Translational competition between globin mRNA and Mengovirus RNA. Areas under the curves of densitometer scans (Fig. 6) are plotted in arbitrary units as total amounts of globin (●) and of Mengovirus RNA-directed products (o). Total ^{35}S-methionine incorporation into protein (X). (B) Co-purification of the activity that relieves translational competition with eIF-2. For purification of eIF-2, see Rosen et al. (1982). The gradient portion of the phosphocellulose column is shown. Aliquots were assayed for binding of ^{35}S-labeled Met-tRNA$_f$ (o) or binding of ^{125}I-labeled Mengovirus RNA (●). Aliquots were added to translation mixtures containing 1 µg of globin mRNA and 0.5 µg of Mengovirus RNA. Incorporation of ^{35}S-methionine into total protein (X) and total amount of globin formed (▲) are shown. Triangles on right indicate amount of globin synthesized in reaction mixtures lacking Mengovirus RNA, incubated with (▲) and without (Δ) material from tube 11. From Rosen et al. (1982).

Since Cl^- or OAc^- ions inhibit translation principally by affecting the binding of mRNA to 40 S subunits carrying Met-tRNA$_f$ (Weber et al., 1977), the data of Fig. 5 suggest that these anions may inhibit the interaction between eIF-2 and mRNA during translation. Indeed, when the effect of Cl^- or

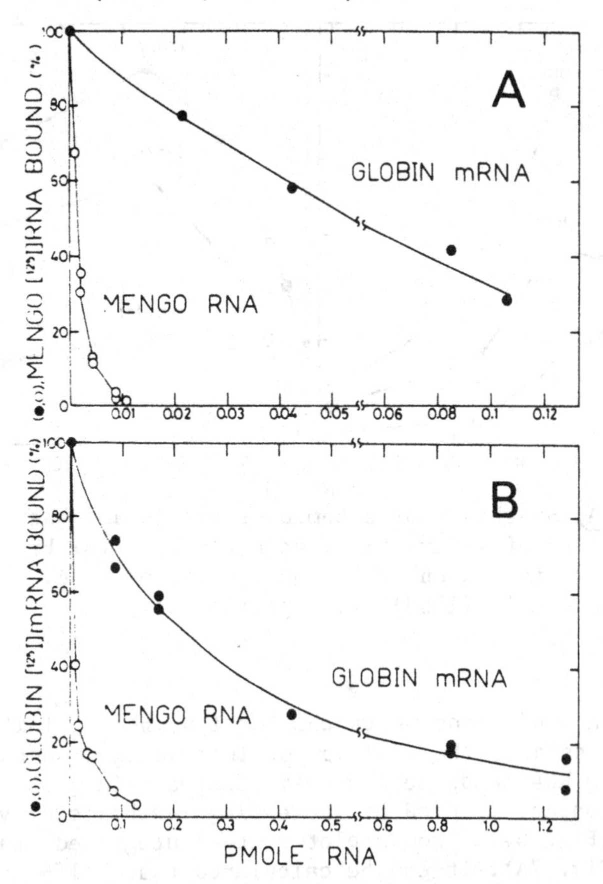

FIG. 8. Competition between globin mRNA and Mengovirus RNA in direct binding to eIF-2. See text. From Rosen et al., (1982).

OAc^- ions on direct binding of globin mRNA to eIF-2 is studied, remarkably parallel results are obtained (Fig. 4C). Moreover, binding of purified α-globin mRNA to eIF-2 is more sensitive to salt than is binding of unfractionated globin mRNA (two thirds α-globin mRNA) (Fig. 4D), attesting to a greater affinity of β-globin mRNA for eIF-2 (Di Segni et al., 1981).

More quantitative evidence for a direct relationship between the affinity of a given mRNA for eIF-2 and its ability to compete in translation came from a study of the competition between globin mRNA and Mengovirus RNA (Rosen et al.,

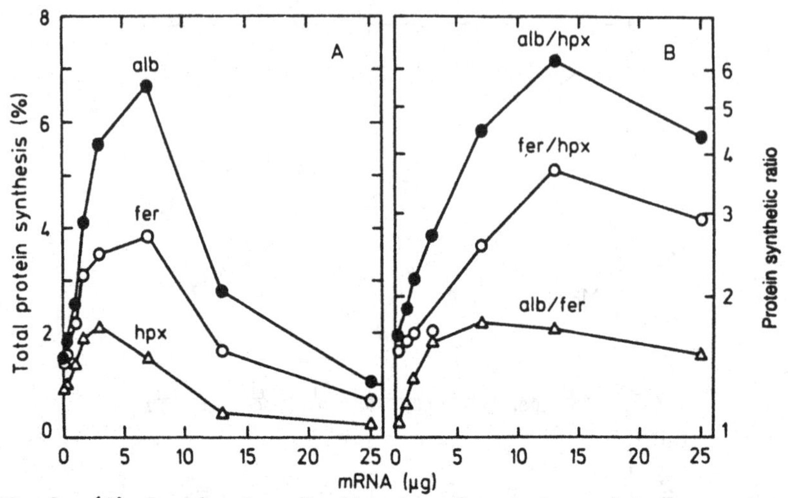

FIG. 9. (A) Synthesis of albumin, ferritin and haemopexin as a function of liver mRNA concentration. (B) Protein synthetic ratios calculated from the values in A. From Kaempfer & Konijn (1983), with permission.

1982). In conditions where the total number of initiations remains constant, the addition of increasing amounts of Mengovirus RNA leads to a drastic decrease in globin mRNA translation,accompanied by increasing synthesis of viral protein (Fig. 6). From a plot of the integrated translation yields (Fig. 7A), it can be calculated that half-maximal inhibition of globin mRNA translation occurs when 35 molecules of globin mRNA are present for every molecule of Mengovirus RNA. Assuming that equal proportions of these RNA species are translationally active, this means that a molecule of Mengovirus RNA competes 35-fold more strongly in translation than does (on average) a molecule of globin mRNA.

Does this competition involve eIF-2? Indeed, in conditions where globin synthesis is greatly depressed by the presence of Mengovirus RNA, the addition of eIF-2 does not stimulate overall translation, yet restores globin synthesis to the level seen in the absence of competing Mengovirus RNA (Fig. 7B). Globin synthesis in controls lacking Mengovirus RNA is not stimulated by the addition of eIF-2 (Fig. 7B, triangles on right). Hence, addition of eIF-2 allows the more weakly competing, but more numerous, globin mRNA molecules to initiate translation at the expense of the more strongly competing, but less numerous, viral RNA molecules.

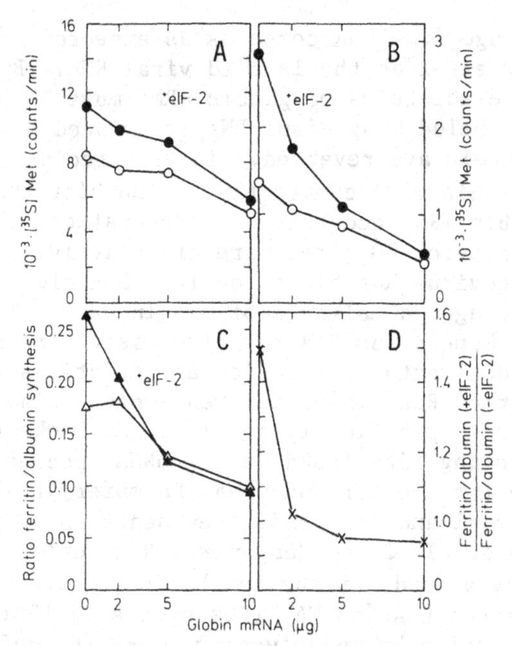

FIG. 10. Effect of initiation factor eIF-2 on synthesis of albumin and ferritin during translational competition by globin mRNA. Synthesis of albumin (A) and ferritin (B) was in the absence (o) or presence (•) of 2 μg purified eIF-2. (C) Ratio of ferritin to albumin synthesized in the absence (Δ) or presence (▲) of eIF-2. (D) Effect of eIF-2 on the ferritin/albumin synthetic ratio, plotted as a function of the amount of globin mRNA present. From Kaempfer & Konijn (1983), with permission.

The fact that eIF-2 acts to shift translation in favor of globin synthesis shows clearly that globin mRNA and Mengovirus RNA compete for eIF-2, but does not eliminate the possibility that eIF-2 could act in a non-specific manner, as by increasing the pool of 40 S/Met-$tRNA_f$ complexes. The results of RNA binding experiments, however, show that Mengovirus RNA and globin mRNA compete directly for eIF-2 with an affinity ratio that matches exactly with that observed in translation competition experiments.

In the experiment of Fig. 8, the only macromolecules present are eIF-2 and mRNA. In A, binding of labeled Mengovirus RNA to a limiting amount of eIF-2 is studied in the presence of increasing amounts of unlabeled, competing RNA.

Unlabeled Mengovirus RNA competes as expected, with the same affinity for eIF-2 as the labeled viral RNA. By contrast, 30 times more molecules of globin mRNA must be present before binding of labeled Mengovirus RNA is reduced by one half. In B, the labels are reversed. In this reciprocal experiment, unlabeled globin mRNA competitively inhibits the binding of labeled globin mRNA according to expectation, while Mengovirus RNA competes 30 times more effectively. Thus, a molecule of Mengovirus RNA binds to eIF-2 30-fold more strongly than (on average) a molecule of globin mRNA. The high affinity of Mengovirus RNA for eIF-2 is not related simply to nucleotide length, for vesicular stomatitis virus (VSV) negative-strand RNA, which is even longer, binds only very weakly and non-specifically to eIF-2 and lacks the high-affinity binding site found in all mRNA species tested, including the far shorter VSV mRNA (Kaempfer et al., 1978a).

Additional and independent evidence for a preferential interaction of eIF-2 and Mengovirus RNA during translation is furnished by a study of the inhibition of translation by double-stranded RNA (dsRNA) (Rosen et al., 1981). DsRNA blocks initiation by inactivating eIF-2 (Kaempfer, 1974; Clemens et al., 1975). Rosen et al. (1981) found that, on one hand, dsRNA binds with higher affinity than globin mRNA to eIF-2 and inhibits globin mRNA translation, while on the other hand, it binds with lower affinity than Mengovirus RNA to eIF-2 and fails to establish translational inhibition when this viral RNA is present. The results of that study indicate that the rate-determining event in the establishment of translational inhibition by dsRNA involves competition between mRNA and dsRNA and that Mengovirus RNA, because of its higher affinity for eIF-2, is able to protect this initiation factor against inactivation by dsRNA. The implications of these findings for translational control during virus infection have been discussed (Kaempfer et al., 1983).

One would predict that the more evenly different species of mRNA are expressed, the less extreme must be the differences in competing ability between them. Turning to the cellular mRNA population, one may ask if individual species of mRNA compete in translation, and if so, whether they differ in competing ability. While this question has been answered affirmatively for the mRNA pair that encodes α- and β-globin, it has not been studied carefully for other, and in particular less differentiated, mRNA populations. Recently, we used competition pressure analysis to study this point for three liver mRNA species, encoding haemopexin, ferritin and albumin (Kaempfer & Konijn, 1983). The application of competition pressure, through the addition of greater than saturating amounts of liver mRNA to micrococcal nuclease-treated reticulocyte lysate, reveals distinct behavior of the three mRNA species, as judged by specific, quantitative immunoprecipitation of their respective translation products (Fig. 9A). At mRNA levels greatly in excess of saturation, there is a general inhibition of translation (Pelham & Jackson, 1976; Di Segni et al., 1979), but at low and intermediate levels of mRNA, the proportion of total protein synthesized as albumin, ferritin and haemopexin exhibits characteristic differences. Synthesis of haemopexin (hpx) declines well before that of ferritin (fer) and albumin (alb), while ferritin synthesis levels off earlier than albumin synthesis. The protein synthetic ratios plotted in Fig. 9B show that with increasing mRNA concentration, the alb/fer ratio increases 1.6-fold, the fer/hpx ratio 2.4-fold, and the alb/hpx ratio 3.8-fold before reaching essentially a plateau value. These results can be analyzed and interpreted in analogy to the α/β globin system (Fig. 3A) to mean that the three mRNA species compete, and do so in the order hpx < fer < alb of ascending competing ability. By comparison to α- and β-globin, the relative abilities of haemopexin, ferritin and albumin mRNA to compete in translation seem to differ far less extensively. Yet, these small differences can be magnified considerably by the application of competition pressure in vitro: translation of the more effectively competing mRNA species is progressively increased at the expense of the translation of more weakly competing ones. Thus, haemopexin, ferritin and albumin mRNA could be rank-ordered in this respect, independent of their relative abundance in the total liver mRNA population, and in spite of the fact that they were present in a complex mixture of mRNA species.

Addition of increasing amounts of globin mRNA to translation mixtures containing a constant amount of liver mRNA causes a progressive decrease in synthesis of both albumin (Fig. 10A, lower curve) and ferritin (Fig. 10B, lower curve), but synthesis of ferritin diminishes more readily, leading to a decline in the ferritin/albumin synthetic ratio (Fig. 10C, lower curve). This is precisely the result predicted by the data of Fig. 9. Addition of a constant amount of eIF-2 does not cause any change in overall protein synthesis, yet it stimulates synthesis of both albumin and ferritin (Figs. 10A and B, upper curves). Translation of ferritin mRNA is stimulated more by eIF-2 than is translation of albumin mRNA (Fig. 10C, upper curve). As the concentration of globin mRNA is increased, the stimulatory effect of eIF-2 on synthesis of ferritin and albumin diminishes progressively (Fig. 10A, B) and concomitantly, the stimulatory effect of eIF-2 on the ferritin/albumin synthetic ratio disappears (Fig. 10C, D). The response of ferritin mRNA is that expected for an mRNA species that competes more weakly for eIF-2 than does albumin mRNA.

DISCUSSION

The concept that mRNA species differ in their efficiency of translation, apparently because of a different affinity for one or more critical components in the initiation step, was first suggested by Lodish (1971, 1974) who showed that initiation of protein synthesis on a molecule of α-globin mRNA occurs with lower frequency than that on a molecule of β-globin mRNA. Lawrence and Thach (1974) observed translational competition between host mRNA and encephalomyocarditis virus RNA in a cell-free system from mouse ascites cells and concluded that the competition was at the level of initiation. Hackett et al. (1978a, b) made similar observations for Mengovirus RNA. In searching for a possible target of competition, Ray et al. (1983), using a reconstituted cell-free system, suggested that it is initiation factor eIF-4A, or a complex of factors, CBP-II, able to bind to the 5'-terminal cap structure. The results reported here, showing that eIF-2 is able to relieve translational competition between α- and β-globin mRNA, or Mengovirus RNA and globin mRNA, and that these mRNA species compete directly in binding to eIF-2, cannot be explained by assuming a contamination. First, eIF-2 prepared by our procedure is about 98% pure (Rosen & Kaempfer, 1979). Second, the purification procedure yields an eIF-2 preparation that contains only a single mRNA-binding component, eIF-2, since binding of mRNA to this preparation is completely sensitive to competitive inhibition by Met-$tRNA_f$, provided GTP is present, but not by uncharged tRNA (Rosen & Kaempfer, 1979). Third, independent verification that eIF-2 binds 30-fold more tightly to Mengovirus RNA is provided by the observation that Mengovirus RNA is 30 to 40 times more effective than globin mRNA as a competitive inhibitor of ternary complex formation between Met-tRNA$_f$, GTP and eIF-2 (Rosen et al., 1981). While it is conceivable that other initiation factors may also be a target for mRNA competition in translation, it is nevertheless clear from our studies that the addition of eIF-2 is sufficient to overcome such competition.

It is worth noting that in our translation experiments, the added eIF-2 was always used immediately after purification. Storage of purified eIF-2 leads to a loss of activity. Even though stored preparations can be active in ternary complex formation with Met-tRNA$_f$ and GTP, they contain a considerable proportion of inactive eIF-2 molecules that can competitively inhibit active ones during translation. This may explain the failure of other groups to obtain more than marginal relief of translational competition by eIF-2, although the addition of too few molecules of eIF-2 could also explain the results (Ray et al., 1983). Moreover, our translation experiments were done in the micrococcal nuclease-treated reticulocyte lysate. This system offers several advantages over reconstituted cell-free systems. It responds to translational control signals, it is capable of extensive and efficient initiation in conditions more likely to be representative of protein synthesis in intact cells, and except for mRNA, it contains all other components for protein synthesis in a proportion much closer to that of the intact cell.

The striking ability of eIF-2 to overcome the Cl^--induced inhibition of α-globin mRNA translation (Figs. 4A, B and Fig. 5), coupled with the remarkably parallel effect of Cl^- and OAc^- ions on translation of globin mRNA on one hand (Fig. 4A) and on binding of globin mRNA to eIF-2 on the other (Fig. 4C), strongly suggest that a direct interaction between mRNA and eIF-2 occurs during protein synthesis. The mRNA competition studies lead to an identical conclusion.

In the translation competition experiments, relief of competition by added eIF-2 was not accompanied by any stimulation of protein synthesis. This explains why, when Mengovirus RNA was also present, the increase in globin mRNA translation caused by addition of eIF-2 was concomitant with a decrease in Mengovirus RNA translation (Fig. 7B and Rosen et al., 1982). The fact that eIF-2 acts to shift translation in favor of globin synthesis shows clearly that globin mRNA and Mengovirus RNA compete for eIF-2, but does not eliminate the possibility that eIF-2 could act in a non-specific manner, as by increasing the pool of 40S/Met-tRNA$_f$ complexes. The results of the mRNA binding experiments (Fig. 8), however, showing that Mengovirus RNA and globin mRNA compete directly for eIF-2 with an affinity ratio (30-fold) that matches exactly with that observed in translation competition experiments (Figs. 6, 7A), provide strong evidence for a direct competition of these mRNA species for eIF-2 during translation. While it is conceivable that globin mRNA and Mengovirus RNA compete for free eIF-2 molecules, it is more likely that they compete for eIF-2 molecules located in 40S/Met-$tRNA_f$ complexes. (See Di Segni et al., 1979; Rosen et al., 1982, for a fuller discussion of this point.) The data with α- and β-globin mRNA (Figs. 3-5) reinforce these conclusions. Moreover, the greater ability of albumin mRNA to compete in translation, as compared to ferritin and haemopexin mRNA (Figs. 9, 10) also appears to be associated with a more favorable competition for eIF-2.

These results point to mRNA affinity for eIF-2 as a critical element in translational control. They support the conclusion that translation of different mRNA species is regulated to a large extent by their relative affinities for eIF-2.

The concept of differential gene expression by mRNA competition for eIF-2 becomes even more convincing with the finding that eIF-2 by itself recognizes a specific sequence in mRNA, and that, at least in the two cases examined, for STNV and Mengovirus RNA, this sequence has turned out to be virtually identical with the ribosome binding site sequence (Figs. 1, 2; for a fuller treatment, see Kaempfer et al., 1981, and Perez-Bercoff & Kaempfer, 1982). These results strongly suggest a leading role for eIF-2 in guiding the 40 S ribosomal subunit to its binding site in mRNA. The fact that eIF-2 must be seated on the 40 S subunit before mRNA can be bound fits well with this concept. It is tempting to suggest that eIF-2 recognizes the sequence and conformation around the initiation site in mRNA, while eIF-4A (Grifo et al., 1982; Ray et al., 1983) and/or cap-binding proteins (Sonenberg, 1981; Lee et al., 1983) may unwind secondary structure in mRNA and anchor it at the 5'-terminal cap. As we have shown by nuclease sensitivity analysis (Fig. 2B), the binding of eIF-2 to STNV RNA results in a considerable conformational change in mRNA sequences located just downstream from the ribosome binding site.

This work was supported by grants from the National Research Council of Israel, the Gesellschaft fuer Strahlen- und Umweltsforschung (Muenchen) and the Israel Academy of Sciences.

REFERENCES

Barrieux, A. & Rosenfeld, M.G. (1977). J. Biol. Chem. 252, 3843-3847.

Barrieux, A. & Rosenfeld, M.G. (1978). J. Biol. Chem. 253, 6311-6315.

Browning, K.S., Leung, D.W. & Clark, J.M., Jr. (1980). Biochemistry 19, 2276-2282.

Clemens, M.J., Safer, B., Merrick, W.C., Anderson, W.F. & London, I.M. (1975). Proc. Natl. Acad. Sci. USA 72, 1286-1290.

Darnbrough, C.H., Legon, S., Hunt, T. & Jackson, R.J. (1973). J. Mol. Biol. 76, 379-403.

Di Segni, G., Rosen, H. & Kaempfer, R. (1979). Biochemistry 18, 2847-2854.

Grifo, J.A., Tahara, S.M., Leis, J.P., Morgan, M.A.,Shatkin, A.J. & Merrick, W.C. (1982). J. Biol. Chem. 257,5246-5252.

Hackett, P.B., Egberts, E. & Traub, P. (1978a). Eur. J. Biochem. 83, 341-352.

Hackett, P.B., Egberts, E. & Traub, P. (1978b). Eur. J. Biochem. 83, 353-361.

Kaempfer, R. (1974). Biochem. Biophys. Res. Commun. 61, 591-597.

Kaempfer, R., Hollender, R., Abrams, W.R. & Israeli R. (1978a). Proc. Natl. Acad. Sci. USA 75, 209-213.

Kaempfer, R., Hollender, R., Soreq, H. & Nudel, U. (1979a). Eur. J. Biochem. 94, 591-600.

Kaempfer, R., Israeli, R., Rosen, H., Knoller, S., Zilberstein, A., Schmidt, A. & Revel, M. (1979b). Virology 99, 170-173.

Kaempfer, R. & Konijn, A.M. (1983). Eur. J. Biochem. 131, 545-550.

Kaempfer, R., Rosen, H., Di Segni, G. & Knoller, S. (1983). In Mechanisms of Viral Pathogenesis (Kohn, A., ed.) Martinus Nijhoff, The Hague, in press.

Kaempfer, R., Rosen, H. & Israeli, R. (1978b). Proc. Natl. Acad. Sci. USA. 75, 650-654.

Kaempfer, R., van Emmelo, J. & Fiers, W. (1981). Proc. Natl. Acad. Sci. USA 78, 1542-1546.

Lawrence, C. & Thach, R. (1974). J. Virol. 14, 598-610.

Lee, K.A.W., Guertin, D. & Sonenberg, N. (1983). J. Biol. Chem. 258, 707-710.

Leung, D.W., Browning, K.S., Heckmann, J.E., RajBhandary, U. L. & Clark, J.M. Jr. (1979). Biochemistry 18, 1361-1366.

Levin, D.H., Kyner, D. & Acs, G. (1973). Proc. Natl. Acad. Sci. USA 70, 41-45.

Lodish, H.F. (1971). J. Biol. Chem. 246, 7131-7138.
Lodish, H.F. (1974). Nature (London) 251, 385-388.
Pelham, H.R.B. & Jackson, R.J. (1976). Eur. J. Biochem. 67, 247-256.
Perez-Bercoff, R. & Kaempfer, R. (1982). J. Virol. 41, 30-41.
Ray, B.K., Brendler, T.G., Adya, S., Daniels-McQueen, S., Miller, J.K., Hershey, J.W.B., Grifo, J.A., Merrick, W.C. & Thach, R.E. (1983). Proc. Natl. Acad. Sci. USA 80, 663-667.
Rosen, H., Di Segni, G. & Kaempfer, R. (1982). J. Biol. Chem. 257, 946-952.
Rosen, H. & Kaempfer, R. (1979). Biochem. Biophys Res. Commun. 91, 449-455.
Rosen, H., Knoller, S. & Kaempfer, R. (1981). Biochemistry 20, 3011-3020.
Schreier, M.H. & Staehelin, T. (1973). Nature (London), New Biol. 242, 35-38.
Sonenberg, N. (1981). Nucleic Acids Res. 9, 1643-1656.
Trachsel, H., Erni, B., Schreier, M.H. & Staehelin, T. (1977). J. Mol. Biol. 116, 755-767.
Weber, L.A., Hickey, E.D., Maroney, P.A. & Baglioni, C. (1977). J. Biol. Chem. 252, 4007-4010.

INITIATION OF PROTEIN BIOSYNTHESIS

Kjeld A. Marcker

Department of Molecular Biology
University of Aarhus
DK-8000 Århus C, Denmark

Initiation of translation of mRNA into protein consists of a series of discrete reactions during which the ribosomal subunits interact with mRNA and with initiator tRNA in such a way that the ribosome binds at a specific site of an mRNA that contains the initiator codon AUG or GUG corresponding to the beginning of a cistron. During this process the initiator tRNA is also positioned on the ribosome by base-pairing between the anticodon of the initiator tRNA and the initiation codon of mRNA. Several specific proteins called initiation factors and GTP direct the various steps leading to the formation of a functional ribosomal chain initiation complex.

In the prokaryotic system a series of discrete reactions can be defined in which mRNA, ribosome, initiator tRNA and initiation factors interact to give a functional ribosomal initiation complex that can engage in peptide bond formation. This is in contrast to the situation in the eukaryotic cell, where the mechanism of polypeptide chain initiation is much more complicated than in prokaryotes. One reason for this complexity is undoubtedly the unusual structure of eukaryotic mRNAs. Thus there is a 7-methyl guanosine 5' triphosphate 2-O methylated residue known as the cap structure present at the 5' end of most eukaryotic mRNAs. In addition many eukaryotic mRNAs are terminated by a poly A tail. Furthermore eukaryotic mRNAs do not contain a Shine-Delgarno sequence, in contrast to prokaryotic mRNAs. An important still unsolved problem in this field is therefore how the eukaryotic ribosome selects the correct initiation site on a mRNA molecule with such precision that translation starts at the AUG codon corresponding to the beginning of a cistron. Another peculiarity of eukaryotic mRNA is that although they may contain more than one initiation site which all are used at some stage during the translation of the mRNA, only one of these sites is functional at any given time during translation. Unfortunately we still do not know why the eukaryotic ribosome only can handle one initiation site at the time while the procaryotic ribosome can initiate at several initiation sites simultanously.

The complexity of eukaryotic initiation is further illustrated by the large number of initiation factors required for forming a functional initiation complex. It is still not clear how many factors participate in this process, but there seems to be general agreement that at least nine such factors are involved. The extraordinary number of factors is probably necessary because some of them interact with the modified structures present in eukaryotic mRNAs. Furthermore the multitude of factors most likely also reflect the need for translational control of eukaryotic protein synthesis resulting from the stability of eukaryotic mRNAs. It is clear that the mode of action of various regulatory elements controlling eukaryotic initiation cannot be elucidated until the details of the molecular mechanisms of the initiation factors are understood. Unfortunately the functions of these factors are very difficult to study primarily because few if any covalent bonds are formed or broken during the formation of the initiation complex.

The papers presented at this session of the symposium all deal with the role of the initiation factors in the formation of the eukaryotic ribosomal initiation complex in addition to their involvement in translational control at the point of initiation. I am sure that the reader will agree with me that even if this is very difficult to study, substantial progress is being achieved in this area.

2. PROTEIN SYNTHESIS ON ENDOPLASMIC RETICULUM

TRANSPORT OF PROTEINS AND SIGNAL RECOGNITION

T.A.Rapoport[1],A.Huth[1],S.Prösch[1], P. Bendzko[1], L.Kääriäinen[2], R.Bassüner[3], R.Manteuffel[3] and A.Finkelstein[4]

1) Zentralinstitut f.Molekularbiol. d. AdW d.DDR,Berlin-Buch,2) Dept.Virology University of Helsinki,3) Zentralinst. f.Genetik u. Kulturpflanzenforschung, 4)Institute of Protein Research, Poustchino, USSR

ABSTRACT

We provide support for the belief that the mRNA for a given protein contains the entire information for its final localization in a cell or its secretion out of a cell. If mRNA coding for carp preproinsulin is injected into Xenopus oocytes, proinsulin is found in the medium. Similarly, if the 26S-RNA coding for the structural proteins of the Semliki-Forest-Virus is injected, the envelope proteins are found on the outer surface of the oocytes. Transport and processing of the membrane proteins are not dependent on glycosylation. However, if plant storage globulin mRNAs are injected into oocytes, the translation products are found to be secreted rather than stored. Obviously, in this case the destination of the proteins is determined in part by the cellular apparatus.

Carp preproinsulin, like most other secretory proteins, contains a N-terminal, cleavable signal peptide. Recombinant plasmids were constructed which contained the complete coding sequence of carp preproinsulin placed into the coding region of prokaryotic genes so that fused translation products are expected. In many cases, the eucaryotic signal peptide was found to transport the insulin antigen out of E.coli cells even if its sequence was internal. Cleavage of the signal peptide is apparently correct since a proinsulin-like material is found in the periplasm. The results support our hypothesis according to which the first stretch of hydrophobic amino acids, following an unfolded part of the polypeptide chain or a complete folding domain with a hydrophilic surface, operates as signal for translocation.

INTRODUCTION

Most polypeptides are synthesized in the cytoplasm of a cell but they are finally located at many different sites. How does a cell "know" where the protein should be transported to? Obviously, there must be signals for the intracellular transport of proteins and devices of the cell to decode them. Are the signals and their receptors ubiquitous in nature or are signals and cells specifically adapted to each other? We shall provide evidence for the general belief that the mRNA for a given protein contains the entire information for its final destination. However, we also have an example, where the cell, in addition to the mRNA, determines the localization of a polypeptide.

The overwhelming evidence suggests that the signals for intracellular traffic of proteins are only decoded at the protein level. Signal sequences direct proteins across the endoplasmic reticulum membrane or across the inner bacterial membrane (Blobel and Dobberstein, 1975; Blobel, 1980). How are these signals recognized given that they vary greatly in length and sequence? An explanation is offered on the basis of theoretical considerations. The hypothesis predicts that the first hydrophobic sequence, following an unfolded part of the polypeptide chain or a complete folding domain with a hydrophilic surface, will operate as signal for translocation. Experimental support will be provided by the demonstration that a signal peptide placed at an internal position, is functional.

RESULTS AND DISCUSSION

Does the mRNA Contain the Entire Information for the Spatial Destination of a Protein?

The hypothesis that the mRNA for a given polypeptide contains the entire information for its final localization may be tested by introducing the mRNA into a foreign cell and asking for the fate of the synthesized protein. We have microinjected mRNAs into oocytes of Xenopus leavis which have been shown to translate them faithfully (for review see Marbaix and Huez,1980).

If mRNA coding for carp preproinsulin was injected into oocytes, proinsulin was found to be synthesized and thereafter contained within membranes inside the oocyte (Rapoport et al,1978; Rapoport,1981). With a lag period of about 8 h, proinsulin appeared in the medium surrounding the oocytes. Globin synthesized in response to coinjected globin mRNA, remained entirely inside the cell, also indicating that no leakage had occurred. Obviously, Xenopus oocytes which normally do not secrete proteins to a large extent, are nevertheless able to do so if the appropriate mRNA is introduced. The result supports the general belief that the mRNA contains the entire information for secretion of a protein and that the cellular apparatus for this process is ubiquitous in nature. In agreement with this notion, the signal sequence of carp preproinsulin appears to be correctly cleaved in Xenopus oocytes (Rapoport et al,1978). It should be noted, however, that no insulin was found inside or outside the oocytes. Thus, processing of proinsulin to insulin might require more specific enzyme(s) not present in oocytes (see,however,below).

Similar results were obtained by others with a variety of secretory proteins which were found to be exported from Xenopus oocytes (Lebleu et al, 1978; Colman and Morser,1979; Lane et al,1980).

The only exception is bee promellitin which apparently remains inside the cell (Lane et al,1981).

In order to test whether the correct transport of foreign proteins in oocytes is a general phenomenon, we have injected mRNA coding for plasma membrane proteins. As a model, the 26S-RNA was used which codes for the structural proteins of the Semliki-Forest-Virus (SFV).

This RNA codes for a cytosolic capsid protein (C) and for three envelope polypeptides (E1,E2 and E3) (for review see Garoff et al,1980). E2 and E3 are initially constituents of a precursor (p62) which is later cleaved into the individual polypeptides.All the envelope proteins are glycosylated at Asn-residues during their passage to the plasma membrane of the virus-infected cell.

After injection of 26S-RNA into Xenopus oocytes, the synthesis of all expected polypeptides was observed (Fig.1). The identity of the envelope proteins was proved by immunological means. The immunoprecipitated material had the expected molecular mass but displayed multiple bands. These are probably due to uncomplete glycosylation of the polypeptides in oocytes (see also Lane et al, 1979).

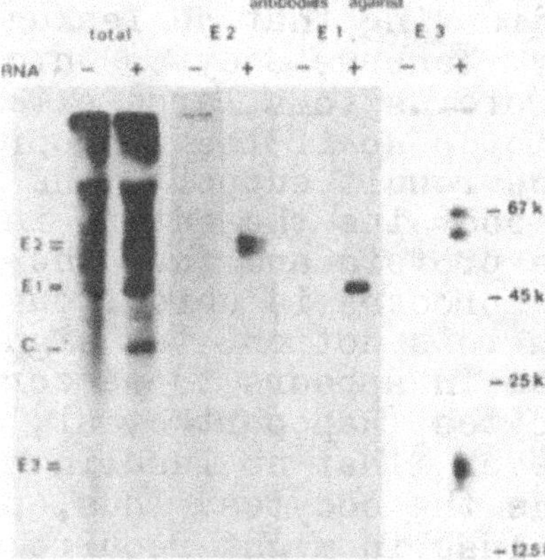

Fig.1. Synthesis of SFV-proteins in oocytes. Oocytes were injected with 26S-RNA or water,cultered overnight and incubated with ^{35}S-Met for 24 h. Products were either applied directly to a SDS-gel or first immunoprecipitated with antisera against the envelope proteins.

If the glycoproteins coded by the viral RNA are correctly handled by the Xenopus oocytes,they should appear on its surface. In order to test the localization of these proteins, oocytes were treated shortly with antiserum directed against the 29S-complex of the envelope proteins, washed and further incubated with ^{125}J-labelled protein A, and the bound radioactivity was determined (Table 1). Significantly more radioactivity was bound to oocytes injected with 26S-RNA as compared to water-injected ones. Other controls also indicate that the SFV-glycoproteins were localized on the oocyte surface (Table 1). Table 2 shows that all three envelope proteins (E1,E2,E3) were found on the plasma membrane of oocytes injected with 26S-RNA. The amount of SFV-glycoproteins on the surface of oocytes depended on the amount of 26S-RNA injected (not shown).

TABLE 1
SFV-glycoproteins appear on the surface of Xenopus oocytes injected with 26S-RNA

Oocytes incubated with	^{125}I-protein A bound to oocytes injected with	
	water (cpm/oocyte)	26S-RNA (cpm/oocyte)
antiserum against 29S-complex of envelope proteins	109 ± 31(9)	634 ± 238(9)
normal serum	40 ± 19(9)	39 ± 13(9)
antiserum against 29S-complex followed by unlabelled protein A(4.5 ,ug/ml)	n.d.	98 ± 48(8)

Oocytes were defolliculated, injected and cultured for 24 h. They were then incubated as shown in the Table before incubation with ^{125}I-protein A. In parenthesis the number of experiments.

The localization of the SFV-glycoproteins was also proved by a different method. Injected oocytes were labelled for 2 h with ^{35}S-Met. Some of the oocytes were analyzed imediately, others incubated further for 4 h with unlabelled Met. The intact cells were treated with antiserum at 0°C, washed and homogenized. The antigen-antibody complexes were precipitated by Staph.aureus bacteria and the supernatant was again treated in sequence with antiserum and Staph. bacteria. In this manner, a separation should be possible of envelope proteins located on the surface and those within the cell. The result of such an experiment is shown in Fig.2. After 2 h of labelling, only very small amounts of E1 and E2 were found on the surface whereas there was a large pool intracellularly. After a 4 h chase period, however, significant amounts of the envelope proteins were observed on the surface. It may be doubted whether the method indeed determines polypeptides located on the surface, particularly since the intracellular pool was always larger than that presumed to be outside. An indication for the validity of the separation procedure is that the ratio of E2 and E1 was always well below 1:1 intracellularly (see also Figs.3 and 4) but 1:1 on the surface, as determined with different antisera. It should be 1:1 since equimolar amounts are synthesized and both proteins contain about equal numbers of Met-residues. The reason for the low intracellular ratio of E2:E1 may be the existence of the p62-precursor protein and/or preferred degradation of E2-containing polypeptides. In any case, the findings suggest that only 1:1 complexes of E2 and E1 (and probably E3) are transported to the plasma membrane and/or are stable there.

The rate of processing of the p62-precursor in Xenopus oocytes was also determined. The pulse-chase labelling experiments shown in Fig.3 indicate a half-life of 1-2 h. If one assumes a temperature coefficient $Q_{10^{o}C}$ of about 2, this would correspond to a half-life of about 20-40 min at 37°C which fits with the rate of conversion found in SFV-infected cells.

How does the oocyte "know" that the SFV-en-

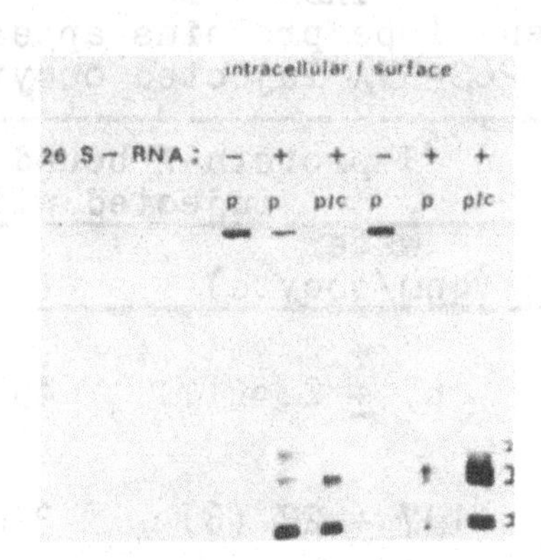

Fig.2.Appearence of SFV-proteins on the oocyte surface.p-2h pulse,p/c-additional 4h chase.p62,E2 and E1 are indicated by brackets.See also text.

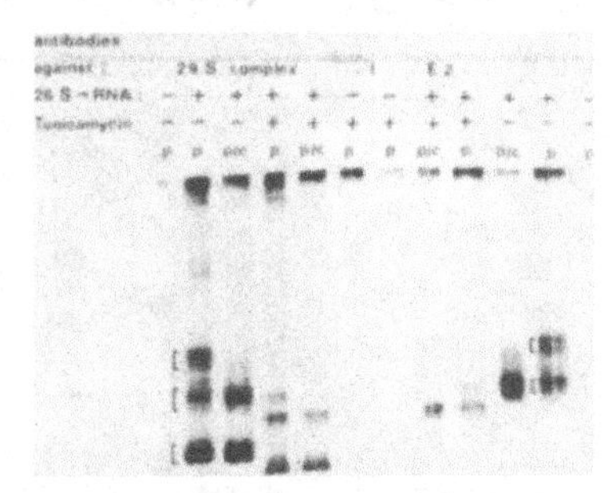

Fig.3.Pulse-chase labelling of oocytes injected with 26S-RNA (± tunicamycin).See also Fig.2.

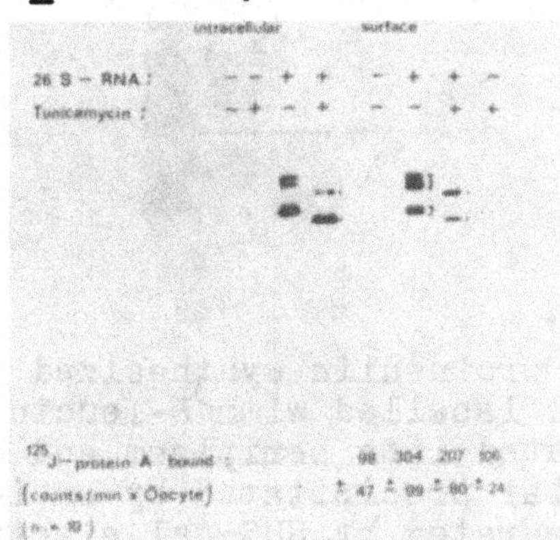

Fig.4. Unglycosylated SFV-envelope proteins are transported to the oocyte surface.See also Fig.2 and Table 1.

TABLE II
All three SFV-envelope proteins appear on the surface of 26S-RNA injected oocytes

Antiserum against	^{125}I-protein A bound to oocytes injected with	
	water (cpm/oocyte)	26S-RNA (cpm/oocyte)
29S-complex of envelope proteins	67 ± 22 (11)	455 ± 97(10)
E1-protein	137 ± 27 (6)	390 ± 108(7)
E2-protein	75 ± 11 (13)	384 ± 161(12)
E3-protein	23 ± 10 (12)	149 ± 18(7)

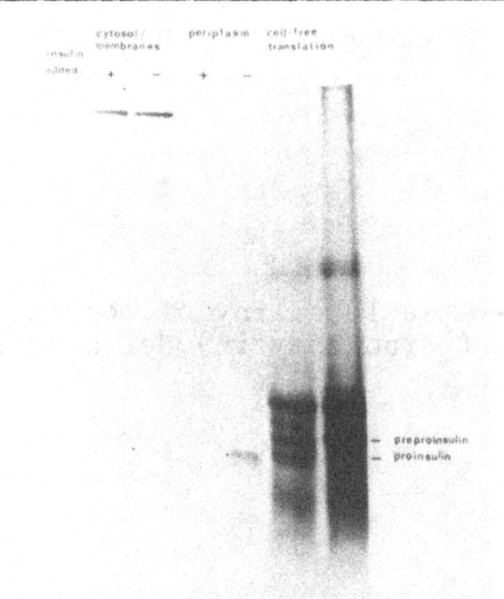

Fig.11. Carp proinsulin synthesized in E.coli. Bacteria were labelled with ^{3}H-leucine for 30 min and fractionated into periplasm and cytosol/membranes.Material precipitated by anti-carp insulin serum was separated by SDS-gel electrophoresis. The products are coded by p(24-S_{PPI}-PI)(Fig.8). Markers were synthesized in a wheat germ system programmed with carp islet RNA (± membranes).

velope proteins should be transported from the endoplasmic reticulum membrane to the plasma membrane? Such a sorting signal might be provided by carbohydrate residues attached to the polypeptide. This hypothesis, though excluded for other plasma membrane proteins (see Gibson et al,1981), could explain the fact that tunicamycin-treated SFV-infected cells accumulate unglycosylated p62-precursor and E1 in the endoplasmic reticulum membrane (Leavitt et al,1977). The unglycosylated polypeptides are apparently poorly soluble. The results shown in Figs.3 and 4 indicate that both cleavage of the p62-precursor and transport of the envelope proteins to the plasma membrane occur in oocytes treated with tunicamycin. The great shift in the mobility of all virus-coded polypeptides indicates that glycosylation was effectively inhibited. Thus, carbohydrate residues cannot be the sorting signal for the SFV-envelope proteins. The differing results obtained in infected cells and oocytes might be due to the different temperatures employed (for a precedent see Gibson et al,1979) or to different properties of the cells.

From these data one may conclude that the 26S-RNA contains the complete information for the final destination of the plasma membrane proteins and that it can be decoded in an unrelated cell. Again, as with secretory proteins, it appears that the cell does not contribute any further information.

A different observation was made when plant storage globulin mRNA was injected into Xenopus oocytes (Bassüner et al,1983). The globulin polypeptides of leguminoses are synthesized in membrane-bound polysomes of the mesophyllum, probably contain signal peptides and some are glycosylated. They are finally found in membrane-bounded vesicles in the seeds. Thus, their pathway is similar to proteins secreted to the outside. What happens if the mRNA coding for storage globulins is injected into oocytes? Are the plant proteins stored or secreted? We have shown for globulins of three different species (vicilin and legumin of vicia faba L., vicilin of pivum sativum L. and phaseolin of phaseolus vulgaris L.) that they are efficiently secreted from Xenopus oocytes (Bassüner et al,1983).Again, as a control, globin cosynthesized remained intracellular. One may conclude that first, the storage of the globulins is homologous to extracellular secretion and has step(s) in common with it, and secondly, that the cell, in addition to the mRNA, determines the final destination of these proteins in an unspecified manner.

Recognition of Signal Sequences- a Hypothesis

How are signal sequences recognized by a single receptor (Signal Recognition Particle, Walter and Blobel, 1981) given their large differences in lengths and amino acid sequence (for review see e.g. Kreil,1981)? By use of a molecular theory (Finkelstein and Ptitsyn,1977) we have recently confirmed previous results that a common secondary structure, which might be recognized by the receptor, cannot be predicted for all signal sequences (Finkelstein et al,1983).We have therefore proposed that the continuous stretch of hydrophobic amino acids present in all signal peptides, is deeply immersed into a hydrophobic pocket of the receptor. Identical backbone conformations are then induced for all sequences owing to the requirement of saturation of all H-bond-donors and acceptors.Quantitative considerations indicate that a minimum of about 6-7 contiguous hydrophobic residues is required for strong binding, in agreement with the actual findings. Any strong interaction of the signal peptide with other parts of the protein would lead to significant weakening of the receptor binding. We have therefore predicted that the first (from the N-terminus) continuous stretch of hydrophobic residues, which is free in solution and follows either an unfolded part of the polypeptide chain or a complete folding domain with a hydrophilic surface, will be recognized by the receptor and function as signal for translocation across the membrane. It should be noted that the proposed mechanism differs from other known ligand-receptor interactions for which side chains of the amino acids play an important role.In the present case the immersion is so deep- and this is only possible for entirely apolar sequences- that side chain interactions are unimportant.This explains the lack of any sequence homology among different signal peptides.

An Internal Eucaryotic Signal Sequence Functions in E.Coli

If our proposal is correct then a signal sequence may not be necessarily located at the N-terminus of an exported polypeptide. In fact, there is evidence that internal signal sequences exist for ovalbumin (Meek et al,1982), for the band III-protein of erythrocytes (Braell and Lodish,1982) and for the E1-protein of SFV and Sindbis virus (see Garoff et al,1980) to mention a few.However, no internal signal sequence has been positively identified as yet. It appears therefore of interest to try the opposite approach, to place a known signal sequence,which is normally found at the N-terminus, at an internal position.What features are required for the preceding parts of a polypeptide chain so that a signal peptide can function as internal sequence?

The process of translocation of proteins across the inner bacterial membrane appears to be similar to the transport across the eucaryotic endoplasmic reticulum membrane (see Michaelis and Beckwith,1982). Thus, it is possible to study the function of internal signal sequences in bacteria. Such investigations might also lead to the utilization of signal peptidase and signal peptide peptidases of bacterial cells in cases where a cloned gene of an eucaryotic protein is to be expressed which is normally processed by specific enzymes only present in the eucaryotic cell.

We have chosen carp preproinsulin as a model for a secretory protein. A cDNA-copy of the mRNA has been cloned (Liebscher et al,1980) and its nucleotide sequence determined (Hahn et al,1983). The sequence completely agrees with the previously established amino acid sequence of carp insulin (Makower et al,1982).

By use of in vitro recombination several

constructions were made which contained the complete coding region of carp preproinsulin including its signal sequence (Figs.5-10). For each construction the level of expression and the extent of export to the periplasm were determined by means of a radioimmunoassay for carp insulin. It should be noted that the cytoplasmic β-galactosidase and the periplasmic β-lactamase were found in their expected compartments (3-6% and 70-90% export,respectively).

In the construction p(S_{bla}-169-S_{PPI}-PI)which contains two signal sequences, an insulin-like peptide was found to be exported (Fig.5).The product in the periplasm appears to correspond in size to proinsulin (not shown). It is surprising that in a similar construction with pre-growth hormone, the protein was found membrane associated rather than secreted (Seeburg et al,1978). It is likely that in our construction there was synthesized initially a fusion product of 181 amino acid residues of β-lactamase and preproinsulin. If a single dG-residue is added in the linker region between the two genes so that they are no longer in a continuous reading frame, no expression was found (Fig.5).

By removing a HpaII-fragment from the β-lactamase gene a frame-shift was obtained which led to termination of protein synthesis in front of the preproinsulin gene (Fig.6). Synthesis of an insulin-like polypeptide was nevertheless observed regardless of the length of the linker region between the two genes. This result indicates that initiation of translation in these cases occurs at the AUG-codon of the preproinsulin gene (there are no other initiator codons in frame with preproinsulin following the stop codon in the β-lactamase gene).The lack of expression in the construction p(S_{bla}-169-disframe-PPI) may be explained by the assumption that ribosomes coming from the β-lactamase part of the mRNA interfere with the reinitiation at the preproinsulin site. The fact that the plasmids p(S_{PPI}-PI) (Fig.6) code for an exported insulin-like peptide shows that the eucaryotic signal sequence functions in E.coli. Our results agree with those of Talmadge et al (1980 a,b)for rat preproinsulin which has a signal

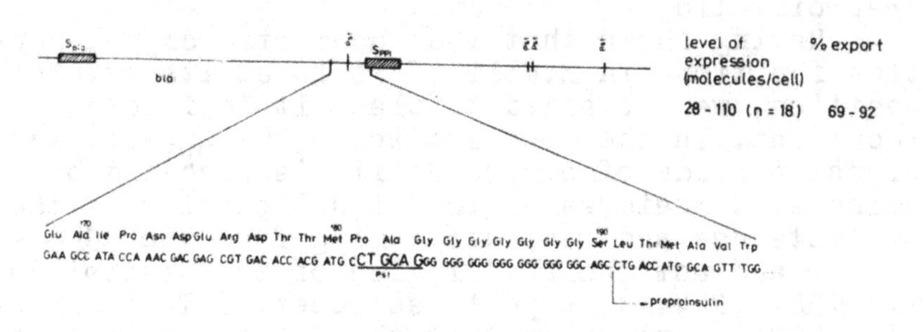

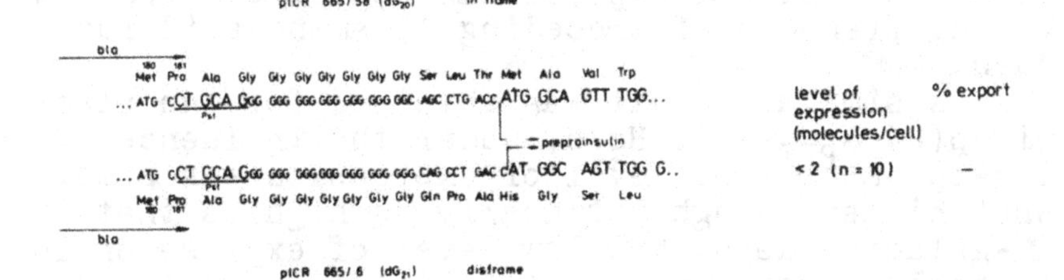

Fig.5. Construction with two signal sequences (S_{bla} and S_{PPI}). A frame-shift variant was also constructed (lower part).

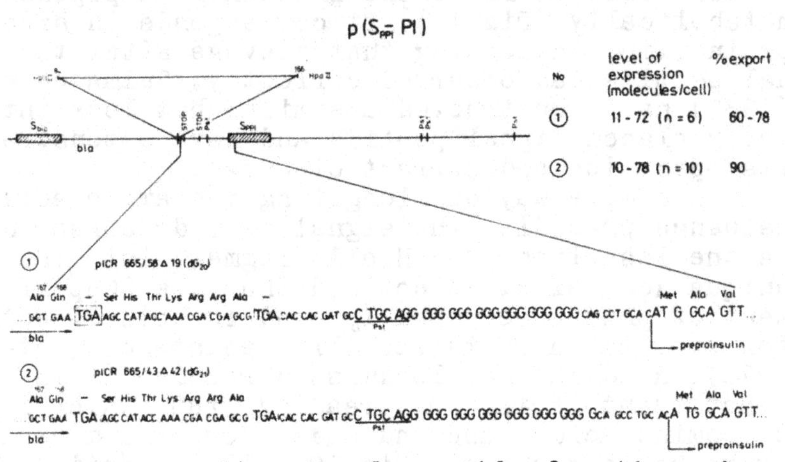

Fig.6. A eucaryotic signal peptide functions in E.coli. A frame-shift in the β-lactamase gene was made by removal of a HpaII-fragment(position 84-166). Reinitiation of translation occurred.

peptide with little similarities to that of carp preproinsulin.

Having shown that the eucaryotic signal peptide functions in E.coli if it is at its natural position, we proceeded to place it to internal locations. In the construction p(19-S_{PPI}-PI) the signal peptide of preproinsulin is preceded by 7 amino acid residues of the signal peptide of the β-lactamase and residues coming from the dG-dC-linker and untranslated region of the preproinsulin mRNA (Fig.7). Export was observed.Talmadge et al (1980a) have shown that the first 7 amino acid residues of β-lactamase are unsufficient to serve as signal peptide. Thus, in our construction a eucaryotic signal peptide was not interfered with by the presence of preceding 19 amino acid residues.

A similar result was obtained for the plasmid p(24-S_{PPI}-PI). Here, under the influence of the lac-operator, the level of expression was somewhat higher though not nearly as high as that of β-galactosidase. The low level of expression is probably caused by inefficient transcription (results not shown). The important result is,however, that again the insulin-like material was exported (Fig.8). Owing to the elevated level of synthesis we were able to lable the insulin-like peptide metabolically (Fig.11). It corresponds in size to proinsulin, indicating that clevage after the signal peptide had occurred correctly. Talmadge et al (1981) have constructed a similar but less internally placed signal peptide and have demonstrated its operation and correct cleavage.

A simple way of elongating the amino acid sequence preceding the signal peptide appeared to be the insertion of a HpaII-fragment into the unique AccI-site. We noticed that the 90bp-fragment of pBR322 is probably read in both directions for expression of tetracyclin resistance (Sutcliff, 1978). Read in the clockwise direction it is a constituent of the open reading frame which is 338 amino acids long and codes for predominantly hydrophobic residues (Fig.10). In the anti-clockwise direction it codes for a highly charged amino acid sequence which is contained in a hypothetical protein with 159 amino acid residues(Fig.9). It

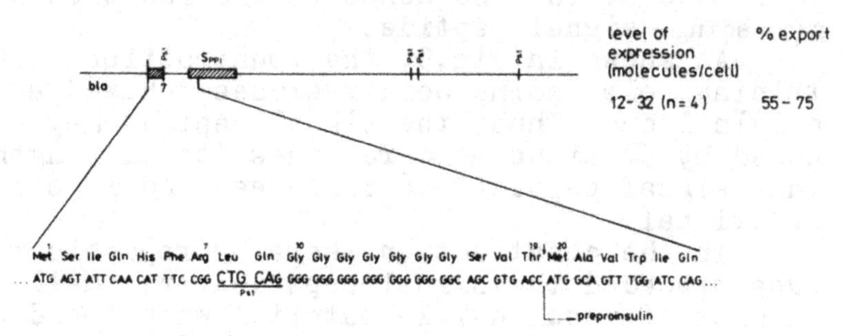

Fig.7.A signal peptide preceded by 19 amino acid residues functions.

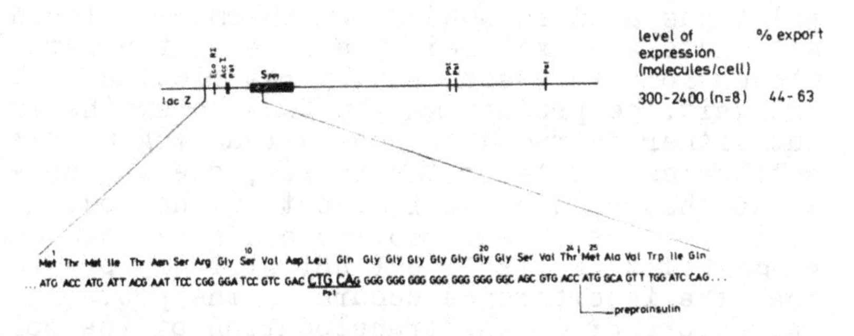

Fig.8. A signal peptide preceded by 24 amino acid residues functions.

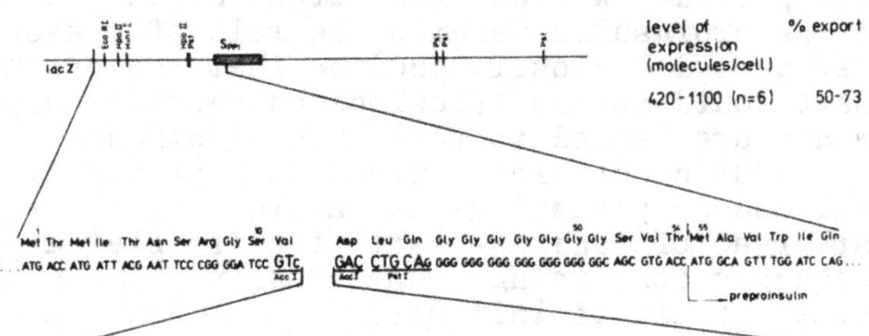

Fig.9. A signal peptide preceded by 54 amino acid residues, including many charged ones, functions.

was therefore of particular interest to test the influence of this sequence on the function of the succeeding signal peptide.

As shown in Fig.9, the construction containing polar amino acid residues behaved as the original one. Thus, the signal peptide may be preceded by 54 amino acid residues (total length of the "signal peptide" 76 residues) and is still functional.

In the construction where hydrophobic residues preceded the signal peptide very small amounts of insulin-like material were found and the export could not be ascertained. Also, in several instances more insulin-like peptides were found inside than outside the cell.

The low level of expression was unexpected since the inserted DNA is normally transcribed and translated in pBR322. Furthermore, cloning artefacts are excluded by analysis of 5 parallel clones for expression and by restriction enzyme analysis.The product may in fact be synthesized but either degraded or poorly reacting with the antibodies. Whatever the reason, one may conclude that a sequence in front of the coding region of the signal peptide has prevented efficient export .It is likely but not strictly proved that the interference occurs on the protein level before or during translocation of the polypeptide across the membrane. This assumption is supported by the fact that it is the hydrophobic variant which prevents secretion.One may speculate that the hydrophobic sequence acts as a signal peptide, whereas the actual signal sequence of preproinsulin assumes the role of a stop-transfer peptide (Blobel,1980) so that the proinsulin part would not be translocated. Further experiments are needed to test this hypothesis.

Our conclusion hinges on the presumption that all codons preceding the natural signal sequence are read off and are part of a translated peptide sequence containing preproinsulin. It would become invalid if initiation occurred in all cases at the AUG-codon of preproinsulin. Such an assumption is contradicted by the contrasting results with the clone p(S_{bla}-169-S_{PPI}-PI) and its frame shift variant (Fig.5) and also by the greatly varying expression levels of p(54-S_{PPI}-PI)a and p(54-S_{PPI}-PI)b. However, we cannot exclude the possibility that the primary translation products are first cleaved to yield a functional, N-terminally located, signal peptide. Of course, such a limited proteolysis would have to be very specific yet independent of the exact amino acid sequence of the N-terminal extension. We therefore favour the idea that the signal peptide is still operative as internal sequence.

In summary, our results support the assumption that the first hydrophobic sequence following an unfolded (in this case) part of the polypeptide chain will function as signal for translocation across the membrane.

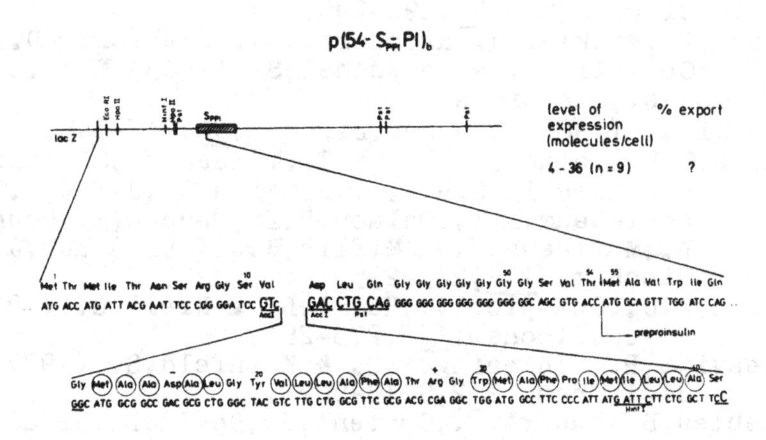

Fig.10. Low synthesis of insulin-like material by a plasmid similar to that in Fig.9 but with hydrophobic amino acid residues preceding the signal peptide.

REFERENCES

Bassüner,R.,Huth,A.,Manteuffel,R. & Rapoport,T.A. (1983) Eur.J.Biochem.,in press
Blobel,G. (1980) Proc.Nat.Acad.Sci.77,1496-1500.
Blobel,G. & Dobberstein,B.(1975) J.Cell Biol. 67, 852-862.
Braell,W.A. & Lodish,H.F. (1982) Cell 28,23-31.
Colman,A. & Morser,J.(1979) Cell 17,517-526.
Finkelstein,A. & Ptitsyn,O.B. (1977)Biopolymers 16,525-529.
Finkelstein,A.,Bendzko,P. & Rapoport,T.A.(1983) FEBS Lett., submitted
Garoff,H.,Frischauf,A.-M.,Simons,K.,Lehrach,H. & Delius,H. (1980) Nature (Lond.)288,236-241.
Gibson,R.,Schlesinger,S. & Kornfeld,S.(1979) J. BiolChem. 254,3600-3607.
Gibson,R.,Kornfeld,S. & Schlesinger,S.(1981)Trends Biochem.Sci.5 ,290-293.
Hahn,V.,Winkler,J.,Rapoport,T.A.,Liebscher,D.H., Coutelle,C. & Rosenthal,S. (1983) Nucl.Acids Res., in press.
Kreil,G. (1981) Ann.Rev.Biochem.50,317-348.
Lane,C.D.,Colman,A.,Mohun,T.,Morser,J.,Champion,J., Kourides,I.,Craig,R.,Higgins,S.,James,T.C., Applebaum,S.W.,Ohlson,R.I.,Paucha,E.,Houghton, M.,Matthews,J. & Miflin,B.J.(1980) Eur.J. Biochem. 111,225-235.
Lane,C.D.,Champion,J.,Haiml,L. & Kreil,G.(1981) Eur.J.Biochem.113,273-281.
Leavitt,R.,Schlesinger,S. & Kornfeld,S. (1977) J. Biol.Chem. 252,9018-9023.
Lebleu,B.,Hubert,E.,Content,J.,DeWit,L.,Braude,I.A. & LeClerq,E. (1978)Biochem.Biophys.Res.Commun. 82,665-673.
Liebscher,D.H.,Coutelle,C.,Rapoport,T.A.,Hahn,V. Rosenthal,S.,Prehn,S. & Williamson,R. (1980) Gene 9,233-246.
Makower,A.,Dettmer,R.,Rapoport,T.A.,Knospe,S., Behlke,J.,Prehn,S.,Franke,P.,Etzold,G. & Rosenthal,S.(1982) Eur.J.Biochem.122,339-345.
Marbaix,G. & Huez,G. (1980)in: Transfer of Cell Constituents into Eucaryotic Cells(Celis,ed) pp.347-381,Plenum Press,New York,London
Meek,R.L.,Walsh,A. & Palmiter,R.D.(1982)J.Biol. Chem. 257,12245-12251.

Michaelis,S. & Beckwith,J.(1982) Ann.Rev.Microbiol. 36,435-465.
Rapoport,T.A.(1981) Eur.J.Biochem.115,665-669.
Rapoport,T.A.,Thiele,B.,Prehn,S.,Marbaix,G., Cleuter,Y.,Hubert,E. & Huez,G. (1978) Eur.J. Biochem. 87,229-233.
Seeburg,P.H.,Shine,J.,Martial,J.A.,Ivarie,R.D., Morris,J.A.,Ullrich,A.,Baxter,J.B. & Goodman, H.M.(1978) Nature (Lond.) 276,795-798.
Sutcliff,J.(1978) Cold Spring Harb.Symp.quant.Biol. 49,77-90.
Talmadge,K.,Stahl,S. & Gilbert,W. (1980a) Proc. Nat.Acad.Sci. 77,3369-3373.
Talmadge,K.,Kaufman,J. & Gilbert,W. (1980b) Proc. Nat.Acad.Sci. 77,3988-3992.
Talmadge,K.,Brosius,J. & Gilbert,W. (1981)Nature (Lond.) 294,176-178.
Walter,P. & Blobel,G. (1981) J.Cell Biol.91,557-561.

BIOSYNTHESIS OF PEPTIDES IN HONEYBEE VENOM GLANDS AND FROG SKIN: STRUCTURE AND MULTI-STEP ACTIVATION OF PRECURSORS

G. Kreil, W. Hoffmann, A. Hutticher, I. Malec, C. Mollay, K. Richter, U. Vilas and R. Vlasak

Institute of Molecular Biology
Austrian Academy of Sciences
Salzburg, Austria

The biosynthesis of peptides invariably proceeds via pre-pro-forms, which are cleaved and modified by a series of enzymatic reactions to ultimately yield the final product. The concept of limited proteolysis, originally developed for the conversion of zymogens to the active enzymes, has amply been documented in studies on the liberation of peptides from their respective precursors. One type of posttranslational reaction, the cleavage at pairs of lysine/arginine residues, as was first detected for proinsulin (1), has frequently been observed in precursor-product conversions. In the past years, the occurrence of other processing reactions has been documented which show apparently a more restricted distribution. In our laboratory, the biosynthesis of peptides in honeybee venom glands and, more recently, in amphibian skin has been investigated and the types of translational and posttranslational reactions involved in the liberation of these peptides from larger precursors are described in this communication.

I) FROM PREPROMELITTIN TO THE FINAL PRODUCT

The main constituent of honeybee venom is melittin, a lytic peptide comprised of 26 amino acids (2). The primary translation product of melittin mRNA isolated from venom glands of young queen bees is prepromelittin, which contains 70 residues (3). Starting from the amino end, four different parts can be discerned in the structure of this molecule: a) a pre- or signal-peptide of 21 mostly hydrophobic residues; b) an acidic pro-region of 22 amino acids with either proline or alanine at evennumbered positions; c) the 26 amino acids of melittin; and d) an extra glycine residue at the carboxyl end (see Fig. 1). The sequence analysis of cloned melittin mRNA (4) has corroborated the amino acid sequence determined previously and it has, moreover, shown that the 70 amino acids of prepromelittin represent the total coding capacity of the mRNA. This is the smallest pre-secretory precursor yet found in nature and its small size actually raises several questions about the detailed mechanism of vectorial discharge of the nascent chain into the lumen of the endoplasmic reticulum. It has been estimated that the complex between the ribosome and signal recognition particle (5) covers 60-80 amino acids. By the time docking protein releases the translational arrest of this particle (6), the synthesis of prepromelittin would be virtually completed and a type of post-translational transfer to the lumen would have to be envisaged.

The liberation of melittin from its precursor is a multistep process which involves three types of enzymatic reactions. These are the endoproteolytic cleavage of the signal-peptide, the formation of the carboxy-terminal amide with concomitant loss of the terminal glycine and lastly the stepwise activation of promelittin by an exoprotease.

a) Cleavage of Prepromelittin by Signal Peptidase from Rat Liver and E. coli:

Cleavage of signal peptides in animal cells is

NH_2.Met-Lys-Phe-Leu-Val-Asn-Val-Ala-Leu-Val-
-Phe-Met-Val-Val-Tyr-Ile-Ser-Tyr-Ile-Tyr-Ala-
-ALA-Pro-Glu-Pro-Glu-Pro-Ala-Pro-Glu-Pro-Glu-
-Ala-Glu-Ala-Asp-Ala-Glu-Ala-Asp-Pro-Glu-Ala-
-GLY-Ile-Gly-Ala-Val-Leu-Lys-Val-Leu-Thr-Thr-
-Gly-Leu-Pro-Ala-Leu-Ile-Ser-Trp-Ile-Lys-Arg-
-Lys-Arg-Gln-Gln-Gly.COOH

Fig. 1. Amino acid sequence of honeybee prepromelittin. The first residue of the pro-peptide and of melittin are underlined.

an early reaction which occurs prior to the completion of polypeptide chains (7,8). The enzyme which catalyzes this reaction has been termed signal peptidase, a membrane-bound protease which is not species specific. This is exemplified by the fact that injection of melittin mRNA into frog oocytes yields promelittin with the same amino-terminal sequence as the precursor from honeybee venom glands (9). Several laboratories have described the post-translational processing of presecretory polypeptides in the presence of microsomes and detergent (10-16), which could not be inhibited by a variety of protease-inhibitors. Signal peptidase as present in isolated microsomes is a cryptic membrane enzyme and detergents or other membrane-perturbing agents are required to unmask its activity. With rat liver rough microsomes we have previously shown that in the presence of e.g. 0.2 % deoxycholate a partial conversion of prepromelittin to promelittin with the correct amino end can be observed. A partial purification of signal peptidase from this starting material has been achieved (17) and only three major bands, as judged by SDS polyacrylamide gel electrophoresis, are present in these preparations. This purification involves a reconstitution of dispersed membrane proteins into phospholipid vesicles. Even after this procedure, signal peptidase is still a cryptic enzyme, whose activity apparently depends on the protein/phospholipid/

detergent ratio. Addition of melittin to the vesicles yields a more reproducible activation of the peptidase.

After incubation of labeled prepromelittin with the partially purified signal peptidase, samples are usually fractionated by extraction with n-butanol. Under these conditions, prepromelittin accumulates at the interface, while promelittin liberated during the reaction can be purified from the aqueous layer. From the organic layer we have succeeded in isolating the intact signal peptide of prepromelittin (17). This clearly shows that rat liver microsomes contain a membrane-bound endoprotease which cleaves prepromelittin at a single peptide bond - the prepro-junction. More recently, similar experiments have been performed with signal peptidase from E. coli obtained from W. Wickner. This enzyme also cleaves prepromelittin correctly and signal peptide and promelittin have been isolated and characterized as the cleavage products (18). The mammalian and bacterial signal peptidase may also be immunologically related as an antibody against the E. coli enzyme precipitates the protease activity from our partially purified rat liver enzyme (19). This antibody interacts with two proteins with molecular weights of about 35.000 and 58.000. It is currently not known which of these proteins is the rat liver signal peptidase; alternatively, the smaller polypeptide may be a fragment of the larger one.

b) Formation of the carboxy-terminal amide:

From the sequence of prepromelittin it seems likely that the extra glycine plays a role in the formation of the carboxy-terminal amide found in melittin (2). These blocked termini are found in a wide variety of physiologically active peptides of diverse origin. In recent years, the precursors of some of these peptides have been elucidated and, without exception, a glycine residue has been found adjacent to the amino acid which becomes amidated in the final product. Examples are the precursors for mammalian calcitonin (20), gastrin

(21), vasopressin (22) and oxytocin (23), caerulein from frog skin (24), corticotropin releasing hormone (25) etc. The mechanism of the amidation reaction is currently not known. It is probably a redox-reaction as the nitrogen of the terminal glycine has been shown to remain in the amidated peptide (26).

c) Stepwise activation of promelittin:

The pro-part of promelittin has a very unusual sequence in that every other residue is either proline or alanine. In extracts from venom sacs of honeybees, an enzyme is present which catalyzes the stepwise cleavage of promelittin (27,28). This enzyme has been shown to be a dipeptidylaminopeptidase, a member of a group of enzymes widely distributed in nature (29-31). Types II and IV of this group show a high activity with substrates having proline or alanine as the second amino acid. The former is a soluble, lysosomal enzyme with a pH optimum around pH 5.5, while the latter group is represented by membrane-bound enzymes found in high amounts in mammalian kidney and intestinal brush borders (30,31). The enzyme from honeybee venom sacs resembles the type II in several respects.

To study the conversion of promelittin to melittin, we have mostly used a fragment prepared from prepromelittin, labeled in vitro with radioactive proline, by digestion with chymotrypsin and a subsequent step of Edman degradation. This yields an acidic peptide which contains the entire pro-region and the amino-terminal hexapeptide of melittin (residues 22-49 of prepromelittin, see Fig. 1). The evidence that promelittin is activated by stepwise cleavage of dipeptides can be summarized as follows: i) venom sac extracts completely hydrolyze the pro-part but cleavage stops at the amino end of melittin; ii) in the presence of mersalyl, which inhibits a dipeptidase also present in these extracts, dipeptides are the only cleavage products which can be detected; iii) the time-course is in agreement with a stepwise hydrolysis in that the first dipeptide Ala-Pro is released earlier than the tenth dipeptide Asp-Pro; iv) inhibition of the dipeptidylpeptidase by diisopropyl-fluorophosphate also abolishes the cleavage of the pro-part; v) fragments containing an extra amino acid at the amino end which, consequently, have the wrong "reading frame" for the dipeptidylamino-peptidase, are not hydrolyzed by these extracts (27,28).

These results demonstrate that the activation of promelittin proceeds by a new mechanism which involves the stepwise cleavage of dipeptides. Such an exoproteolytic activation mechanism solely governed by the nature of the second amino acid can proceed with exquisite specificity even in the presence of many other polypeptides. It has, moreover, an inherent time scale and may thus offer additional safeguards against the premature release of the final product. This is of obvious importance in case of melittin, a lytic peptide which strongly interacts with membranes and, therefore, must be liberated late in the venom sac which has a chitin wall.

Dipeptidylaminopeptidases have been discovered more than ten years ago but it has largely been a group of enzymes in search of a biological role. Our findings assign an important function to one of these enzymes. Recently, it has been shown that stepwise processing is also taking place during the liberation of yeast alpha-mating factor from its precursor. The last step in this activation is the cleavage of three dipeptides all terminating with alanine (32,33). Judging from the amino acid sequence of the pro-region, a similar mechanism could be involved in the liberation of an antifreeze protein of an arctic fish from its precursor (34). In our studies on the biosynthesis of caerulein in frog skin we have probably found another case of stepwise processing (see later).

II) Honeybee Prepro-Secapin

Among the cDNA clones obtained from venom gland mRNA of queen bees, several were found which do not hybridize with the nick translated insert of a prepromelittin clone. Two of these were found to contain an insert coding for the precursor of secapin. This peptide was discovered some years ago as a minor constituent of worker bee venom (35) and its sequence has been determined by two groups (36,37). The amounts available have been insufficient to determine the physiological properties of secapin; it is remarkably non-toxic and its role in bee venom is not known (35).

Figure 2 shows the nucleotide sequence of part of the insert of clone Cla1. A second clone, pBM9, has an identical sequence except that it contains about 100 additional bases in the 3'-untranslated region. Both clones have only one open reading frame which codes for a polypeptide of 77 amino acids, the secapin sequence being at the carboxyl end. The predicted polypeptide starts with a sequence of 10 mostly hydrophilic residues, followed by a stretch rich in amino acids with hydrophobic side chains characteristic for signal peptides. The pro-region, presumably starting at serine-28 or alanine-32, is very hydrophilic and terminates with a single arginine residue. The characteristic structure in the sequence of the pro-part of promelittin, where even numbered residues are either proline or alanine, is not found in prosecapin. Consequently, the activation of this precursor must proceed by a different mechanism ultimately involving cleavage at a single arginine residue. It is surprising that in the same gland, two different activation mechanisms are present, one proceeding via exoproteolytic cleavage of dipeptides and the other involving hydrolysis after a single arginine residue which may be particularly exposed. It is noteworthy that bee venom also contains small amounts of melittin-F (36), a fragment of melittin generated by hydrolysis after lysine-7. The enzyme which activates prosecapin may also cleave melittin at a slow rate. This enzyme must be very different from trypsin, which cleaves melittin much more readily near the carboxyl end in the Lys-Arg-Lys-Arg-sequence. It is also likely that

..AAGAATTATGAAGAACTATTCAAAAAATGCAACACACTTAATTACG
MetLysAsnTyrSerLysAsnAlaThrHisLeuIleThr

GTTCTTCTATTCAGCTTTGTTGTTATACTTTTAATTATTCCATCAAAA
ValLeuLeuPheSerPheValValIleLeuLeuIleIleProSerLys

TGTGAAGCCGTTAGCAATGATAGGCAACCATTGGAAGCACGATCTGCT
CysGluAlaValSerAsnAspArgGlnProLeuGluAlaArgSerAla

GATTTAGTCCCGGAACCAAGATACATTATTGATGTTCCTCCTAGATGT
AspLeuValProGluProArgTyrIleIleAspValProProArgCys

CCTCCAGGTTCTAAATTCATTAAGAACAGATGTAGAGTCATAGTGCCT
ProProGlySerLysPheIleLysAsnArgCysArgValIleValPro

TAAATTCGTATGAACT(1oo bases)poly(A)
end

FIGURE 2. Nucleotide sequence of the insert of clone Cla1 and amino acids sequence of the open reading frame. The insert contains 58 additional bases at the 5'-end which are not shown. The secapin sequence is underlined.

the enzyme activating prosecapin is not related to the one involved in typical prohormone-hormone conversions at pairs or multiples of lysine/arginine residues (1). Processing at single arginine residues have been reported for several other precursors of physiologically active peptides (38-40).

III) PRECURSORS OF FROG SKIN PEPTIDES AND THEIR PROCESSING

From amphibian skin, a variety of peptides have been isolated which were found to be similar or identical to mammalian hormones or neuropeptides (41). The biological role of these peptides and other constituents of frog skin secretions, which have played a prominent role in the folk medicine of many areas, is still obscure. It is noteworthy, however, that the amounts of different peptides in skin is often far greater than those found in any mammalian tissue. We thus consider the biosynthesis of frog skin peptides a promising model system for the formation of

```
 ACCTCAACAACGAGAAGCCAATGACGAACGTCGCTTTGCTGATGGA
...ProGlnGlnArgGluAlaAsnAspGluArgArgPheAlaAspGly

CAACAAGACTACACAGGTTGGATGGATTTTGGCCGCCGTGATGATGAA
GlnGlnAspTyrThrGlyTrpMetAspPheGlyArgArgAspAspGlu

GATGATGTAAACGAACGAGATGTCCGAGGATTTGGCTCTTTCCTAGGT
AspAspValAsnGluArgAspValArgGlyPheGlySerPheLeuGly

AAAGCTTTAAAGGCTGCTTTAAAAATTGGTGCAAATGCGCTGGGAGGA
LysAlaLeuLysAlaAlaLeuLysIleGlyAlaAsnAlaLeuGlyGly

TCACCTCAACAACGAGAAGCCAATGACGAACGTCGCTTTGCTGATGGA
SerProGlnGlnArgGluAlaAsnAspGluArgArgPheAlaAspGly

CAACAAGACTACACAGGTTGGATGGATTTTGGCCGCCGCAATGGTGAA
GlnGlnAspTyrThrGlyTrpMetAspPheGlyArgArgAsnGlyGlu

GATGATTAATATTCTTCTTGAAAAC(60 bases)poly(A)
AspAsp///
```

FIGURE 3. Nucleotide sequence of the insert of clone pUF37 and amino acid sequence of the open reading frame. Clone pUF48 starts at the A marked (•) and differs by only two point mutations near the 3'-end. Caerulein sequences are underlined, pairs of arginine residues are emphasized. The direct repeat TGATG is marked above the sequence.

homologous peptides in the mammalian gastrointestinal tract and nervous system.

In the past years we have mainly concentrated on the biosynthesis of caerulein, a decapeptide first isolated from the skin secretion of the Australian frog *Hyla caerulea* (42). Caerulein has the same carboxy-terminal sequence as cholecystokinin and gastrin and it may actually be the common ancestor of these hormones (43). This peptide has subsequently been detected in several other frog species, including *Xenopus laevis* (44). Using total mRNA prepared from the skin of *X. laevis*, a cDNA library has been constructed with the inserts being ligated into the PstI site of plasmid pUC8 (45). This library was screened with radioactive cDNA obtained from skin mRNA primed with d(AGTCCATCCA), an oligodeoxynucleotide complementary to the region of the mRNA coding for the Trp-Met-Asp-Phe - sequence of the caerulein precursor. Close to ten percent of the clones were found to hybridize with this radioactive cDNA and many of these do indeed contain inserts derived from mRNAs of caerulein precursors. The sequence of one of these clones, pU37, is shown in Fig. 3. The insert of this clone codes for a polypeptide of 97 amino acids, where the caerulein sequence occurs twice (24). This represents the carboxy-terminal part of a precursor with an internal duplication as the sequence of the first and the last 33 amino acids is almost identical. Variants of the sequence given in Fig. 3 have been found in other clones, which have either a deletion (24) or an insertion of 45 nucleotides. These variants could have arisen through unequal crossing over at the direct repeat TGATG which flanks the deleted and inserted segments (see Fig. 3). This yields precursors containing one or three copies of the caerulein sequence, respectively.

From the sequences flanking caerulein in these precursors, it is evident that the liberation of the final product must involve several post-translational processing steps. Excision of the pairs of arginine residues, the typical yet still poorly understood prohormone-hormone conversion, would yield an intermediate form with four additional residues at the amino end and an extra glycine at the carboxyl end of the caerulein sequence. As discussed earlier, the glycine is required for the formation of the terminal amide. The further maturation the resulting amidated peptide starting with the extension Phe-Ala-Asp-Gly apparently proceeds by the same mechanism as originally detected for promelittin. The skin secretion of Xenopus laevis contains large amounts of a dipeptidylaminopeptidase which readily cleaves Ala-Pro- and Ala-Ala-pNitroanilide. To test whether it would also cleave after glycine residues, as required by the nature of the peptide extension, the tetrapeptide Ala-Gly-Ala-Ala and the reference peptides Ala-Ala-Ala-Ala and Ala-Pro-Ala-Ala, all labeled in the terminal alanine, were synthesized. At pH 5.5, the relative rates of hydrolysis of these tetrapeptides by frog skin dipeptidylaminopeptidase was 15:23:100, respectively. It thus appears quite likely that stepwise cleavage of dipeptides also plays a role in the liberation of caerulein from its precursors.

The caerulein generated in this sequence of processing reactions would still contain an amino-terminal glutamine, which presumably cyclizes spontaneously to form the pyroglutamic acid found in the final product. The tyrosin-O-sulphate present in caerulein represents a post-translational modification also found in gastrin and cholecystokinin. This modification probably occurs early, in the Golgi region (46) and it may be present in quite a few other polypeptides (47). No details about the biosynthesis of tyrosine-O-sulphate have been published as yet.

Several clones hybridizing with the cDNA primed with synthetic oligonucleotide derived from the caerulein sequence were found to contain an insert for an unknown peptide (48). The largest insert found in clone pUF38 could code for a polypeptide of 64 amino acids which starts precisely with a codon for methionine. The first twenty amino acids are largely hydrophobic ones as is typical for signal peptides. It appears thus likely that this insert actually codes for a complete pre-pro-peptide, which would be the smallest of its kind. In this predicted precursor, the carboxy-terminal sequence is ...Lys-Arg-Tyr-Val-Arg-Met-Ala-Ser-Lys-Ala-Gly-Ala-Ile-Ala-Gly-Lys-Ile-Ala-Lys-Val-Ala-Leu-Lys-Ala-Leu-Gly-Arg-Arg-Asp-SerCOOH (48). Processing at the pairs of basic residues and amidation with concomitant loss of the terminal glycine would then yield a highly basic peptide comprised of 24 amino acids. Following a suggestion of Tatemoto and Mutt (49), this postulated peptide has been termed PYL^{a}. When arranged into an alpha helix, PYL^{a} can form a perfect amphipathic structure. Several peptides with high affinity for cell membranes have been shown to form such helices. Examples are melittin and a melittin-like synthetic peptide (50), delta hemolysin (51) and cecropins (52). We are currently trying to isolate this peptide from skin secretions of *X. laevis.*

These studies on the structure and activation of precursors of peptides synthesized in honeybee venom glands and frog skin have revealed unusually complex conversion mechanisms involving several discrete steps. However, the similarities encountered in the two diverse model systems are encouraging in that it could be taken as an indication that the majority of processing reactions have been discovered by now.

ACKNOWLEDGEMENTS

This work has been supported by grants no. 3810, 4174 and 4907 from the Austrian "Fonds zur Förderung der wissenschaftlichen Forschung", by grant no. 1725 from the "Jubiläumsfonds" of the Austrian National Bank and by a grant from the "Politzer Stiftung" of the Austrian Academy of Sciences.

REFERENCES

1. Docherty, K., and Steiner, D.F., Ann. Rev. Physiol. 44, 625-638 (1982).
2. Habermann, E., Science 177, 314-318 (1972).
3. Suchanek, G., Kreil, G., Hermodson, M.A., Proc.Natl. Acad.Sci. USA 75, 701-704 (1978).
4. Vlasak, R., Unger-Ullmann, C., Kreil, G., and Frischauf, A.M., Eur. J. Biochem. submitted (1983).
5. Walter, P., Ibrahim, I., and Blobel, G., J. Cell Biol. 91, 545-550 (1981).
6. Meyer, D.I., Krause, E., and Dobberstein, B., Nature 297, 647-650 (1982).
7. Blobel, G., Dobberstein, B., J. Cell Biol. 67, 835-851 (1975).
8. Kreil, G., Ann. Review Biochem. 50, 317-348 (1981).

9. Lane,C.D., Champion, J., Haiml, L., Kreil, G., Eur. J. Biochem. 113, 273-281 (1981).

10. Kreil, G., Suchanek, G., Kaschnitz, R., Kindas-Mügge, I., FEBS Symp. 47, 79-88 (1978).

11. Jackson, R.C., Blobel, G., Proc. Natl. Acad. Sci. USA 74, 5598-5602 (1977).

12. Kaschnitz, R., Kreil, G., Biochem. Biophys. Res. Commun. 83, 901-907 (1978).

13. Strauss, A.N., Zimmermann, M., Boime, I., Ashe, B., Mumford, R.A., Alberts, A.W., Proc. Natl. Acad. Sci. USA 76, 4225-4229 (1979).

14. Thibodeau, S.N., Walsh, K.A., Ann. N.Y. Acad. Sci. 343, 180-191 (1980).

15. Jackson, R.C., Blobel, G., Ann. N.Y. Acad. Sci. 343, 391-404 (1980).

16. Jackson, R.C., White, W.R., J. Biol. Chem. 256, 2545-2550 (1981).

17. Mollay, C., Vilas, U., Kreil, G., Proc. Natl Acad. Sci. USA 79, 2260-2263 (1982).

18. Mollay, C. and Vilas, U., unpublished experiments (1982)

19. Mollay, C., Vilas, U. and Wickner, W., unpublished experiments (1983).

20. Amara, S.G., David, D.N., Rosenfeld, M.G., Ross, B.A., and Evans, R.M., Proc. Natl. Acad. Sci. USA 77, 4444-4448 (1980).

21. Yoo, O.J., Powell, C.T., and Agarwal, K.L., Proc. Natl. Acad. Sci. USA 79, 1049-1053 (1982)

22. Land, H., Schütz, G., Schmale, H., and Richter, D., Nature 295, 299-303 (1982).

23. Land, H., Grez, M., Ruppert, S., Schmale, H., Rehbein, M., Richter, D., and Schütz, G., Nature 302, 342-344 (1983).

24. Hoffmann, W., Bach, T.C., Seliger, H. and Kreil, G., EMBO J. 2, 111-114 (1983).

25. Furutani, Y., Morimoto, Y., Shibahara, S., Noda, M., Takahashi, T., Hirose, T., Asai, M., Inayama, S., Hayashida, H., Miyata, T., and Numa, S., Nature 301, 537-540 (1983).

26. Bradbury, A.F., Finnie, M.D.A., and Smyth, D.G., Nature 298, 686-688 (1982).

27. Kreil, G., Mollay, Ch., Kaschnitz, R., Haiml, L., Vilas, U., Ann. N.Y. Acad. Sci. 343, 338-346 (1980).

28. Kreil, G., Haiml, L., Suchanek, G., Eur. J. Biochem. 111, 49-58 (1980).

29. Hopsu-Havu, V.K., Ekfors, T.O., Histochemie 17, 30-38 (1969).

30. David, R., MacNair, C., Kenny, A.J., Biochem. J. 179, 379-395 (1979).

31. McDonald, J.K., Schwabe, C. In, "Proteinases in Mammalian Cells and Tissues" (Barrett, Ed.) Elsevier/North Holland Biomedical Press, 1977, p. 311-391.

32. Kurjan, J., and Herskowitz, I., Cell 30, 933-943 (1982).

33. Julius, D., Blair, L., Brake, A., Sprague, G., and Thorner, J., Cell 32, 839-852 (1983).

34. Davies, P.L., Roach, A.H., Hew, C.L.: Proc. Natl Acad. Sci. USA 79, 335-339 (1982).

35. Gauldie, J., Hanson, J.M., Rumjanek, F.D., Shipolini,R.A., and Vernon, C.A., Eur. J. Biochem. 61, 369-376 (1976).

36. Gauldie, J., Hanson, J.M., Shipolini, R.A., and Vernon, C.A., Eur. J. Biochem. 83, 405-410 (1978).

37. Kudelin, A.B., Martynov, V.I., Kudelina, I.A., and Miroshnikov, A.I., Abstr. 15th Europ. Peptide Symp., Gdansk (1978) p. 84.

38. Schwartz, T.W., and Tager, H.S., Nature 294, 589-591 (1981).

39. Scheller, N.G., Jackson, J.F., McAllister L.B., Rothman, B.S., Mayeri, E., and Axel, R., Cell

32, 7-22 (1983).

40. Land, H., Schütz, G., Schmale, H., and Richter, D., Nature 295, 299-303 (1982).

41. Erspamer, V., and Melchiorri, P., Trends in Pharmacol. Sci. 1980, p. 391-395.

42. Anastasi, A., Erspamer, V., and Endean, R., Arch. Biochem. Biophys. 125, 57-68 (1968).

43. Rehfeld, J.F., Americ. J. Physiol. 240, G255-G266 (1981).

44. Anastasi, A., Bertaccini, G., Cei, J.M., De Caro, G., Erspamer, V., Impicciatore, M. and Roseghini, M., Brit. J. Pharmacol. 38, 221-228 (1970).

45. Vieira, J., and Messing, J., Gene 19, 259-268 (1982).

46. Young, R.W., J. Cell Biol. 57, 175-189 (1973).

47. Huttner, W.B., Nature 299, 273-275 (1982).

48. Hoffmann, W., Richter, K., and Kreil, G., EMBO J. 2, 711-714 (1983).

49. Tatemoto, K., and Mutt, V., Nature 285, 417-418 (1980).

50. DeGrado, W.F., Kézdy, F.J., and Kaiser, E.T. J.Amer.Chem.Soc. 103, 679-681 (1981).

51. Freer, J.H., and Birkbeck, T.H., J. Theor. Biol. 94, 535-540 (1982).

52. Merrifield, R.B., Vitioli, L.D., and Boman, H.G., Biochemistry 21, 5020-5031 (1982).

ON THE ATTACHMENT OF RIBOSOMES TO HEAVY ROUGH AND LIGHT ROUGH ENDOPLASMIC RETICULUM IN EUKARYOTIC CELLS.

Ian F. Pryme

Department of Biochemistry,
University of Bergen,
Årstadveien 19, N-5000 Bergen,
Norway.

A 48 hr period of starvation resulted in an almost complete conversion of rough endoplasmic reticulum (ER) membranes to the smooth (S) type. The order of re-appearance of the light rough (LR) and heavy rough (HR) ER fractions after feeding a starved culture was dependent on the nutritional conditions. Inhibition of protein synthesis by cycloheximide resulted in a conversion of HR to S and/or LR membranes. The HR fraction re-appeared within $\frac{1}{2}$ hr after removal of the block. RNase treatment of isolated HR membranes resulted in a release of ribosomes sufficient to lower their density causing them to migrate to the position of LR membranes upon re-centrifugation. The migration of LR membranes was unaffected by this treatment. The migratory behaviour of HR membranes was also affected by raising the salt concentration of the gradient buffer from 25mM to 100mM KCl: the membranes appearing in the LR region of the gradient. The results add further support to the suggestion that the HR and LR fractions represent distinct compartments of the rough ER. In addition differences in the mechanisms of association of ribosomes with HR and LR membranes are indicated.

1. INTRODUCTION

The ER membranes of MPC-11 cells, L-cells and Krebs II ascites cells have been separated into HR, LR and S subfractions by discontinuous sucrose gradient centrifugation (Svardal and Pryme, 1978, Svardal et al 1981, Pryme et al, 1981, Fjose and Pryme, 1983). Appreciable amounts of both LR and S membranes were observed in normal tissues such as liver, pancreas, kidney and bone marrow (Pryme et al 1981, Pryme et al, 1982) but the HR fraction, however, was absent.

The HR and LR subfractions of MPC-11 cells differ with respect to RNA/protein and RNA/phospholipid ratios, polysome profiles and marker enzymes though no differences were observed in the electron microscope (Svardal et al, 1981). Based on the finding that four times as many nascent light chain immunoglobulin polypeptide chains were observed in association with the polysomes of HR membranes compared to LR membranes, a functional compartmentalization of the rough ER has been suggested (Svardal and Pryme, 1980a). The respective amounts of the HR and LR subfractions vary according to the phase of cell cycle, the major amount of HR membranes being found in cells late in G1/early S phase (Svardal and Pryme, 1980b) coinciding with the maximal *in vivo* synthesis of the protein in synchronized cells (Garatun-Tjeldstö et al 1976).

We have recently shown that Krebs II ascites cells contain only negligible amounts of HR membranes after $\frac{1}{2}$ hr *in vitro* incubation following removal from the mouse peritoneal cavity, LR membranes being the dominant fraction (Fjose and Pryme, 1983). Large amounts of HR membranes, however, were observed after 12 hr of incubation. It was thus obvious that the step-up conditions represented by the exposure of 'starved' cells to rich culture medium resulted in profound differences in the relative amounts of rough ER subfractions.

In light of the results obtained with Krebs II ascites cells a series of experiments was designed in order to examine the effects of nutritional conditions on the yields of ER membrane subfractions isolated from transformed cells adapted to grow in suspension culture. In addition the effects of various parameters on ER membrane gradient profiles were studied with a view to obtaining an insight into the form of attachment between ribosomes/polysomes and HR and LR membranes.

2. RESULTS

Cultures of L-cells can be maintained under linear growth conditions in roller suspension culture by replacing half the culture volume every 24 hr with fresh medium. Omission of feeding cell cultures for 48 hr results in an arrest of cell growth reflected by an incomplete doubling of the cell population. Under these conditions L-cells, as opposed to MPC-11 cells, show no decrease in viability.

In L-cells grown under linear conditions the distrib-

being the most marked. The total recovery of radioactivity in gradients of 24 hr and 48 hr membranes was identical (46,700 cpm and 46,000 cpm respectively) while the period of $2\frac{1}{2}$ hr incubation resulted in an almost two-fold increase (87,900 cpm). The reason for this is not known.

A further experiment was performed to explore the effects of nutritional differences on ER membrane profiles. After 48 hr incubation a labeled culture was harvested in two equal portions, one was resuspended in a volume of medium obtained by centrifugation of a culture after 24 hr of growth while the other was resuspended in a volume of completely fresh medium. In the culture fed with poor medium the S fraction was by far the most predominant (Fig. 2 A), the profile in fact resembling that in Fig. 1 C though the LR fraction was not so marked. Feeding with

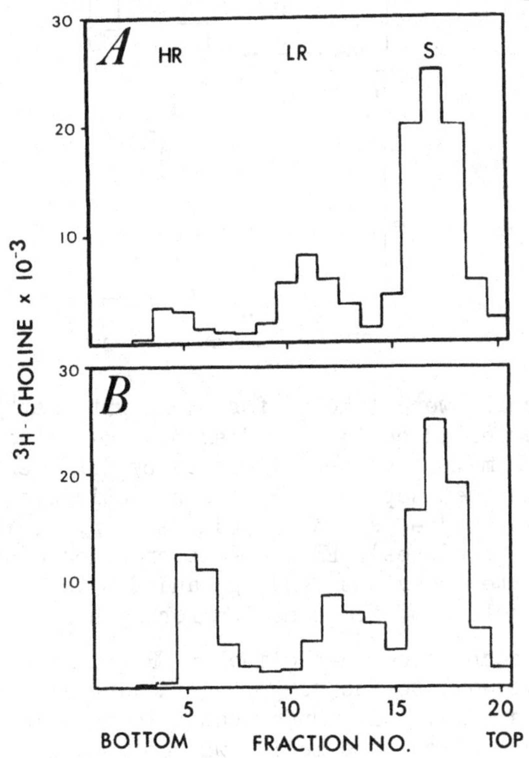

Fig. 2. ER membrane profiles in ^{3}H-choline labeled L-cells. A - 48 hr cells resuspended in '24 hr old' medium, and B - 48 hr cells resuspended in completely fresh medium, both incubated for $2\frac{1}{2}$ hr. (See legend to Fig. 1 for details).

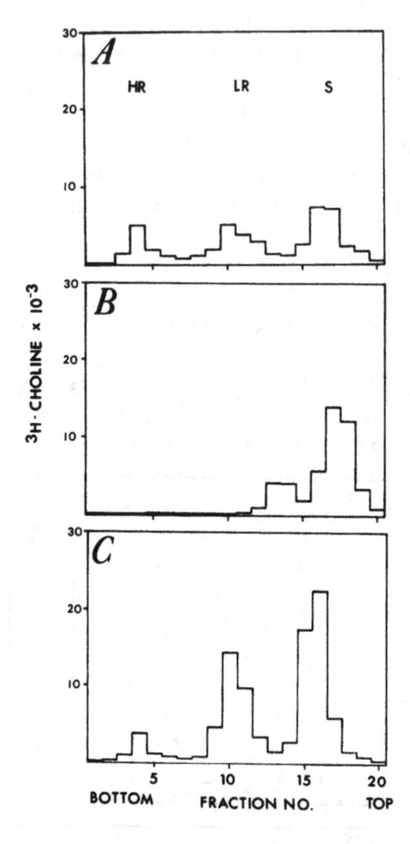

Fig. 1. L-cells were labeled for 24 hr with methyl-^{3}H-choline. Harvested cells were disrupted by nitrogen cavitation and ER membranes were isolated by differential centrifugation and then separated into subfractions on discontinuous sucrose gradients (see Svardal and Pryme, 1978 for details of procedures). ER membrane profiles after A - 24 hr, B - 48 hr after previous feeding, and C - as B then fed with equal volume of fresh medium for 2½ hr.

ution of incorporated ^{3}H-choline in ER membranes is almost equally divided into the rough (HR+LR) and smooth (S) compartments (Fig. 1 A). Under conditions where the routine replenishment of culture medium was omitted for 48 hr then the HR fraction became unidentifiable (Fig. 1 B), while the S fraction accounted for almost 80% of the radioactivity in the ER subfractions. Feeding of a 48 hr starved culture for 2½ hr resulted in the re-appearance of both HR and LR subfractions (Fig. 1 C), the increase in the LR fraction

completely fresh medium resulted in a decrease in the amount of radioactivity in the S fraction (75,200 cpm reduced to 65,800 cpm) while that in the HR fraction increased from 8,900 cpm to 29,600 cpm and that in the LR fraction was identical in both cultures. The feeding of a 48 hr starved culture with completely fresh medium thus promoted a more rapid re-appearance of the HR fraction than the LR fraction (Fig. 2 B). From these nutritional experiments it was apparent that the actual yields of HR and LR membranes differed quite considerably according to the culture conditions. It was thus of interest to investigate how an inhibitor of protein synthesis would affect ER membrane distribution profiles under good growth conditions. The inhibitor chosen for these experiments was cycloheximide, which, at a concentration of 20µg/ml, was known to inhibit the incorporation of radioactive amino acids into protein in MPC-11 cells by more than 95%.

An MPC-11 culture was harvested, resuspended in completely fresh medium and divided into two identical aliquots. ^{3}H-choline was added to both while cycloheximide (20µg/ml) was only added to one of them. After a 2½ hr incubation cells were harvested, disrupted by nitrogen cavitation and ER membranes were separated on discontinuous sucrose gradients. The results shown in Fig. 3 indicate that incubation with cycloheximide causes a selective decrease in the yield of HR membranes (Fig. 3 A) which were abundant in the non-treated cells (Fig. 3 B). The percentage distributions of radioactivity in HR, LR and S subfractions in the cycloheximide and non-treated control cultures were 1.4, 35.2 and 63.4%, and 22, 33 and 45% respectively. It appeared that the radioactivity in the HR fraction in non-treated cells was recovered in the S fraction in the cycloheximide treated cells, while the amount of radioactivity in the LR fraction was unaffected by the treatment. Thus the major observation in this experiment was the reduction in the HR fraction and the increase in the S fraction.

In order to determine whether or not the effect of cycloheximide could be reversed a further experiment was performed. An MPC-11 culture, labeled for 24 hr with ^{3}H-choline, was harvested and resuspended in completely fresh medium containing cycloheximide (20µg/ml). After a ½ hr incubation half of the culture volume was removed and the cells were harvested for disruption and subcellular fractionation. The remainder of the culture was diluted tenfold with fresh medium (without cycloheximide) and incubated

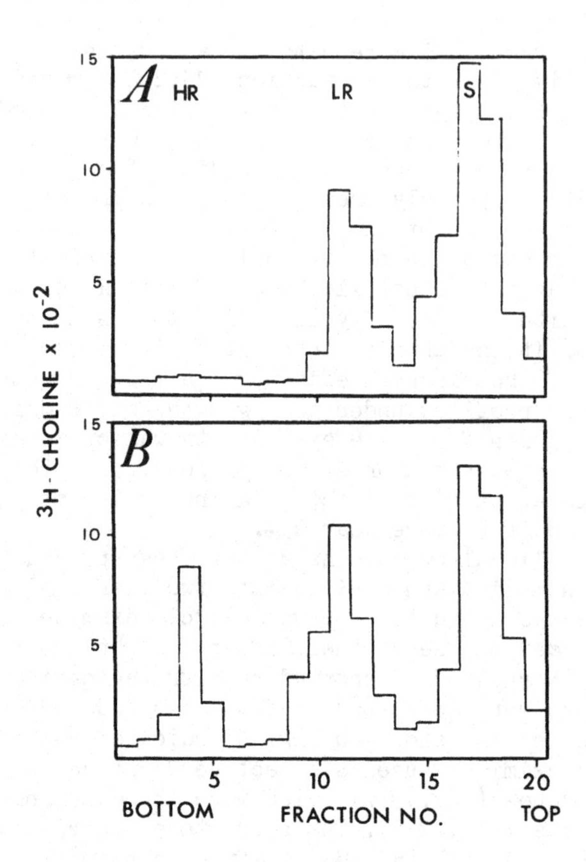

Fig. 3. An MPC-11 culture was harvested, resuspended in completely fresh medium and divided into two identical aliquots. ^{3}H-choline was added to both while cycloheximide (20μg/ml) was added to only one. After a 2½ hr incubation both cultures were harvested and cells were disrupted by nitrogen cavitation. ER membranes were separated on discontinuous sucrose gradients (Svardal and Pryme, 1978). ER membrane profiles in A - cycloheximide treated cells, and B - non-treated cells.

for a further period of 1 hr before being harvested. In the cycloheximide inhibited culture more than 60% of the ER membranes were of the S type, 27.3% were LR and 8.9% HR membranes (Fig. 4 A). A removal of the blocking effect of cycloheximide by performing a tenfold dilution with fresh medium followed by a 1 hr incubation resulted in a reduction in the amount of S membranes and a threefold increase in that of the HR fraction (Fig. 4 B).

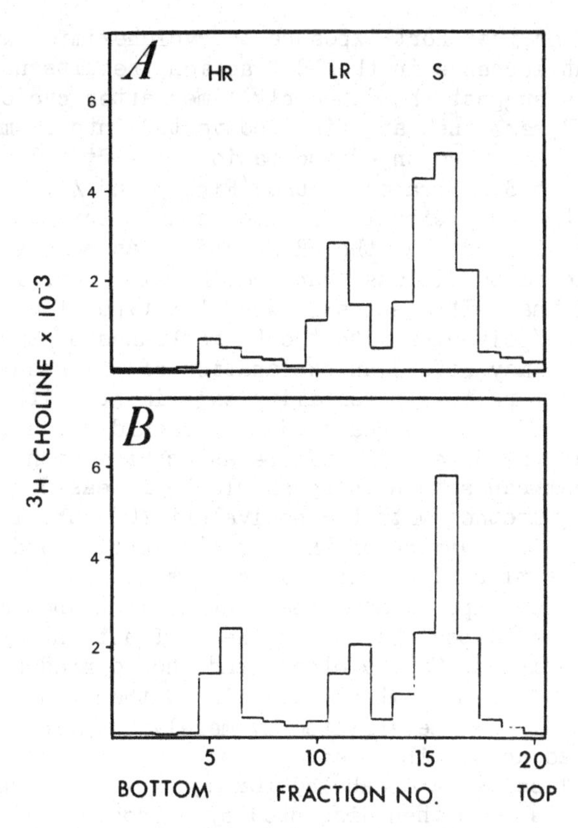

Fig. 4. An MPC-11 culture was labeled for 24 hr with ^{3}H-choline. The culture was harvested and resuspended in completely fresh medium containing cycloheximide (20µg/ml). After a $\frac{1}{2}$ hr incubation half of the culture was removed and the cells disrupted by nitrogen cavitation. The remainder of the culture was diluted tenfold with fresh medium and incubated for 1 hr before being harvested. ER membranes were separated on discontinuous sucrose gradients (Svardal and Pryme, 1978). ER membrane profiles in A - cells treated with cycloheximide for $\frac{1}{2}$ hr, and B - cells treated as in A then diluted tenfold with fresh medium and incubated 1 hr.

In an experiment to determine how rapidly cycloheximide exerts an effect on ER membrane profiles a 5 min. study with the inhibitor was performed. Cycloheximide (20µg/ml) was added to one of two identical MPC-11 cultures. Both were harvested after 5 min. incubation and cells were disrupted by nitrogen cavitation. ER membrane profiles showed that there was a marked decrease in the amount of HR membranes already after the short exposure to cycloheximide with a concomitant increase in the LR fraction (results not shown). The results suggest that at early times after cycloheximide addition HR membranes are first converted into LR membranes while after more prolonged incubation ($\frac{1}{2}$ - $2\frac{1}{2}$ hr) a further conversion to S membranes occurs (Fig. 3 and 4).

Both the nutritional experiments and those with cycloheximide indicated that the HR fraction was more sensitive to the altered conditions than the LR membranes. It was considered that differences between the type of association of ribosomes/polysomes with the ER membranes of the two subfractions may afford an explanation of the results. In order to explore this possibility experiments with RNase were performed. It was reasoned that should not all 60S subunits at the 5' end of mRNA be associated with either HR or LR membranes then these should be released upon RNase treatment (together with the equivalent 40S subunits) resulting in a lowering of the vesicle density and thereby altering the migration behaviour on discontinuous sucrose gradients. Thus upon RNase treatment HR membranes may appear in the LR position while the LR fraction may appear in the S position. On the other hand should predominantly all 60S subunits be membrane-associated then the vesicle density should not be affected by merely cleaving the mRNA inbetween adjacent ribosomes.

MPC-11 cells were labeled for 24 hr with ^{3}H-choline and the cells were then disrupted by nitrogen cavitation. Gradient fractions containing HR and LR membranes respectively were pooled and the membranes pelleted. The HR and LR membrane pellets were resuspended in buffer using gentle homogenization and each preparation was divided into two identical aliquots. RNase (5μg/ml) was added to one aliquot of each while the parallel remained untreated. After a 10 min. incubation at room temperature membranes were again pelleted, resuspended in buffer and applied to discontinuous sucrose gradients. The results presented in Fig. 5 show that HR membranes after RNase treatment migrated to the position of LR membranes while the migration of LR membranes was not affected by treatment with the enzyme. These observations are concomitant with a release of ribosomes from the HR membranes resulting in lowered density of the membrane vesicles. From experiments where ribosomal subunits in MPC-11 cells were labeled with ^{3}H-uridine prior to RNase treatment of isolated ER membrane preparations, it was found that almost 30% of label was released from HR membranes

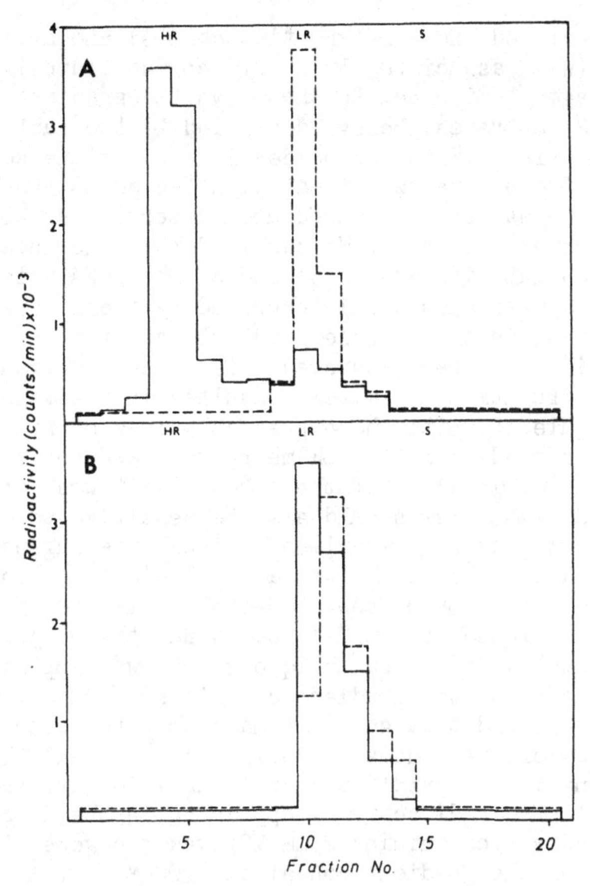

Fig. 5. HR and LR membranes were isolated from 24 hr ^{3}H-choline labeled MPC-11 cells. Each membrane preparation was divided into two equal aliquots : to one RNase (5µg/ml) was added while the parallel aliquot remained untreated. After a 10 min. incubation at room temperature membranes were pelleted, resuspended in buffer and applied to discontinuous sucrose gradients (Svardal and Pryme, 1978).
A - HR membranes. B - LR membranes.
——— buffer alone. - - - buffer + RNase (5µg/ml).

while less than 5% of radioactivity was released from the LR membranes (results not shown).

We have previously shown that the centrifugation of LR membranes from MPC-11 cells on discontinuous sucrose gradients containing 100mM KCl results in a 'shearing' effect on the polysomes (Svardal and Pryme, 1980c) causing a cleavage of the mRNA molecule producing a 5' mRNA fragment (with associated short polypeptide chains) and a 3' mRNA fragment (with associated long polypeptide chains). The former fragment migrated further down the gradient tube free of LR membranes, being identified by the fact that A_{260nm} material did not coincide with ^{3}H-choline containing material. The 3' fragment remained attached to the membrane vesicles and was only released upon detergent solubilization of the membrane material. We surmised that the interaction of some 60S subunits at the 5' end of the mRNA molecule with the LR membranes was interrupted by the high salt concentration in the sucrose gradients and that the high hydrostatic pressures generated under the conditions of ultracentrifugation ultimately resulted in cleavage of the mRNA molecule. Despite the appreciable loss of ribosomes under these conditions the LR membranes were not converted to the S type. In light of these results it was considered that the HR membranes should also be sensitive to such conditions if, as the experiments with RNase suggested, that the HR fraction contained a population of ribosomes which were not membrane-bound. ^{3}H-choline labeled ER membranes were isolated from MPC-11 cells and the preparation divided into two identical aliquots. One was applied to a discontinuous sucrose gradient containing 25mM KCl while the other was applied to a gradient made up with 100mM KCl. Examination of the gradient profiles showed that the HR was not present in the gradient containing 100mM KCl (results not shown). The radioactivity appearing in the HR region of the gradient containing 25mM KCl was recovered in the LR region of the gradient containing 100mM KCl. A mere suspension of the ER membranes in buffer containing 100mM KCl before ultracentrifugation on 25mM KCl gradients did not cause any alteration in the migration behaviour of either HR or LR membranes thus the salt concentration of the gradient buffer was the important factor in causing a conversion of HR membranes to those of a density equivalent to LR membranes during ultracentrifugation.

3. DISCUSSION

Two major populations of polyribosomes exist in the majority of eukaryotic cells hitherto studied : the free polysomes present in the cell sap and the membrane-bound polysomes associated with the rough ER. It was earlier considered that only secretory proteins were synthesized on membrane-bound polysomes, however, it has now been shown that a variety of other proteins which cannot be classified as secretory are also synthesized on membrane-bound poly-

the membrane through binding proteins independent, therefore, of a signal sequence and the length of the nascent polypeptide chain.
somes, for example, Sindbis and VSV glycoproteins (Wirth et al, 1977, Grubman et al, 1975), integral membrane proteins (Lodish and Small, 1975) and luminal proteins of membrane containing organelles (Sabatini and Kreibich, 1976). With such a wide spectrum of proteins being synthesized on membrane-bound polysomes a variety of post-translational modification enzymes must also be available in the ER. This then raises the question as to whether all regions of the rough ER seen in the electron microscope contain the same mRNA species and the same complement of necessary enzymes, or are specific mRNA species sequestered into separate domains of the ER where the correct range of modification enzymes is present. It was proposed some years ago that the rough ER may serve to segregate topographically or compartmentalize the synthesis of some intracellular protein species in particular regions of the cell (Tata, 1971,1973). Some evidence for the synthesis of light chain immunoglobulin (Pryme, 1974) and mitoplast proteins (Shore and Tata, 1977a,b) on specific compartments of rough ER membranes has been provided. We have demonstrated that it is possible to separate the rough ER in MPC-11 cells into nuclear associated ER, HR ER and LR ER (Svardal et al, 1981), each compartment having a characteristic polysome profile.

Three mechanisms have been described suggesting possible forms of interaction between polyribosomes and endoplasmic reticulum membranes : (a) via the 60S subunit, (b) via the growing polypeptide chain, and (c) via the mRNA molecule (see review, Svardal and Pryme, 1980a). Evidence is strongly in favour that 60S subunits closest to the 5' end of mRNA molecules coding for secretory proteins are not initially membrane-bound (Blobel and Dobberstein, 1975). As translation proceeds, however, and sufficient of the signal sequence has been synthesized then attachment between peptide and membrane can occur ultimately followed by a binding of the 60S subunit to the membrane surface. Thus many 60S ribosomes could be membrane-bound (peptide chains of adequate length to interact with membranes) whilst others in the same polysome complex would be membrane-free (signal sequence of insufficient length to cause interaction with membranes). On the other hand one can envisage that all 60S ribosomes of a polysome complex could be attached to

If one assumes that the 60S and 40S subunits are released from membranes upon completion of the translation of mRNA's coding for polypeptides initially bearing a signal sequence, then exposure of cells to poor growth conditions or the addition of an inhibitor of protein synthesis would be expected to result in a breakdown of the ribosome/membrane complex. The results of the nutritional experiments and those with cycloheximide indicated that the HR fraction was sensitive to these conditions. Reversal of the blocking conditions resulted in both cases in the re-appearance of the HR fraction. RNase treatment of isolated membranes had a specific effect on the migration properties of HR membranes in that radioactivity initially appearing in the HR region of the gradient was recovered in the LR region after treatment with the enzyme. The reduced density of HR vesicles is attributed to a loss of ribosomes promoted by RNase. This would then mean that not all 60S subunits at the 5' end of mRNA are membrane-bound. Such an arrangement is concomitant with the requirements of the signal hypothesis.

If there exists a more permanent form of attachment between 60S subunits and the membrane surface such that the subunits are not released from their binding sites upon completion of a round of translation then such membranes should be more refractory to nutritional conditions and to conditions where protein synthesis is inhibited. Based on the evidence that the migration properties of LR membranes were unaffected by RNase treatment, and that only a minor fraction of ribosomes was released from membranes by the enzyme, and in addition the fact that cycloheximide treatment did not affect the yield of LR membranes, then one can surmise that the majority of 60S subunits contained in the polysomes of the LR fraction are membrane-associated in a relatively permanent manner independent of nascent polypeptide chain/membrane interaction. Such an arrangement is not in accordance with the requirements of the signal hypothesis.

The observations described here support the concept that there are differences in the manner of association of ribosomes/polysomes with HR and LR ER membranes in MPC-11 and L-cells grown in suspension culture.

REFERENCES

Blobel, G. and Dobberstein, B. (1975) J. Cell Biol. 67, 835-851.

Fjose, A. and Pryme, I.F. (1983) Mol. Cell. Biochem. (in press).

Garatun-Tjeldstö, O., Pryme, I.F., Weltman, J.K. and Dowben, R.M. (1976) J. Cell. Biol. 68, 232-239.

Grubman, M.J., Moyer, S.A., Banerjee, A.K. and Ehrenfeld, E. (1975) J. Cell Biol. 74, 43-57.

Lodish, H.F. and Small, B. (1975) J. Cell Biol. 65, 51-64.

Pryme, I.F. (1974) FEBS Lett. 48, 200-203.

Pryme, I.F., Svardal, A.M. and Skorve, J. (1981) Mol. Cell. Biochem. 34, 177-183.

Pryme, I.F., Svardal, A.M., Skorve, J. and Lillehaug, J.R. (1982) In : Methodological Surveys in Biochemistry - Cancer Cell Organelles (E. Reid, G.M.W. Cook and D.J. Morre, eds.) vol. 11, pp 293-298, Horwood, Chichester, U.K.

Sabatini, D.D. and Kreibich, G. (1976) In : The Enzymes of Biological Membranes (Martonosi, A. ed.) vol. 2, pp 531-579, Plenum, U.S.A.

Shore, G.C. and Tata, J.R. (1977a) J. Cell Biol. 72, 714-725.

Shore, G.C. and Tata, J.R. (1977b) J. Cell Biol. 72, 726-743.

Svardal, A.M. and Pryme, I.F. (1978) Anal. Biochem. 89, 332-336.

Svardal, A.M. and Pryme, I.F. (1980a) Subcell. Biochem. 7, 117-170.

Svardal, A.M. and Pryme, I.F. (1980b) Mol. Cell. Biochem. 29, 159-171.

Svardal, A.M. and Pryme, I.F. (1980c) Mol. Biol. Rep. 6, 105-110.

Svardal, A.M., Pryme, I.F. and Dalen, H. (1981) Mol. Cell. Biochem. 34, 165-175.

Tata, J.R. (1971) Subcell. Biochem. 1, 83-89.

Tata, J.R. (1973) In : Reproductive Endocrinology, 6th Symposium, Karolinska Symposia on Research Methods (E. Diczfalusy, ed.) pp 192-224, Karolinska Institute, Stockholm.

Wirth, D.F., Katz, F., Small, B. and Lodish, H.F. (1977) Cell 10, 253-263.

BIOSYNTHESIS AND INTRACELLULAR TRANSPORT OF Ia ANTIGENS

B. Dobberstein, J. Lipp, W. Lauer & P. Singer*

European Molecular Biology Laboratory,

Heidelberg and *Max-Planck-Institute für

Biologie, Tübingen.

Cellular organelles play a major role in the structural and functional organisation of eucaryotic cells. How these organelles are formed, how their specific membrane proteins are inserted and why these proteins are mainly found in one organelle and not in another is the subject of intensive research (Simons and Warren, 1983; Brown et al., 1983; Sabatini et al., 1982; Dobberstein et al., 1982; Kondor-Koch et al., 1982; Bretscher et al., 1980). Proteins destined for the plasma membrane such as the histocompatibility antigens or the glycoprotein of Semliki Forest or Vesicular stomatitis virus, are first inserted into the membrane of the endoplasmic reticulum. They are subsequently transported to the Golgi complex where their carbohydrate side chains are modified and ultimately reach the plasma membrane (Rothman, 1981; Garoff et al., 1982; Ploegh et al., 1981; Krangel et al., 1979; Owen et al., 1980; Dobberstein et al., 1982). Proteins destined for other locations such as lysosomal or secretory proteins are first translocated across the membrane of the endoplasmic reticulum as well and then transported to lysosomes or to the extracellular space (see Hasilik, 1980). The endoplasmic reticulum might furthermore be the site of assembly of endogenous proteins of the endoplasmic reticulum, the Golgi complex and other cellular organelles. The question thus arises as to how these proteins are sorted from one another and what the signals are that direct a protein to its destination.

For some proteins the destination is one specific membrane (resident proteins), yet for others, and these might represent the majority, the site of destination is a group of membranes through which proteins can cycle or migrate (migrant proteins). Receptors, for instance, have the ability to shuttle between different organelles (Brown et al., 1983). It appears that cellular organisation of membranes is a very dynamic one. This is also reflected by the fact that membrane proteins are often found predominantly in one organelle, but also, at lower concentration, in an other one. An approach to elucidate the biosynthetic and dynamic properties of membrane proteins would be to follow a specific marker and determine its time of residence in the different organelles it passes through. A second goal would be to identify the factors which determine this residency in a specific organelle.

Histocompatibility antigens are a very attractive system in which to study such questions. They are assembled from different subunits in the membrane of the endoplasmic reticulum, pass through the Golgi complex and finally reach the plasma membrane. A unique feature of class II histocompability antigens is that one of their subunits dissociates from the oligomeric complex at an intracellular site, i.e. prior to expression at the cell surface. The location of this protein, in a specific membrane, is determined by its assembly with the other subunit proteins. Thus, for this protein oligomeric assembly and disassembly is a means to route it to a specific membrane.

A. Structure

Class II antigens called HLA-DR antigens in man and Ia antigens in mouse are composed of two polymorphic polypeptide chains, α (MW 33,000) and ß (MW 26,000) and an invariant chain Ii or γ (MW 31,000) (see Shakelford et al., 1982; Uhr et al., 1979; Winchester and Kunkel, 1979; Jones et al., 1978). The α and ß chains are encoded by genes located in the major histocompatibility complex, the Ii chain is coded by a gene outside this complex on a different chromosome (Uhr et al., 1979; Koch et al., 1982; Day and Jones, 1983).

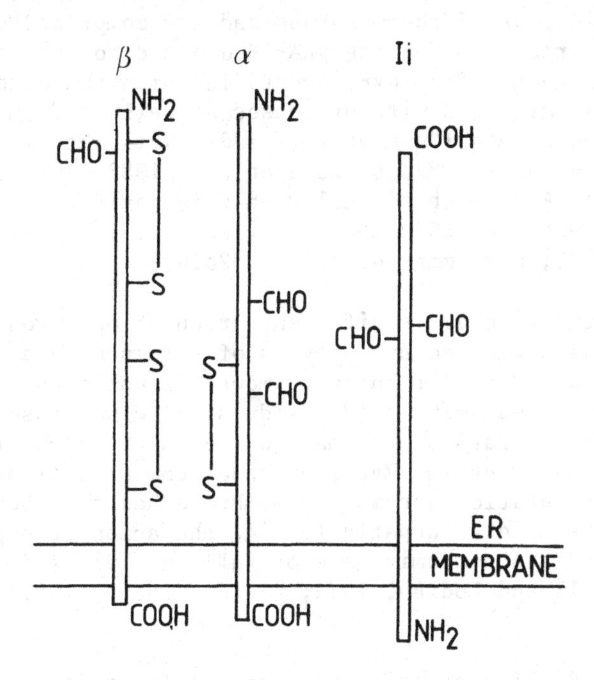

Figure 1. Structural features of Ia antigens.
In the ER they are composed of α, ß and Ii chains. All three chains span the membrane. α and ß chains expose the COOH terminus on the cytoplasmic side whereas Ii chain, most likely exposes the NH_2 terminus on this side. The structures are deduced from Auffray et al., 1982; Larhammar et al., 1982a and b; Lee et al., 1982; Long et al., 1983; Singer et al., 1983. CHO: carbohydrate side chain linked to asparagine; S-S: indicates intramolecular disulphide bonds.

All three chains span the membrane (Fig.1). The COOH terminal segments of the α and ß chains are located on the cytoplasmic side of the membrane and are comprised of 15 amino acid residues for the HLA-DR α and 8 for the ß chains. This is suggested from experiments in which the cytoplasmic portion was digested with proteases as well as from amino acid sequence data (Owen et al., 1981; Kvist et al., 1982; Larhammar et al., 1982a,b; Long et al., 1983; Auffray et al., 1982). A stretch of uncharged amino acid residues is found close to the COOH terminus (Lee et al., 1982; Long et al., 1983; Larhammar et al., 1982a,b).

For the Ii chain a different orientation across the membrane is suggested by analysis of sequence data from cDNA clones coding for human or mouse Ii antigens. No stretch of uncharged amino acid residues is found close to the COOH terminus, rather proximal to the NH_2-terminus (Singer et al., 1983; Long and Mach, personal communication) (Fig.1). Such an orientation across the membrane has also been suggested for the chicken hepatic lectin, the anion transport protein and the neuraminidase of influenza virus (Drickamer, 1981; Braell and Lodish, 1982; Fields et al., 1981).

B. Biosynthesis and Membrane Insertion

α, ß and Ii chains are translated from separate mRNAs and inserted co-translationally into the membrane of the endoplasmic reticulum (ER) (Kvist et al., 1982; Lipp et al., 1983; Owen et al., 1981). There they become N-glycosylated. Insertion of α and ß chains into the membrane is preceded by the synthesis of a signal sequence which initiates the interaction of the nascent polypeptide chain to the membrane (Blobel and Dobberstein, 1975; Korman et al., 1980). Signal sequences have been determined for α, as well as for ß chains by sequence analysis of cDNAs or genes, by comparison of α and ß chains synthesized in vitro with the mature unglycosylated forms (Lee et al., 1982; Long et al., 1983; Korman et al., 1980). They have a length between 20 and 25 amino acid residues with a central core of uncharged amino acid residues. This sequence is cleaved from the nascent polypeptide chain after insertion into the membrane and is thus not found on the mature chain. Interestingly no such cleavable signal sequence has been found for the Ii chain (Lipp et al., 1983).

Insertion of α, ß and Ii chains into the membrane proceeds in the manner recently demonstrated for the light chain of immunoglobulin and human prolactin (Meyer et al., 1982; Meyer, 1982; Walter et al., 1981; Walter and Blobel, 1981; Walter and Blobel, 1982; Ullu et al., 1982). For these proteins two components have been characterized which are required for translocation across the membrane, Signal Recognition Particle (SRP) and Docking Protein (DP). Their characterization has been made possible by the use of a cell free system in which translocation of proteins across a membrane is reconstituted from separate components. SRP is a ribonucleoprotein complex consisting of at least 6 polypeptide chains with MWs of 72, 68, 54, 19, 14 and 9Kd and a 7S RNA. This particle interacts with the NH_2-terminal portion of nascent secretory polypeptide chains, probably with the signal sequence and arrests elongation after about 60 amino acid residues have been polymerised. This would just be sufficient to allow the signal sequence to emerge from the ribosome. The arrest in elongation is suspended when contact is made at a specific side of the ER membrane, the DP. This is a membrane protein of 72Kd of which a 60Kd domain is exposed on the cytoplasmic side. As secretory proteins can only be translocated across the membrane of the ER during their synthesis this arrest in elongation guarantees that no precursor of secretory proteins accumulates in the cytoplasm.

The elongation of all three subunit chains, α, ß and Ii is blocked by SRP and this block is released by microsomal membranes containing DP (Lipp et al., 1983). Insertion into the membrane can therefore be depicted as in: Fig.2A for α and ß chains, and as in 2B for Ii chains. SRP and DP mediate specific binding of all three chains to the membrane (1). As Ii chain lacks a cleavable signal sequence it remains to be shown which segment in this case performs the signal function, it might be the membrane spanning segment (fig.2B). According to figure 2, it is assumed that a loop like structure first interacts with the membrane, as first suggested by Inouye et al. (1977). In such a case the membrane spanning segment of Ii, located close to the NH_2 terminus, might also function as signal sequence without being cleaved by signal peptidase (fig.2B, steps 1 and 2).

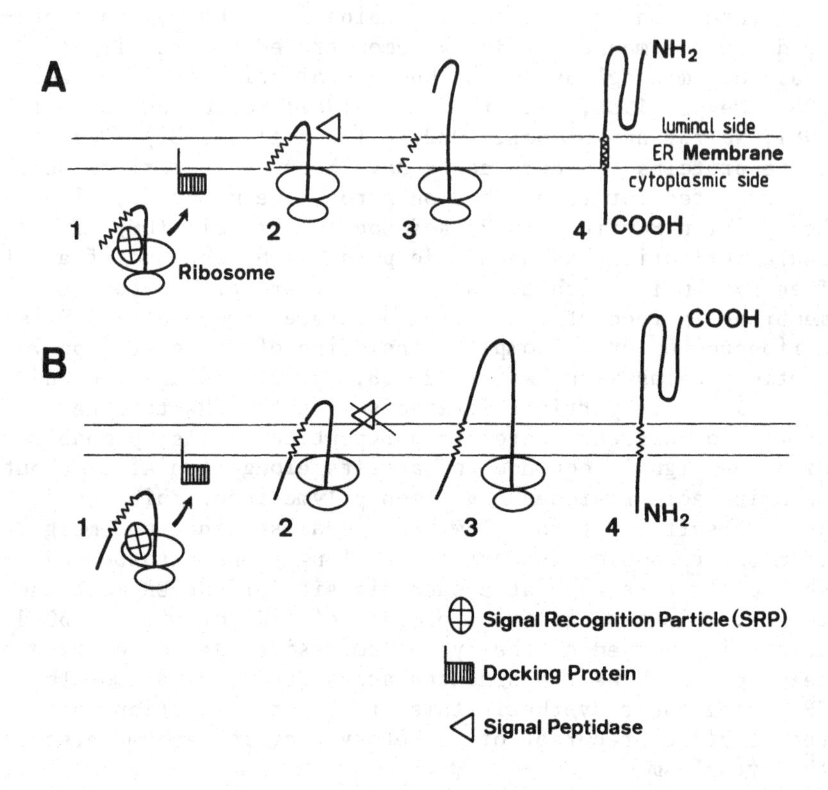

Figure 2. Membrane insertion of α and ß chains (A) and Ii chains (B).
Signal recognition particle (SRP) interacts with nascent α, ß and Ii chains and arrests further translation (step 1). After contact has been made with the docking protein (DP) at the ER membrane translation continues now coupled to membrane translocation (step 2). The signal sequence is cleaved from nascent α and ß chains, but not from Ii chains (step 3). Cleaved signal sequence is probably degraded (A step 3). Translocation proceeds until a stretch of uncharged amino acid residues anchors α and ß chains, close to the COOH terminal end, to the membrane (A step 4). The COOH terminal portion of the Ii chain does not contain a stretch of uncharged amino acid residues and is therefore most likely traversing the membrane (Singer et al., 1983) (B step 4).

Chains are then elongated and extruded across the membrane. For α and ß chains translocation would stop when the stretch of uncharged amino acid residues close to the COOH terminal end emerges from the ribosome, and anchors these proteins in the membrane (fig.2, step 4).

C. Oligomeric Assembly

As the subunit chains of class II histocompatibility antigens are translated from separate mRNAs, the question arises, at which stage of intracellular transport does assembly into the oligomeric complex occur. When the human cell line Raji or mouse spleen cells were pulse labelled for 5min with ^{35}S-methionine and HLA-DR or Ia antigens were precipitated, they were found already assembled into an oligomeric complex consisting of α, ß and Ii chains (Jones et al., 1978; Kvist et al., 1982; Owen et al., 1981). This rapid assembly after biosynthesis would suggest that oligomeric assembly occurs already in the endoplasmic reticulum. This notion is supported by in vitro experiments in which α, ß and Ii chains were found complexed in microsomal membranes (Kvist et al., 1982). Assembly could also be obtained after injection of mRNA derived from Raji cells into frog oocytes (Long et al., 1982; Finn and Levy, 1982).

With genes available for α, ß and Ii antigen, assembly can also be tested in cells after transfection. Rabourdin-Combe and Mach (1983) could demonstrate that genes coding for the human α and ß chains are synthesized in mouse L cells and an oligomeric complex composed of α and ß chains is expressed on the cell surface. Whether the Ii chain plays a role in this surface expression remains unclear as L cells themselves express Ii chains.

D. Intracellular Transport and Surface Expression

Intracellular transport of Ia and HLA-DR antigens can be followed by pulse-chase labelling. In the case of HLA-DR antigens modification of carbohydrate side chains from high mannose to the complex-type can be observed on α, ß and Ii chains about 30min after synthesis. Shortly after this modification has occurred, Ii chains dissociate from the α, ß complex and become rapidly degraded.

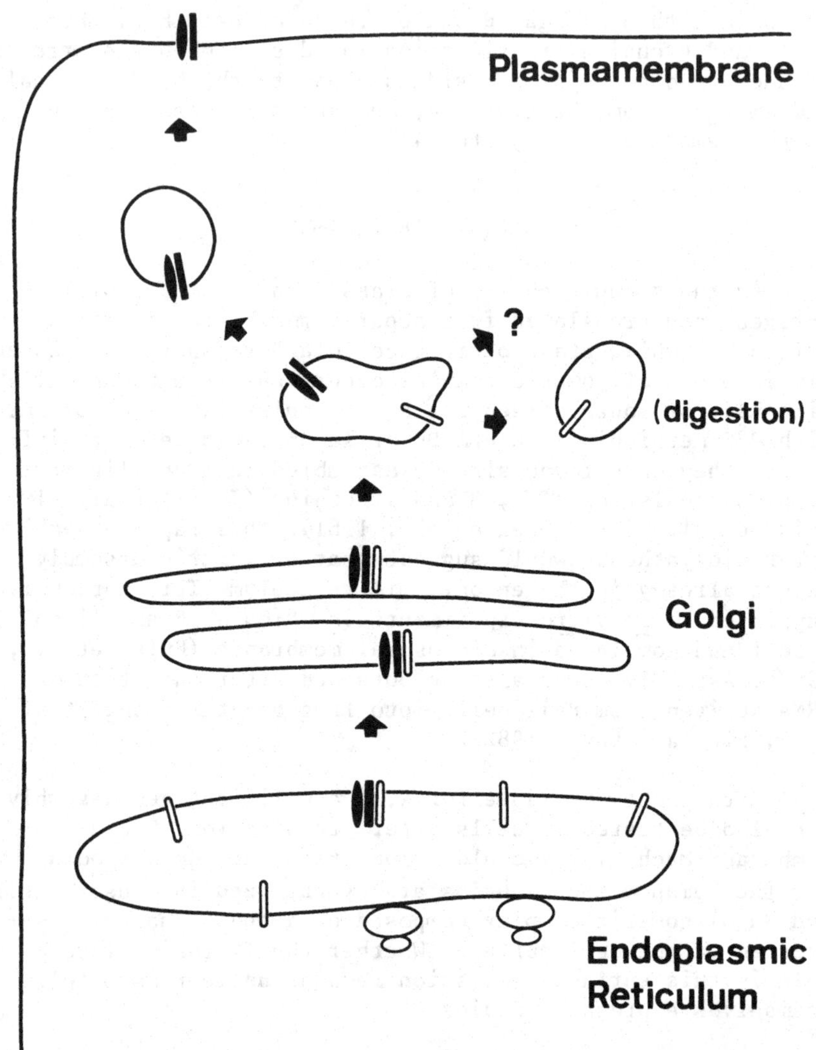

Figure 3. Intracellular transport of Ia antigens.
α, ß and Ii chains assemble in the ER into an oligomeric complex. Ii chains not assembled with α and ß remain in the ER. α, ß and Ii chains pass through Golgi and Ii chains dissociate from α and ß in an yet unidentified organelle. α and ß chains appear on the cell surface. Ii is rapidly digested. Whether it also appears on the cell surface is unclear.
α chain , ; ß chain, ; Ii chain, .

As the addition of complex type oligosaccharide side chains is thought to occur in Golgi, it can be assumed that dissociation occurs at a post-Golgi stage (fig.3). Compartments involved in transport of membrane proteins from the Golgi complex to the plasma membrane are ill defined. There is, however, a compartment, the endosome, in which ligands such as low density lipoprotein or transferrin, dissociate from their receptors during endocytosis (Brown et al., 1983). This dissociation is mediated by the low pH found in this compartment. It is conceivable that membrane proteins like the class II antigens enter such a compartment and dissociate into an α,ß complex and free Ii chains. α and ß chains would then be transported further to the cell surface. The fate of the Ii chains is uncertain. Some authors but not all claim that it can be detected on the cell surface (Koch et al., 1982b; Sung and Jones, 1981).

Intracellular transport of Ia antigens reveals that oligomeric assembly is one of the means by which protein transport within the cell is regulated. Manipulation of the levels of individual subunits within cells will enable the elucidation of the requirements for assembly and cellular transport routes of Ia. Moreover, with genes available from all 3 subunits, it will be possible to determine the protein sequences required for proper assembly and transport by introducing modified nucleic acid sequences into the appropriate cells.

ACKNOWLEDGEMENTS

We are grateful to Drs. D. Meyer and I. Ibrahimi for critical reading of the manuscript and thank A. Steiner for typing the manuscript. The work was supported by the Deutsche Forschungsgemeinschaft.

LITERATURE

1. Auffray, C., Korman, A.J., Roux-Dosseto, M., Bono, R. and Strominger, J.L. 1982. Proc. Natl. Acad. Sci. USA 79, 6337-6341.
2. Blobel, G. 1980. Proc. Natl. Acad. Sci. USA 77, 1496-1500
3. Blobel, G. and Dobberstein, B. 1975. J. Cell Biol. 67, 852-862.
4. Braell, W.A. and Lodish, H.F. 1982. Cell 28, 23-31
5. Bretscher, M.S., Thomson, J.N. and Pearse, B.M.F. 1980. Proc. Natl. Acad. Sci. USA 77, 4156-4159.
6. Brown, M.S., Anderson, R.G.W. and Goldstein, J.L. 1983. Cell 32, 663-667.
7. Day, C.E. and Jones, P.P. 1983. Nature 302, 157-159
8. Dobberstein, B., Kvist, S. and Roberts, L. 1982. Phil. Trans. R. Soc. Lond. B 300, 161-172.
9. Drickamer, K. 1981. J. Biol. Chem. 256, 5827-5839
10. Engelman, D.M. and Steitz, T.A. 1981. Cell 23, 411-422
11. Fields, S., Winter, G. and Brownlee, G.G. 1981. Nature 290, 213-217
12. Finn, O.J. and Levy, R. 1982. Proc. Natl. Acad. Sci. USA 79, 2658-2662
13. Garoff, H., Kondor-Koch, C. and Riedel, H. 1982. In Current Topics in Microbiology and Immunology, vol. 99 (ed. M. Cooper et al.), Springer Verlag, Berlin & Heidelberg, pp. 1-50.
14. Hasilik, A. 1980. Trends Biochem. Sci. 5, 237-240.
15. Inouye, S., Wang, S., Sekizawa, J., Halegoua, S. and Inouye, M. 1977. Proc. Natl. Acad. Sci. USA 74, 1004-1008.
16. Jones, P.P., Murphy, D.B., Hewgill, D. and McDevitt, H.O. 1978. Immunochem. 16, 51-60.
17. Koch, N., Hämmerling, G.J., Szymura, J. and Wabl, M.R. 1982a. Immunogenetics 16, 603-606.
18. Koch, N., Koch, S. and Hämmerling, G.J. 1982b. Nature 299, 644-645
19. Kondor-Koch, C., Riedel, H., Söderberg, K. and Garoff, H. 1982. Proc. Natl. Acad. Sci. USA 79, 4525-4529
20. Korman, A.J., Ploegh, H.L., Kaufman, J.F., Owen, M.J. and Strominger, J.L. 1980. J. exp. Med. 152, 65s-82s.
21. Krangel, M.S., Orr, H.T. and Strominger, J.L. 1979. Cell 18, 979-991.
22. Kvist, S., Wiman, K., Claesson, L., Peterson, P.A. and Dobberstein, B. 1982. Cell 29, 61-69.

23. Larhammar, D., Gustafsson, K., Claesson, L., Bill, P., Wiman, K., Schenning, L., Sundelin, J., Widmark, E., Peterson, P.A. and Rask, L. 1982. Cell 30, 153-161
24. Larhammar, D., Schenning, L., Gustafsson, K., Wiman, K., Claesson, L., Rask, L. and Peterson, P.A. 1982a. Proc. Natl. Acad. Sci. USA 79, 3687-3691.
25. Lee, J.S., Trowsdale, J., Travers, P.J., Carey, J., Grosveld, F., Jenkins, J. and Bodmer, W.F. 1982. Nature 299, 750-752.
26. Lipp, J., Hämmerling, G. and Dobberstein, B. 1983. Manuscript in preparation.
27. Long, E.O., Gross, N., Wake, C.T., Mach, J.P., Carrel, S., Accolla, R. and Mach, B. 1982. EMBO J. 1, 649-654.
28. Long, E.O., Wake, C.T., Gorski, J. and Mach, B. 1983. EMBO J. 2, 389-394.
29. Meyer, D.I., Krause, E. and Dobberstein, B. 1982. Nature 297, 647-650.
30. Meyer, D.I. 1982. Trends Biochem. Sciences 7, 320-321.
31. Owen, M.J., Kissonerghis, A.-M., Lodish, H.F. and Crumpton, M.J. 1981. J. Biol. Chem. 256, 8987-8993.
32. Owen, M.J., Kissonerghis, A.-M. and Lodish, H.F. 1980. J. Biol. Chem. 255, 9678-9684.
33. Ploegh, H.L., Orr, H.T. and Strominger, J. L. 1981. Cell 24, 287-299.
34. Rabourdin-Combe, C. and Mach, B. 1983. Nature 503, 670-674.
35. Rothman, J.E. 1981. Science 213, 1212-1219
36. Sabatini, D.D., Kreibich, G., Morimoto, T. and Adesnik, M. 1982. J. Cell Biol. 92, 1-22
37. Simons, K. and Warren, G. 1983. Advances in Protein Chemistry, in press.
38. Shakelford, D.A., Kaufman, J., Korman, A.J. and Strominger, J.L. 1982. Immunol. Rev. 66, 129-183.
39. Singer, P., Lauer, W., Hämmerling, G. and Dobberstein, B. 1983. Manuscript in preparation.
40. Sung, E. and Jones, P.P. 1981. Molec. Immunol. 18, 899-913.
41. Uhr, J.W., Capra, J.D., Vitetta, E.S. and Cook, R.G. 1979. Science 206, 292-297
42. Ullu, E., Murphy, S. and Melli, M. 1982. Cell 9, 195-202
43. Von Heijne, G. 1980. Eur. J. Biochem. 103, 431-438.

44. Walter, P., Ibrahimi, J. and Blobel, G. 1981. J. Cell Biol. 91, 545-550.
45. Walter, P. and Blobel, G. 1981. J. Cell Biol. 91, 557-561.
46. Walter, P. and Blobel, G. 1982. Nature 299, 691-696
47. Winchester, R.J. and Kunkel, N.G. 1979. In Advances in Immunology, Academic Press, vol. 28, pp. 221-291.

PROTEIN SYNTHESIS ON ENDOPLASMIC RETICULUM

T. Hultin
Department of Cell Physiology,
Wenner-Gren Institute
University of Stockholm
S-113 45 Stockholm, Sweden

This section deals primarily with the molecular mechanisms by which newly synthesized proteins are selected for different intracellular destinations or for secretion to the extracellular medium. Much of the data center around the so called signal hypothesis, which implies that in typical cases the assignation mechanism is cotranslational. This would mean that not only the structure of nascent proteins, but also their destination, is genetically encoded in the translated mRNA.

A summation of recent data from G. Blobels group, clarifying the mechanism of mRNA-directed translocation of proteins across endoplasmic membranes,was presented by B. Dobberstein (G. Blobel was unable to attend the Conference as scheduled). So far two components required for translocation have been characterized: (1) The signal recognition particle (SRP), an 11 S ribonucleoprotein consisting of six polypeptide chains and a 7 S RNA. SRP interacts with the mRNA-encoded signal sequence of the nascent pre-protein as it emerges from the ribosome. Further elongation is thereby temporarily arrested. The structure of SRP has been studied in some detail after cleavage with micrococcal nuclease, which produces two unequal SRP fragments with two and four proteins, respectively. (2) The SRP receptor ("docking protein"). This protein has been purified from detergent extracts of endoplasmic membranes by affinity chromatography on SRP-Sepharose. Membrane-bound or purified docking protein interacts with SRP-arrested ribosomes, thereby releasing the translation block. In the former case the nascent protein chain with its signal sequence is passed through the membrane into the endoplasmic vesicle. Probably the SRP/docking protein interaction is transient and does not persist throughout the entire chain translocation.

One might expect the recognition of the signal sequence by SRP to involve a specific amino acid sequence similar for all preproteins. This is not the case. According to a model for the SRP/signal sequence recognition presented by T.A. Rapoport the interaction involves the immersion of a continuous stretch of 6-7 hydrophobic amino acids, constituting the recognition domain, into a deep hydrophobic pocket within the SRP. Consistent with this model, the mechanism of signal recognition is shown to be phylogenetically highly unspecific. Thus, carp preproinsulin is correctly translated, translocated and processed in bacterial cells (E. coli). It is also of little importance for the translocation whether the signal sequence has a terminal or internal location in the nascent chain. Of the underlying experiments it is of particular interest to learn that frog oocytes injected with either carp insulin mRNA or Semliki forest virus mRNA discriminate correctly between the encoded destinations of the translation products. Thus, proinsulin is secreted to the medium, while the virus envelope polypeptides remain attached to the plasma membranes as is the case in the normal target cells.

The experiments reported by G. Kreil deal in part with a closely related topic. Prepromelittin, as translated from melittin mRNA (prepared from bee venom glands) is a small protein for being translocated. It contains only 70 amino acids, of which 21 belong to the signal peptide. Promelittin can be formed in vitro by removal of this peptide by endoproteolytic cleavage by a special "signal peptidase", a membrane-bound protease without phylogenetic specificity. The enzyme has been extensively purified from rat liver microsomes. A signal peptidase with exactly the same restricted action on prepromelittin has been prepared from E. coli. Signal peptidases are cryptic enzymes, and their activity in vitro depends on an appropriate protein/phospholipid/detergent ratio.

As a rule, translocation involves the insertion of nascent preprotein chains into the endoplasmic reticulum with the N-terminal end on the luminal side. The paper presented by B. Dobberstein deals with the assembly complex of class II histocompatibility antigens. The γ-component of the complex is cotranslationally inserted into the endoplasmic membrane with the reverse orientation, i.e. with the C-terminal end on the luminal side. Still, the insertion depends on SRP and docking protein interactions. During the further intracellular transport the γ-chain dissociates from the normally oriented α- and β-chains, and is rapidly degraded.

The heterogeneity of the endoplasmic reticulum has been studied by I.F. Pryme. Polysomes from heavy rough microsomes from MPC-11 cells contain four times as many immunoglobuline light chain polypeptides as polysomes from light rough microsomes. The experimental data indicate that the two fractions represent distinct compartments of the rough endoplasmic reticulum. On this premise the non-random distribution of the nascent light chains would suggest a discrimination among the various nascent polypeptides carrying signal sequences for being further translated within different domains of the endoplasmic reticulum.

The several aspects of the protein destination mechanisms dealt with at the Conference pose more questions for the future than those that can be answered at present. For instance: How and where does SRP interact and interfere with the mRNA-translating ribosome ? Is the 7 S SRP-RNA directly involved in the arrest of translation ? How is the ultimate location of the various translocated proteins determined with reference to the different cell organelles, cell surface and extracellular medium ? Are other, more specific carriers and receptors involved in the transport of individual proteins to their final destinations ?

3. TRANSLATION FIDELITY

SPERMINE ENHANCES THE SPEED AND FIDELITY OF AMINOACYLATION OF TRANSFER RIBONUCLEIC ACID: A CORRELATION OF PRECISION OF CATALYSIS WITH MAXIMAL RATE ENHANCEMENT.

Robert B. Loftfield, Elizabeth Ann Eigner and Peter Simons

University of New Mexico School of Medicine Dept. of Biochem.

Albuquerque, New Mexico 87131, United States

ABSTRACT

We show that polyamines specifically enhance the syntheses of "correct"aminoacyl tRNAs (i. e., using the phenylalanine:tRNA ligase of yeast, one synthesizes Phe-$tRNA^{Phe}$) and specifically inhibit the possible misacylations such as Phe-$tRNA^{Tyr}$ that are encouraged by the commonly used high magnesium reaction conditions. We relate this enhancement of rate and precision to current proposals on the mechanisms of enzymic catalysis and extend the concept to the certainty that some of the pre-transition state "strain" in the enzyme· substrate complex is in the form of uniquely structured water associated with the enzyme. Release of this structured water during the transition state contributes high positive entropy of activation which lowers the free energy of the transition state, thereby enhancing catalysis.

INTRODUCTION

Among biochemists one needs to work with data that are reproducible, where one can be confident that his observations constitute a signal that is well above the noise level. In consequence biochemists tend to begin with a tentative impression and then to refine the experimental conditions to improve the signal/noise ratio. In part this may mean selecting a pH where reaction rates are high and relatively insensitive to the pH. It may mean choosing a temperature where a single product appears at a linear rate over a long period of observation. It may mean the selection of salt concentrations and protein concentrations that protect the enzyme from denaturation. It may involve a selection between glass and plastic reaction vessels and pipettes or a best choice of some exotic buffer substance. While such efforts to improve the signal/noise ratio are essential, they are fraught with danger in the possibility that the investigator may end up examining a phenomenom that is very different from the original interest.

One might cite a few historical precedents. The Coris won a Nobel prize for their work on phosphorylase. Their studies on this system prospered most under conditions of low phosphate buffer concentrations. It became clear that phosphorylase catalyzes an easily reversible reaction in which glucose-1-phosphate adds a glucose residue to a growing glycogen molecule with the release of phosphate OR a phosphate ion attacks the terminal glucose residue of a glycogen molecule to remove the glucose as glucose-1-phosphate. This easy reversibility,under highly un-physiological conditions, long obscurred the need for a glycogen synthetase system and doubtless delayed its discovery for years.

Similarly generations of biochemists had been raised on the dogma of Michaelis-Menten kinetics and the practical difficulties of collecting good data at low concentrations of substrate (and at correspondingly low reaction rates). In 1964 Riggs and Dowd (1) surveyed the literature and found a majority of reported determinations of enzymic parameters included few or no observations of rates where $(S) \ll K_m$ even though all textbooks on enzymology urged against such neglect. The combination of unreliable experimental conditions and faith in the Michaelis-Menten equation led investigators to avoid collecting data at low substrate concentrations or, worse, to ignore the data they did collect, thus delaying by years the discovery of sigmoid kinetics. When Pardee and Gerhardt(2) finally assured themselves that their own data were good and that important information lay in the experimentally difficult low range, they opened a new era in biochemistry; the era of sigmoid kinetics, substrate cooperativity, allosteric control and metabolic regulation. A similar neglect of the obvious and a similar pursuit of the familiar may be responsible for our failure to bring polyamines (spermine, spermidine, putrescine and cadaverine) into the biochemical consciousness until quite recently.

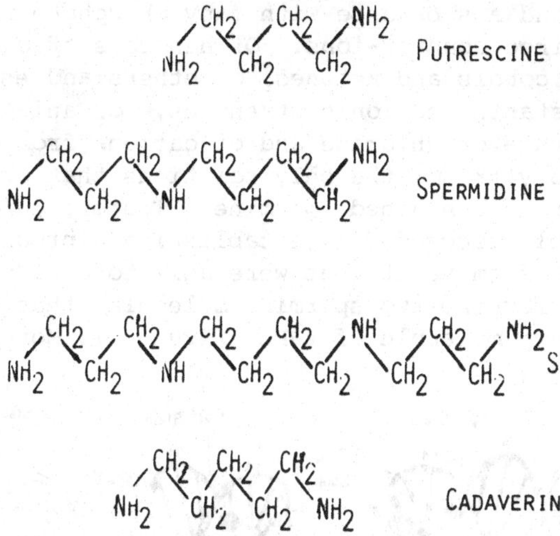

PUTRESCINE

SPERMIDINE

SPERMINE

CADAVERINE

Figure 1. The structures of the four naturally occurring polyamines

POLYAMINES AS CO-FACTORS IN THE AMINOACYLATION OF tRNA

Anton von Leuwenhoek, the inventor of the microscope, also isolated and described one of the polyamines, spermine phosphate, which had crystallized from human semen. Despite this very early discovery and successful efforts by organic chemists to determine the structures and to develop chemical syntheses, biochemists largely ignored these simple substances even though they appeared to be ubiquitious in living tissue. It is understandable that nucleic acids, protein, prostaglandins, gangliosides and the like evaded effective biochemical attack either because of their inherent complexity, sensitivity, rarity or scarcity, and lack of appropriate study techniques. But the polyamines are simple, stable, C-4 to C-10 straight chain compounds whose concentrations may exceed those of amino acids and major metabolites in many cells. It is surprising that their simplicity and ubiquity did not make them interesting to biochemists.

The next hint that polyamines might have great significance was largely ignored as well. Transfer RNA had been isolated, sequenced, its function was largely understood and it is a relatively small molecule (MW= ca 30 000 Dalton). It looked like an ideal target for X-Ray crystallographers and they attacked enthusiastically. Of course good crystals are essential and a substance with many phosphate residues certainly required counter-ions. Of hundreds of combinations of water and alcohols and ketones, of ethers and esters , of dielectric constants and ionic strengths, of anions like sulfate or citrate or chloride and of cations from lithium to strontium to yttrium, the only solutions that ever yielded useful crystals contained spermine. Teeter, Quigley and Rich (3) not only succeeded in establishing a precise structure for $tRNA^{Phe}$ from yeast, but were able to assign unambiguous locations for the two spermine molecules that co-crystalized with every molecule of tRNA (Figure 2). Moreover they

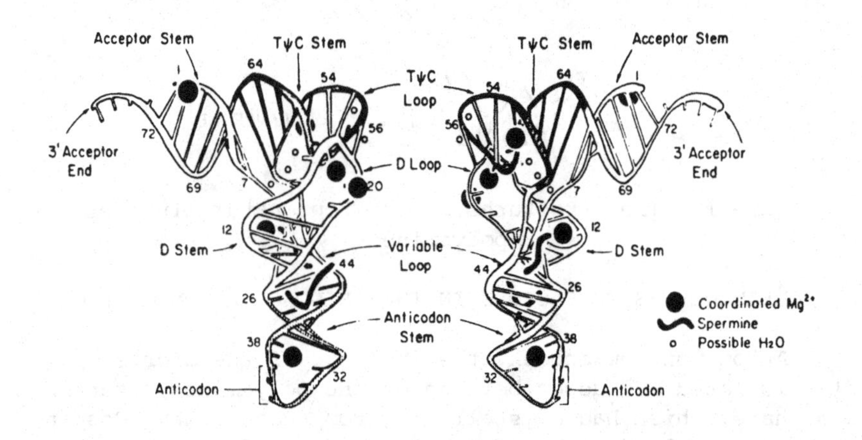

Figure 2. Two views of the structure of $tRNA^{Phe}_{yeast}$ as determined by the Rich group(3). Note the precisely oriented molecules of spermine lying in the grooves of the double helix structure. The drawing is a gift from Dr. Martha Teeter.

showed by experiment and calculation that a tRNA structure containing two twenty-Ångström long spermine molecules lying in the grooves of the RNA helix had a structure different from that of the same RNA paired with two (or more) magnesium or thallium or other cations in the same area. The long extended charge dispersion of the spermine tetra-cation causes a narrowing of the groove and a corresponding bending of the helix axis, an effect not seen with inorganic cations.

Another early hint that polyamines might have significance in this area came from efforts to explore and optimize the aminoacylation of tRNA. It appeared from the earliest experiments that polyvalent cations were essential, certainly for the synthesis of aminoacyl-adenylate and probably for the transfer of the activated aminoacyl residue to tRNA. Inevitably all sorts of cations were tested, Co^{2+}, Zn^{2+}, Mn^{2+}, K^{+}, etc., but also a few organic cations like histamine, putrescine and spermine. Spermine appeared to have some catalytic value in some systems, particularly in the translation of a poly-U message so there were further efforts to establish its role. This led to a lengthy and now pointless controversy about whether spermine "replaces" magnesium in the aminoacylation of tRNA. The fact is that magnesium ion is essential at physiological levels and that spermine is also essential under physiological conditions if an acceptable enzymic activity is to be maintained. Surprisingly, this observation was not made until the late 1970's because of the way biochemists have learned to do research. One begins with whole animals or tissues and learns what one can; protein synthesis requires O_2, it can be inhibited by azide or dinitrophenol, etc. Then one disrupts the tissue and finds that synthesis occurs on ribonucleoprotein particles (now known as ribosomes), but that cytosolic factors are needed. These cytosolic factors include not only ATP and GTP but also "activating enzymes" and a new low molecular weight ribonucleic acid (now known as tRNA). In order to study the latter two, each biochemist selected what appeared to be optimal conditions of ionic strength, temperature, pH, buffer material, ATP concentration, etc. Notably every investigator found that relatively high concentrations of Mg^{2+} (5 to 20 mM) gave stable, rapid and reproducible rates of aminoacylation of tRNA. As Figure 3 shows, the stimulatory and essential participation of spermine in the synthesis of Ile-$tRNA^{Ile}$ under physiological conditions is totally obscurred by the use of 5 mM Mg^{2+} which not only prevents spermine stimulation, but slows the reaction two-fold.

It is now clear that polyamines stimulate the rate of aminoacylation of every tRNA if the magnesium concentration is in the physiological range of 0.5 to 1.0 mM. Far from it being a question of whether spermine replaces Mg^{2+}, it is clear that in many laboratories high concentrations of Mg^{2+} were "replacing" polyamines; concealing the participation of polyamines that is essential under more nearly physiological conditions. Kinetic analysis of the formation

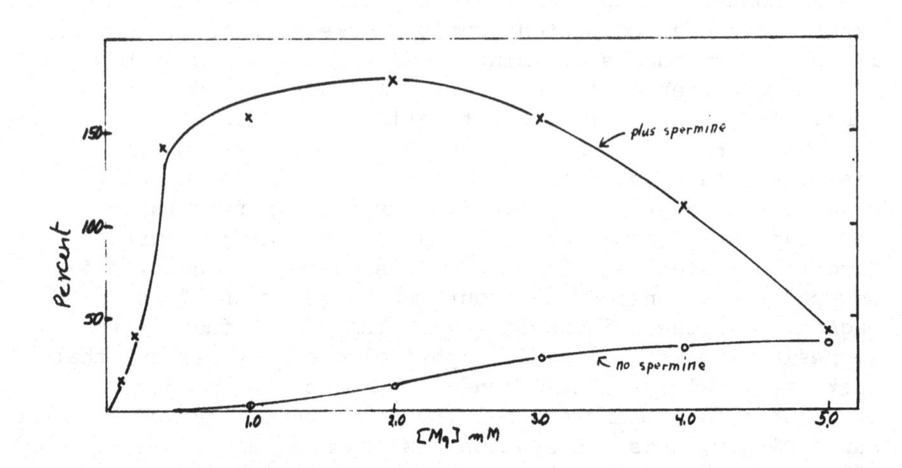

Figure 3. The rate of formation of Ile-$tRNA^{Ile}$ as a function of Mg^{2+} concentration without added spermine (o) and in the presence of 0.2 mM spermine (x).

of isoleucyl-tRNA shows that the transition state contains at a minimum enzyme, tRNA, spermine, ATP, Mg^{2+} and isoleucine, a six component or senary complex (4).

THE ROLE OF POLYAMINES IN ENHANCING THE PRECISION OF AMINOACYLATION OF tRNA

We recognized that those workers who were studying misacylation, the attachment by an amino acid:tRNA ligase of a"correct" amino acid to an "incorrect" tRNA (i.e., the use of the phenylalanine ligase to form Phe-$tRNA^{Tyr}$) had optimized their systems in a number of ways. Organic solvents were sometimes used, ATP concentrations were low, interspecific enzymes and tRNAs were matched, Tu factor might be used to stabilize the product: one universal tool to optimize the misacylation reaction was to use high Mg^{2+} commonly in the range of 20 to 30 mM. It occurred to us that the "optimal" reaction conditions used to study many aminoacylations of tRNA might, in fact, be so unnatural as to also optimize the misacylation reaction.

Table 1 shows the rate of phenylalanination of several tRNAs at low and high Mg^{2+} concentrations and in the presence or absence of polyamine and shows the frequency of misacylation relative to the rate of "correct" or cognate aminoacylation under the same conditions. For the cognate homologous synthesis of Phe-$tRNA^{Phe}$, it will be noted that 1.1 mM Mg^{2+} plus 1.0 mM spermine gives a rate of esterification comparable with the "optimal" rate obtained with 15 mM Mg^{2+}. While spermine accelerates the correct acylation in 1.1 mM Mg^{2+} by a factor of fifteen-fold, neither spermine nor spermidine has a significant effect on the rate of misacylation. On the other hand, in the absence of polyamine an additional 14 mM Mg^{2+} stimulates correct aminoacylation some eighteen-fold BUT also stimulates misacylation 300 to 1000-fold. The net effect is that relative to the rate of synthesis of Phe-$tRNA^{Phe}$, the rates of misacylations are lowest with physiological concentrations of Mg^{2+} and polyamines and are highest with polyamine-free 15 mM Mg^{2+} (where Val-$tRNA^{Phe}$ is formed one-fifth as rapidly as Phe-$tRNA^{Phe}$).

In numerous other studies of non-cognate Enzyme·tRNA interactions it had been shown that tRNAs bind to the non-cognate activating enzymes almost as well as to cognate enzymes. Frequently the association constants are almost the same and they rarely differ by a factor of one hundred. Furthermore it was known that high Mg^{2+} inhibits the binding of non-cognate tRNA much more than the binding of cognate tRNA. Reference to Table 2 will show that the K_m's for the

TABLE 1: RATE OF FORMATION OF Phe-tRNA[a]

	1.1 mM Mg^{2+}		15 mM Mg^{2+}	
	0.0 PA[b]	1.0 mM PA[b]	0.0 PA[b]	1.0 mM PA[b]
$tRNA^{Phe}_{yeast}$	13	216	244	360
$tRNA^{Val}_{E.coli}$	.05	.11	57	18
Error freq.[c]	.004	.0005	.23	.05
$tRNA^{Ala}_{E.coli}$	.015	.03	4.5	4.3
Error freq.[c]	.001	.00014	.018	.012

a) rates in moles of product/mole of enzyme per minute
b) PA = spermine with yeast tRNA, spermidine with E.coli tRNA, other combinations gave similar results.
c) Error Frequency is the rate of non-cognate acylation divided by the rate of cognate acylation under the same conditions.

TABLE 2: ENZYMATIC PARAMETERS OF SYNTHESIS OF Phe-tRNA (5)

	$(Mg^{2+})^c$	$K_m(tRNA)^c$	V_{max}^a	$K_m(tRNA)^c$	V_{max}^a
		no polyamines		0.2 mM polyamine (b)	
$tRNA^{Phe}_{yeast}$	.06	60	.95	40	19.2
	14	20	217	90	350
$tRNA^{Val}_{E.coli}$	.03	220	.05	700	.039
	14	1100	40	1600	24
$tRNA^{Ala}_{E.coli}$	.06	100	.003	100	.003
	14	500	4.5	1000	3.0

a) rates in moles of product formed per mole of enzyme per minute
b) for $tRNA_{yeast}$, spermine; for $tRNA_{E.coli}$, spermidine. other combinations were similar.
c) K_m(tRNA) in nanomolar; concentration of free Mg^{2+} in mM

cognate aminoacylation are essentially constant with or without high Mg^{2+} and with or without polyamine. The K_m's for non-cognate tRNA are increased with either high Mg^{2+} or polyamine such that less misacylation would be expected whereas, in fact, more misacylation is found at high Mg^{2+} concentrations especially in the absence of polyamine. The reason that polyamines reduce the error rate is that under otherwise identical conditions the presence of 0.2 mM polyamine has accelerated the V_{max} of cognate aminoacylation twenty-fold while depressing the V_{max} of non-cognate aminoacylation. The Ebel group, in particular, has observed that discrimination against misacylation lies far more in the V_{max} term than in the K_m term. The present work emphasizes the particular contribution of polyamines to the V_{max} term which is to say that polyamines stabilize the transition states for cognate aminoacylation while Mg^{2+} accelerates misacylation; in other words Mg^{2+} tends to stabilize the transition states of non-cognate aminoacylation of tRNA.

THE NATURE OF ENZYMIC CATALYSIS

Beginning with the Lock and Key concept, a variety of proposals have been made to explain how enzyme molecules enhance the rates of an otherwise thermodynamically possible reaction. By binding one or more substrates the enzyme brings substrates into closer proximity not only with each other but also with reactive groups on the enzyme. It can be estimated that the "effective concentration" of a reagent is as high as 10^6 M, in other words that a particular group in desirable orientation reacts a million times more rapidly than a similar group not appropriately anchored. Enzymes provide the possibilities for general acid and general base catalysis, for the formation of co-valent intermediates, for inducing stress in the substrate, etc. However, as Jencks (6) and Atkinson (7) have pointed out, none of these factors or all taken together are capable of explaining the degree of enhancement that enzymic catalysis displays. Thermodynamically what is necessary is that an enzyme·substrate complex, even though it be tight, is unable to express all of the intrinsic binding energy possible by complete interaction of substrate and protein. It is the expression of this potential binding energy (very negative enthalpy) in the transition state that lowers the free energy of the transition state which is equivalent to saying that the reaction goes more rapidly. One interesting example is Jencks' Succinyl-CoA:Keto Acid CoA Transferase which catalyzes the following reaction: (8)

Suc∼SCoA + Enz-COOH ⟶ SucOH + Enz-CO∼SCoA

Enz-CO∼SCoA + CH_3COCH_2COOH ⟶ Enz-COOH + CH_3COCH_2CO∼SCoA

In the ground state succinyl∼CoA binds only about as well to the enzyme as the equally chemically reactive SucCO∼SCH_3. The ground states of the enzyme complex with correct or incorrect substrate are of comparable energy, but in the transition state the large CoA group develops charge-charge interactions, van der Waal's interactions, hydrogen bonds, etc., of low enthalpy; the intrinsic binding capacity is realized and the free energy of the transition state is reduced by some 16 kilocalories per mol, enough to enhance the reaction rate by 10^{12}. We can only speculate on the way that the forces of this developing binding are coupled to the active site of the enzyme; it is sufficient here to note that the potential for expressing very specific binding forces *ONLY* in the transition state is a *sine qua non* for enzymic catalysis. It may be that some of these forces will be expressed before the transition state is achieved; in other words that a detectable conformational change will occur with a correct substrate that cannot be detected with an equally well bound but incorrect substrate.

THE SPECIFIC CATALYSIS OF THE AMINOACYLATION OF tRNA

We have already noted that the aminoacyl-tRNA synthetases bind a wide variety of non-cognate tRNAs more or less equally well. The binding energy that can be expressed in the ground state of the enzyme·tRNA complex is relatively non-specific. Every tRNA is roughly "L" shaped, each has many ionized phosphate residues which can be attracted by Coulombic forces to appropriately located positive charges on the enzyme. The bringing of two such charged and bulky molecules together releases a great deal of structured water; all such associations show a very high positive entropy of association. (The release of one ice-like structured water molecule is equivalent to about 5 kcal/mol·deg or a factor of 10 in the association constant at 25°C. In these cases it is common to find entropies of association in the range of 30 to 70 cal/mol·deg). These non-specific forces lead to enzyme·tRNA complexes with K_m's in the range of 10^{-6} molar depending largely on pH, ionic strength and solvent as would be expected from the nature of the binding forces.

Krauss and co-workers (9) using Temperature-Jump techniques have been able to establish that only the cognate tRNA is capable of driving a conformational transition in the enzyme·tRNA complex. As one example of several, they found:

$$Enz^{Phe}_{yeast} + tRNA^{Tyr}_{E.coli} \underset{k_{21}= 1600\ s^{-1}}{\overset{k_{12}= 2.4 \times 10^{8}\ M^{-1}s^{-1}}{\rightleftharpoons}} \text{non-cognate } Enz\cdot tRNA \qquad K_a=5 \times 10^5\ M^{-1}$$

while

$$Enz^{Phe}_{yeast} + tRNA^{Phe}_{yeast} \underset{k_{21}= 250\ s^{-1}}{\overset{k_{12}= 2 \times 10^{8}\ M^{-1}s^{-1}}{\rightleftharpoons}} \text{cognate } Enz\cdot tRNA \qquad K_a=8 \times 10^5$$

$$\text{cognate } Enz\cdot tRNA \underset{k_{32}= 750\ s^{-1}}{\overset{k_{23}=420\ s^{-1}}{\rightleftharpoons}} \text{cognate } (Enz^{Phe}\cdot tRNA^{Phe})^* \qquad K_2=0.6$$

Cognate and non-cognate tRNA bind with comparable rates and equilibrium constants to form an E S complex. Only the cognate complex is able to change conformation to a second, thermodynamically unstable complex, E S*. Only this second complex is capable of reacting with ATP and phenylalanine to bring about phenylalanination of the tRNA (entirely comparable results have been obtained with cognate and non-cognate tRNAs and the enzymes specific for $tRNA^{Tyr}$ or $tRNA^{Ser}$).

It should be noted that the entropy change in forming the non-cognate complexes (e.g. $Enz^{Phe} \cdot tRNA^{Tyr}$) is in the range of 75 cal/mol·deg while the entropy of formation of the cognat compexes (e.g. $(Enz^{phe} \cdot tRNA^{Phe})+(Enz^{phe} \cdot tRNA^{Phe})^*$) is generally much lower from which one may infer that the reaction $E \cdot S \longrightarrow E \cdot S^*$ is driven by a high negative enthalpy accompanied by a corresponding negative entrophy change. Note that the equilibrium favors the E·S form rather than the E·S* conformation.

THE ROLE OF WATER IN ENZYMIC CATALYSIS

In 1975 Low and Somero (10) determined volumes of activation ($\Delta V^{\ddagger}$) for a number of enzymic reactions; i.e., the difference between the molar volume of the ground state of the enzyme·substrate complex and the molar volume of the transition state. For each enzyme Low and Somero found a structural or intrinsic volume change that presumably reflects a contraction or expansion of the enzyme·substrate complex as it passed through the transition state. Beyond this there was a much larger "hydration effect" that was detected by conducting the reactions in the presence of Hofmeister (chaotropic) salts. Hofmeister anions break up or discourage the formation of water structures around exposed charged groups and amino acid side chains. To the extent that the transition state is transiently stabilized by newly structured water at the enzyme·water interface, it would be expected that Hofmeister anions would prevent the development of dense water structures (i.e., increase $\Delta V^{\ddagger}$) and in proportion they would inhibit the reaction. Low and Somero found that Hofmeister salts invariably affected both the activation volumes and the reaction rates; an observation that lead to the conclusion that changes in the enzyme·water interface are used to stabilize the transition state.

Bulk water has high volume relative to water structured around aliphatic groups or charged groups. Similarly, bulk water has high entropy. We reasoned that reactions whose transition state is stabilized by the release of structured water would show a high positive entropy and high positive volume of activation. A high positive entropy of activation seems almost an intuitive contradiction. A transition state with its perfectly oriented reacting groups cannot be "disordered". The only source of high positive entropy must be in the "melting" of bound water that is somehow coupled to the action at the catalytic site. One example outside the tRNA field will suffice:

TABLE 3: THE HYDROLYSIS OF ACETYLCHOLINE AND SOME ANALOGS BY ACETYLCHOLINESTERASE

Substrate	K_d	ΔS_{ass}	ΔV_{ass}	k_{cat}	$\Delta S^{\pm}$	$\Delta V^{\pm}$
CH_3-$\overset{CH_3}{\underset{CH_3}{N^{+}}}$-$CH_2CH_2OAc$ (I)	7×10^{-4}	-20	-54	4×10^4	+24	+ 25
CH_3-$\overset{CH_3}{\underset{H}{N^{+}}}$-$CH_2CH_2OAc$ (II)	2×10^{-3}	+2		7×10^3	-8	
CH_3-$\overset{H}{\underset{H}{N^{+}}}$-$CH_2CH_2OAc$ (III)	1×10^{-2}	+4		3×10^3	-9	
H-$\overset{H}{\underset{H}{N^{+}}}$-$CH_2CH_2OAc$ (IV)	2×10^{-2}	+7		2×10^2	-9	
CH_3-$\overset{CH_3}{\underset{CH_3}{N^{+}}}$-$CH_2CH_2NHAc$ (V)	1×10^{-4}			1×10^3	+12	
CH_3-$\overset{CH_3}{\underset{H}{N^{+}}}$-$CH_2CH_2NHAc$ (VI)	2×10^{-4}			3×10^2	-13	
CH_3-$\overset{CH_3}{\underset{CH_3}{C}}$-$CH_2CH_2OAc$ (VII)	3×10^{-3}	+54	+115	3×10^3	-12	-25

K_d in moles/liter; ΔS_{ass} and $\Delta S^{\pm}$ in cal/mol·deg; ΔV_{ass} and $\Delta V^{\pm}$ in ml/mol (11).

It will be noted that only acetylcholine (I) and its isoelectric isoster, trimethylaminoacetamide (V) have negative entropies and negative volumes of association; each of the other analogs lacking only a methyl group or a positive charge (VII) complex with the enzyme, sometimes as well as the natural substrate, but with positive ΔV_{ass} and ΔS_{ass}: correspondingly, only I and V are able to go from the compact, low entropy ground state to a transition state of higher volume and higher entropy. One visualizes a series:

$$E + S \rightleftharpoons \underset{\text{high}\,\Delta V,\Delta S}{E\cdot S} \rightleftharpoons \underset{\text{low}\,\Delta V,\Delta S}{E\cdot S} \rightleftharpoons \underset{\text{high}\,\Delta V^{\pm},\Delta S^{\pm}}{E\cdot S^{\pm}}$$

Of these, E·S is the predominant ground state for compounds II, III, IV, VI, and VII while I and V will be found largely in the ground state E·S*.

If the ground state in predominantly E·S*, the positive entropy of activation becomes a free energy-reducing contribution to the transition state. If, as for the poorer substrates II, III, IV, VI, and VII, the ground state is nearer to E·S, the complex lacks the driving force of organized water and the reaction goes more slowly.

APPLICATION TO THE AMINOACYLATION OF tRNA

Transfer RNA is a relatively small molecule but large enough and complex enough to have a well-defined structure that can be changed according to conditions of temperature, solvent and counter-ions. As noted earlier, one form containing two molecules of spermine and five magnesium ions is so precisely defined in structure that it, and only it, is capable of forming crystals for high resolution X-Ray analysis. Several physico-chemical approaches have demonstrated that in the absence of counter-ions the molecule becomes extended as a result of internal repulsion of the anionic phosphate residues. As the concentration of magnesium ion increases, the tRNA molecule becomes more compact attaining a structure similar to that found in spermine-containing structures but inevitably not the same. In the absence of spermine or at high magnesium concentrations the spermine-binding sites are occupied by magnesium. At this point the structure is somewhat more extended and less well defined. It is easy to suppose that while $tRNA^{phe}$ has lost some of the precise structure conferred on it by its composition <u>and</u> the spermine, other tRNAs will have lost some of the rigidly oriented components that distinguished them from $tRNA^{Phe}$. In other words, at very high magnesium the spermine-maintained distinctions between different tRNAs are blurred and the resulting structures are not only able to form the E·S complex, but to a greater or lesser extent, able to stimulate a conformational change to $E.S^*$. As in the case of the acetylcholinesterase substrate, we find a rough constancy between the sum of the entropy of association and the entropy of activation (Table 4).

This is consistent with the idea that a function of spemine is to define the tRNA structure so precisely that more of the total instrinsic binding energy can be ex pressed only in the formation of an $E{\cdot}S^*$ complex whose free energy approaches that of $E{\cdot}S^{\ddagger}$. This complex presumably has a low

TABLE 4: ENTROPIES OF ASSOCIATION AND ACTIVATION FOR SOME ENZYME:tRNA REACTIONS

	ΔS_{ass}	$\Delta S^{\pm}$
Enz^{Phe}		
+ $tRNA^{Phe}$, low Mg^{2+}, no spermine	+25	no reaction
the same, but high Mg^{2+} (slow Rxn)	+15	+25
same, 2mM Mg^{2+}, .2mM Spm (fast Rxn)	+10	+35
+ $tRNA^{Tyr}$ + Mg^{2+} + spermine	+75	no reaction
Enz^{Ile}		
+ $tRNA^{Ile}$ + 5mM Mg^{2+} (slow Rxn)	+20	+27
same + 2mM Mg^{2+} + .1mM spm(fast Rxn)	+10	+35
$tRNA^{Val}$ + Mg^{2+} + spm	+ 45	no reaction

ΔS_{ass} and $\Delta S^{\pm}$ in Cal/mol·deg

entropy because some amino acid side chains are oriented to organize adjacent water. In turn the water-organizing residues are coupled to the active site is such a way that, as reaction proceeds, water is released--one example might be that two oppositely charged residues move towards each other thus gaining both negative enthalpy and, by charge-cancellation and water release, positive entropy. Both effects would reduce the $\Delta G^{\pm}$ and assist in catalysis.

If, as seems inevitable, the enormous entropy changes of binding and activation are due to the structuring and melting of superficial water, one would expect to find these reactions to be very sensitive to Hofmeister anions. In fact, 16 mM dinitrophenoxide ion reduces the catalytic rate constant by 90%; other Hofmeister anions are similarly inhibitory (12).

A non-cognate tRNA brought into a blurry almost "correct" configuration by high magnesium would not be expected to gain as much advantage from being able to reorganize precisely the enzyme and its surrounding water; in proportion the reaction would show a lower entropy of activation and a lessened sensitivity to Hofmeister anion. Preliminary experiments suggest both to be true even though the E·S complexes are formed.

CONCLUSION

Polyamines do increase the fidelity of aminoacylation of tRNA entirely by their ability to increase V_{max} for acylation rather than by affecting K_m(tRNA). In turn, this effect appears to result from the ability of polyamines to confer a very precise and relatively inflexible structure on all tRNAs such that only cognate tRNA can bring the enzyme from a simple non-specific Enz·tRNA complex of high volume and entropy into a low entropy/low volume complex whose properties and "strain" are derived from structured water. This second complex approaches the free energy of the transition state as schematically shown in Figure 4. (The quantitative data are appropriate for the spermine catalyzed synthesis of Ile-$tRNA^{Ile}$). The complexes of non-cognate spermine·tRNAs with enzymes are unable to form the E·S* conformer while high magnesium blurs the distinctions between all tRNAs and permits many to achieve a conformation similar to, but not as strained as the cognate spermine-containing E·S*.

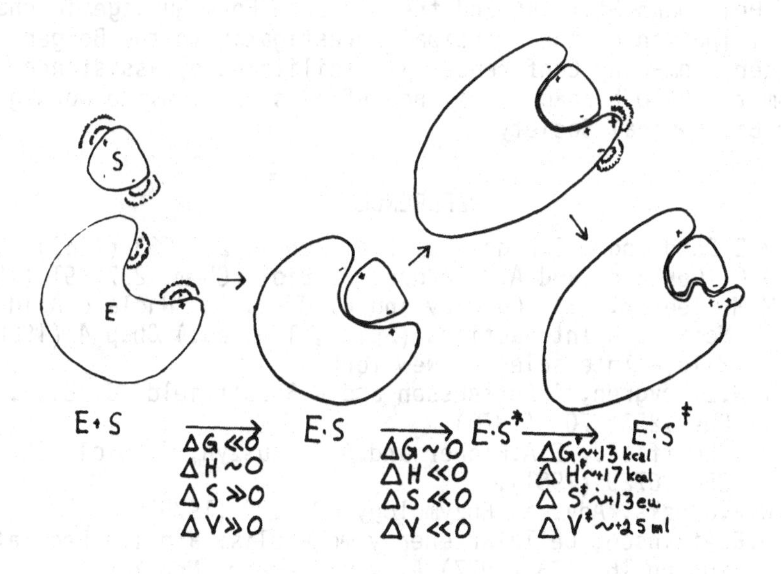

Figure 4; A schematic representation of the thermodynamic changes accompanying the formation of the non-specific Enz·tRNA E·S complex; the unique conversion of the spermine stiffened cognate E·S complex to the strained hydrated E·S* complex; and the passage of this into the transition state.

Although studies of this sort may tell us much of how the precision of synthesis of protein is maintained, they may be even more valuable as a guide to the nature of enzymic catalysis.

Finally there is humble pie. As has happened before, hindsight reveals the importance of events or substances or phenomena that convenience made us overlook. Polyamines are becoming prominent in many fields of biochemistry, physiology, psychiatry and medicine. It is hard to see how they could have been overlooked. How many similarly important phenomena or substances are we ignoring as we design our day's protocol?

ACKNOWLEDGEMENT

This work has been supported by the National Institute of General Medical Sciences. Some of the experiments were done at the Medizinische Hochschule Hannover with the help of Professor Günther Maass and Dr. Claus Urbanke and with financial assistance from the Deutsche Fulbright Kommission, the Heinemann Stiftung and the Deutsche Forschungsgeminschaft. Participation of the Principal Investigator in the Bergen Linderström-Lang Conference was facilitated by assistance from the NATO bureau of Science Affairs and from the Norwegian Biochemical Society.

REFERENCES

1. J.E.Dowd and D.S.Riggs, J. Biol. Chem. 249,863 (1965)
2. J.C. Gerhardt and A.B.Pardee, J. Biol. Chem. 237,891 (1962)
3. M.M. Teeter, G.J. Quigley and A. Rich, in "Nucleic Acid-Metal Ion Interactions"(Spiro, T.G. ed.) Chap 4 (1980) Wiley- Interscience, New York.
4. T.N.E.Lövgren, A.Pettersson and R.B.Loftfield, J. Biol. Chem.253,6702 (1978).
5. R.B.Loftfield, E.A.Eigner and A.Pastuszyn, J. Biol. Chem. 256, 6729 (1981).
6. W.P.Jencks, Adv. in Enzymology 43, 220 (1975)
7. D.E.Atkinson, Cellular energy metabolism and its Regulation.pp 169-173 (1977) Academic Press, New York.
8. H.White and W.P.Jencks, J. Biol. Chem. 251, 1688 (1976)
9. G.Krauss, D.Riesner and G.Maass, Eur. J. Bioch. 68, 81 (1976); see also EJB 68, 71 (1976).
10. P.S.Low and G.N.Somero, Proc. Nat. Acad. Sci. 72, 3014 & 3305, (1975)
11. I.B.Wilson and E.Cabib, J.A.Chem. Soc. 78, 202 (1956) and unpublished work from the authors' laboratory.
12. R.B.Loftfield, E.A.Eigner, A. Pastuszyn, T.N.E.Lövgren and H.Jakubowski, Proc. Nat. Acad. Sci. 77, 3374 (1980)

THE ALLOSTERIC BEHAVIOUR OF THE ELONGATION FACTOR EF-Tu

Barend Kraal, Johannes M. Van Noort and
Leendert Bosch

Department of Biochemistry
State University of Leiden, Wassenaarseweg 64
2333 AL Leiden, The Netherlands

INTRODUCTION

The *E. coli* factor EF-Tu is a multifunctional protein that interacts with guanine nucleotides, tRNA, the elongation factor EF-Ts, the ribosome, the antibiotics kirromycin and pulvomycin, and probably bacteriophage RNA. It thus fulfils an essential function in protein synthesis (Miller & Weissbach, 1977; Kaziro, 1978) and phage RNA replication (Blumenthal & Carmichael, 1979). It is an abundant cytoplasmic protein that, depending on bacterial growth conditions, can amount from five to ten percent of the total intracellular protein (Furano, 1975; Van der Meide *et al.*, 1983). Its amino acid sequence (393 residues) has been determined (Arai *et al.*, 1980; Jones *et al.*, 1980) and considerable progress has been made in the elucidation of its three-dimensional structure (Kabsch *et al.*, 1977; Jurnak *et al.*, 1977; Rubin *et al.*, 1981; Clark *et al.*, 1982).

EF-Tu is encoded by two genes, *tufA* and *tufB*, that are distantly located on the *E. coli* genome. Their nucleotide sequences are remarkably conserved and differ at 13 positions only (Yokota *et al.*, 1980; An & Friesen, 1980). The corresponding gene products, EF-TuA and EF-TuB, are identical except for the C-terminal acid residue. No functional difference between the two proteins has been reported (Miller *et al.*, 1978; Shibuya *et al.*, 1979).
The considerable amount of structural information regarding EF-Tu and its encoding genes, as well as its capacity to interact with so many ligands makes this protein an ideal

object for studies of structure/function relationships. In the two following sections we will subsequently describe our approach with chemical modification and affinity labelling techniques, and our investigations on specific mutants of EF-Tu. The last section is a discussion of the new findings in the light of the available information on the 3D-structure of EF-Tu and with regard to its multifunctional behaviour.

THE INDUCTION OF TWO tRNA BINDING SITES ON THE EF-Tu MOLECULE

The key role of EF-Tu in polypeptide chain elongation is depicted in the classical scheme of Fig. 1. In each step of the EF-Tu cycle the protein molecule undergoes a conformational change upon subsequent interaction with the allosteric effectors such as GTP, aminoacyl-tRNA, etc. A major event is the delivery of an aminoacyl-tRNA to the A-site on the ribosome·mRNA complex, an interaction for which EF-Tu requires a P-site filled with either uncharged or peptidyl-tRNA (De Groot *et al.*, 1971; Lührmann *et al.*, 1979). At the stage of codon-anticodon interaction the GTPase centre on EF-Tu is activated and a proof reading mechanism comes to

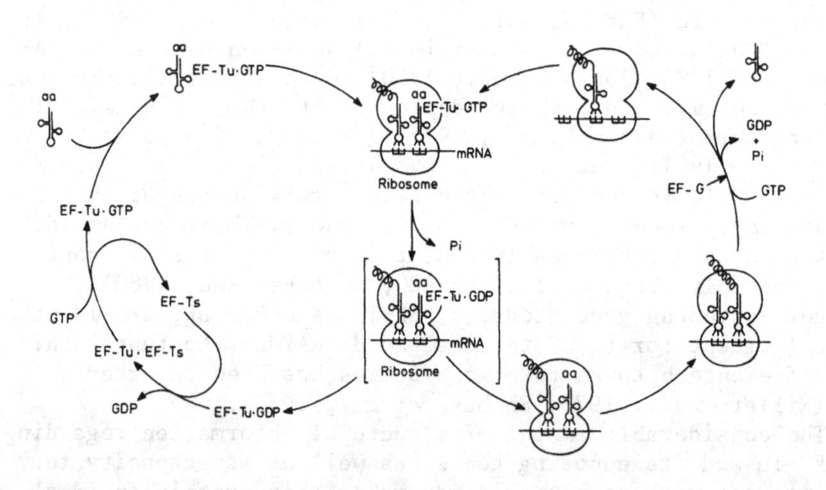

Fig. 1. *The polypeptide chain elongation cycle in E. coli.*

action (Hopfield, 1974; Gavrilova *et al.*, 1981; Thompson & Karim, 1982). The EF-Tu·GDP formed is released from the ribosome as a binary complex, eventually accompanied by the aminoacyl-tRNA in the case of a non-cognate anticodon.

An allosteric effector not shown in Fig. 1 is the antibiotic kirromycin (see Fig. 2) that inhibits peptide elongation. Upon binding to EF-Tu it activates the GTPase centre and induces a "GTP-like" conformation, thus preventing the release of EF-Tu·GDP from the ribosome (Chinali *et al.*, 1977; Wolf *et al.*, 1977; for a review see Parmeggiani & Sander, 1980). As a result of this induced conformation EF-Tu·GDP complexed with kirromycin becomes able, like EF-Tu·GTP, to form a complex with aminoacyl-tRNA. We recently found (Van Noort *et al.*, 1982) that the antibiotic induces an additional tRNA binding site both in EF-Tu·GDP and EF-Tu·GTP, that accepts uncharged tRNA, aminoacyl-tRNA and N-acetylaminoacyl-tRNA.

Evidence Obtained by Differential Chemical Modification

The existence of an additional tRNA binding site was deduced from modification studies of EF-Tu with sulfhydryl reagents like N-tosyl-L-phenylalanine chloromethylketone and N-ethylmaleimide. In both cases one unique site in the polypeptide chain is involved, namely Cys-81. An illustration of the results obtained with (^{3}H)N-ethylmaleimide is shown in Fig. 3.
In the absence of kirromycin (open symbols) the only effect measured concerns the interaction between EF-Tu·GTP and aminoacyl-tRNA: increasing concentrations of the latter eventually lead to a complete protection of Cys-81. Here we are dealing with the classical tRNA binding site (Arai *et al.*, 1974), which we shall call site I. In the presence of kirromycin (closed symbols) the picture is totally different. For

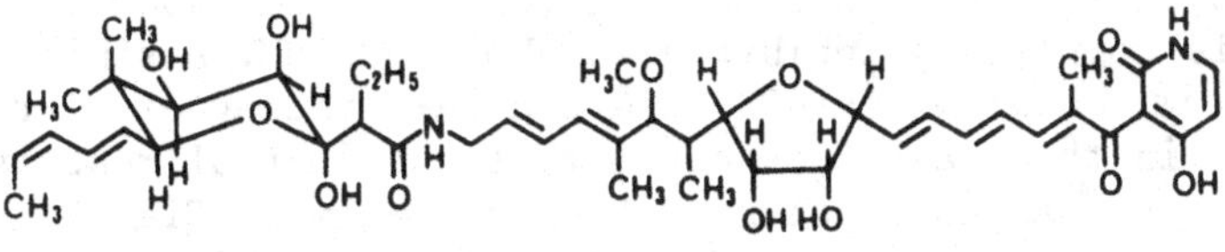

Fig. 2. *The structure of the antibiotic kirromycin.*

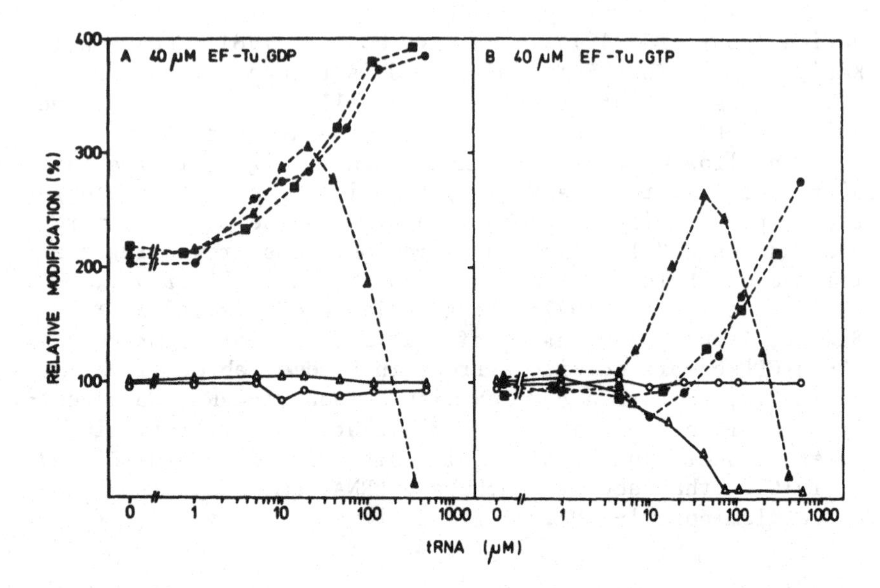

Fig. 3. *The reactivity of Cys-81 in EF-Tu towards modification by (^{3}H)N-ethylmaleimide under the influence of various ligands.* The modification was performed either in the absence (open symbols) or presence (closed symbols) of 180 µM kirromycin, and at varying concentrations of uncharged tRNA (circles), aminoacyl-tRNA (triangles), or N-acetylaminoacyl-tRNA (squares). The reaction mixtures (25 µl) further contained 40 µM EF-Tu, 65 µM (^{3}H)N-ethylmaleimide (22 Ci/mol), 20 mM Tris-HCl pH 7.6, 3.5 mM $MgCl_2$, 75 mM KCl and 10% (v/v) methanol (added as the solvent for kirromycin and N-ethylmaleimide). After 30 min at 0°C the reaction was terminated by the addition of 25 µl 20 mM 2-mercaptoethanol, followed by 10 µl 1% (w/v) bovine serum albumin solution and 3 ml 7% (w/v) trichloroacetic acid. The precipitates were filtered and counted. For further details see Van Noort *et al.* (1982).

EF-Tu·GDP a two-fold enhanced reactivity is found upon kirromycin binding, which is even further enhanced upon addition of increasing amounts of uncharged tRNA (circles), aminoacyl-tRNA (triangles), or N-acetylaminoacyl-tRNA (squares). Only in the case of aminoacyl-tRNA the labelling drops again when the concentration is raised above 20 µM, and is completely inhibited at about 300 µM. A similar picture is obtained for EF-Tu·GTP after binding of kirromycin, which in this case does not affect the initial reactivity of Cys-81. An interesting difference between Fig. 3A and 3B is, that enhancement of Cys-81 modification in EF-Tu·GTP·kirromycin requires higher concentrations of uncharged tRNA and N-acetylaminoacyl-tRNA than that of aminoacyl-tRNA.

Experiments like these strongly suggest that the antibiotic opens up not one but at least two binding sites for aminoacyl-tRNA. Binding to one site of the EF-Tu·nucleotide·kirromycin complex results in the exposure of Cys-81, whereas binding to the second site blocks the reaction with the modifying agent. The latter site is identified with the classical tRNA binding site (site I) on account of the protection of Cys-81 and its selectivity for aminoacyl-tRNA. The former site is disclosed by kirromycin and readily accepts uncharged tRNA, aminoacyl- or N-acetylaminoacyl-tRNA. It is designated site II. In the presence of kirromycin site II displays a higher affinity for tRNA than site I. On the other hand, the shielding of Cys-81 upon binding at site I dominates the exposure caused by the tRNA binding at site II.

Our interpretation of the results exemplified in Fig. 3 is given in the model of Fig. 4. Several alternative models could not provide a satisfactory explanation (Van Noort *et al.*, 1982). Although the data do not permit a quantitative evaluation of dissociation constants, Fig. 4 gives a qualitative impression of the various affinities between the interaction partners. The presence of the additional site II on EF-Tu may have important implications. It means that in the presence of kirromycin EF-Tu·GTP can bind two tRNA molecules simultaneously, one bearing an aminoacyl and the other a peptidyl moiety. On the ribosome this may account for a stable complex of EF-Tu with tRNAs at the A- and P-sites. The stability of this complex will even be enhanced when hydrolysis of GTP occurs, as site II on EF-Tu·GDP·kirromycin has a higher affinity for peptidyl-tRNA than site II on the GTP containing complex. This explains the tenacity with which EF-Tu·GDP·kirromycin binds to the ribosome because EF-Tu becomes anchored both to aminoacyl- and peptidyl-tRNA.

Evidence Obtained by Affinity Labelling

In order to obtain independent support for our model and to characterize the two tRNA binding sites on EF-Tu in a more direct fashion we tried to cross-link tRNA to the EF-Tu·kirromycin complexes. The method employed consisted of periodate oxidation of the 3' end of uncharged tRNA and reduction of the intermediary Schiff's base complexes formed with ε-NH_2 side chains of the protein molecule (Hansske & Cramer, 1979).

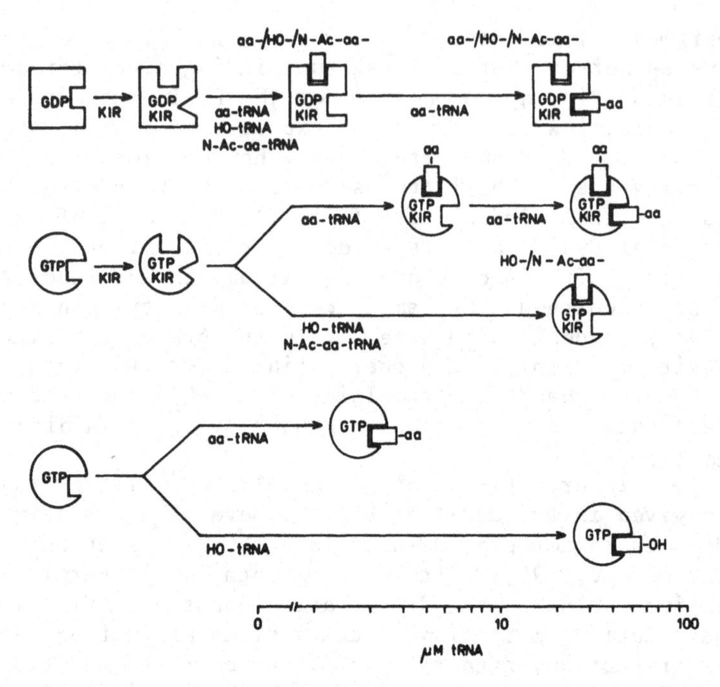

Fig. 4. *A tentative model of complex formation between EF-Tu and tRNA in the presence or absence of kirromycin.* The protein molecule is represented by a circle (EF-Tu·GDP) or a square (EF-Tu·GTP). The indentations in the outlines represent the tRNA binding sites on the elongation factor; the triangular indentations indicate that we do not know whether or not the affinity of aminoacyl-tRNA for site I of the protein·kirromycin complex increases upon occupation of site II by tRNA (see also last section). The presence of a tRNA binding site on EF-Tu·GDP and the binding of uncharged tRNA to EF-Tu·GTP was detected by modification with N-tosylphenylalanine chloromethylketone (Jonák *et al.*, 1980). Aminoacyl-tRNA (aa-tRNA), uncharged tRNA (HO-tRNA) and N-acetylaminoacyl-tRNA (N-Ac-aa-tRNA) are all represented by rectangles. At the bottom of the figure a logarithmic tRNA concentration scale is drawn. The positions of the complexes relative to this scale indicate the lowest tRNA concentrations at which the interactions with EF-Tu (40 μM) could be detected under the conditions of Fig. 3.

A first control, whether $NaIO_4$-oxidized tRNA ($tRNA_{oxi}$) is able to bind at all to EF-Tu·GDP in the presence of kirromycin, is illustrated in Fig. 5. Both $tRNA_{oxi}$ and uncharged tRNA stimulated the modification of Cys-81 with (^3H)N-ethylmaleimide (compare Fig. 3A), albeit not with the same efficiency. The oxidation of tRNA lowers its affinity for EF-Tu·GDP·kirromycin apparently four-fold. Further evidence for the specificity of the binding is presented in Fig. 6.

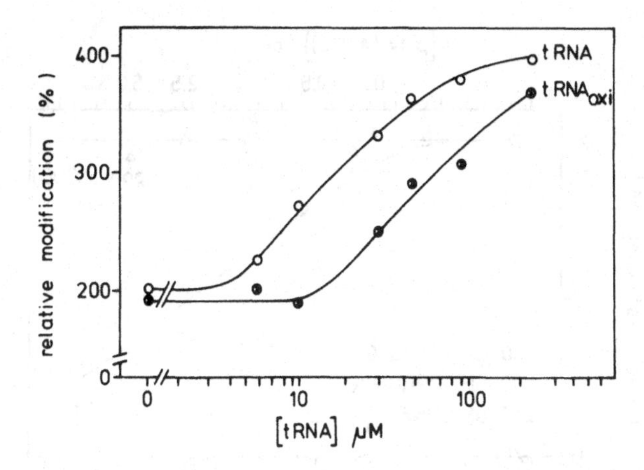

Fig. 5. *The effects of uncharged tRNA* (O) *and 3' oxidized tRNA* (●) *on the modification of EF-Tu·GDP by (^{3}H)N-ethylmaleimide in the presence of 200 μM kirromycin.* Further experimental conditions are described in the legend to Fig. 3.

A mixture of EF-Tu·GDP·kirromycin and $tRNA_{oxi}$ was treated with (^{3}H)borohydride in the presence of increasing amounts of either uncharged $tRNA_{OH}$ or a random polynucleotide poly(AGUC). Taking the net amount of (^{3}H)label incorporated into the protein as a measure of the extent of cross-linking, these data show that $tRNA_{oxi}$ in its interaction with EF-Tu is competed out by uncharged tRNA but not by poly(AGUC). The competition between $tRNA_{oxi}$ and uncharged tRNA again points to a similar ratio of affinities for EF-Tu·GDP·kirromycin as found in Fig. 5.

For identification of the cross-linking site on EF-Tu the reaction with (^{3}H)borohydride (see legend to Fig. 6) was performed on a preparative scale. Cross-linking was stopped by acidification with boric acid to pH 6.0 and extensive dialysis. Free and coupled tRNAwere degraded with RNAse I, the radioactive protein was digested with thermolysin and analyzed by peptide mapping (Duisterwinkel *et al.*, 1981b). Except for Lys-89, all the lysine-containing peptides could be traced. Peptides of interest were detected by fluorescamine staining and autoradiography. The identification of the radioactive peptides by amino acid analysis was confirmed by manual Edman degradation: radioactivity was exclusively recovered at the predicted degradation step for

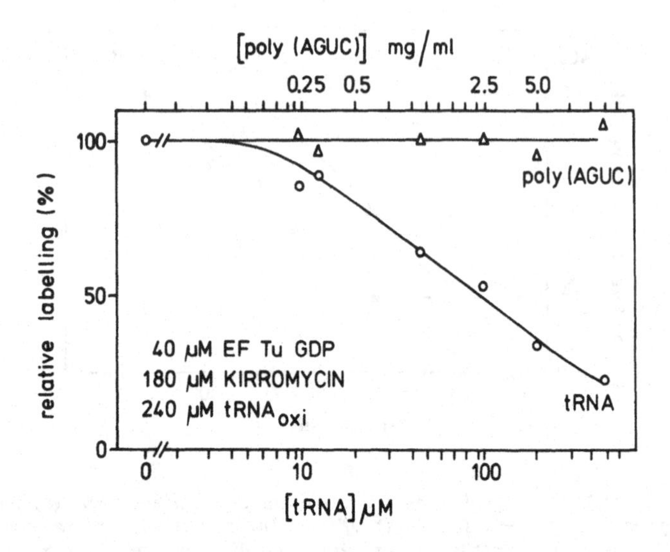

Fig. 6. *The effect of increasing amounts of uncharged tRNA* (O) *or poly(AGUC)* (Δ) *on the cross-linking efficiency of 300 μM 3' oxidized tRNA to 40 μM EF-Tu·GDP in the presence of 200 μM kirromycin.* The buffer system of Fig. 3 was used, and reduction of the Schiff's base adducts took place with 800 μM (^{3}H)$NaBH_4$ at 0°C for 1 hr. The reaction was followed by precipitation with 7% (w/v) trichloroacetic acid, filtration and counting of the filter residues. Values are corrected for background counts of tritiated products in the absence of kirromycin.

a lysine residue. In Fig. 7 the relative cross-linking is plotted against the lysine positions of the EF-Tu polypeptide chain. It shows that cross-linking only occurred in the presence of kirromycin and, with a remarkable specificity, at three positions only: Lys-208, Lys-237 and Lys-357. In the case of EF-Tu·GDP Lys-208 was the major labelled residue, in the case of EF-Tu·GTP both Lys-208 and Lys-237. Omission of $tRNA_{oxi}$ in a control experiment obliterated labelling of Lys-208 and Lys-237 but, unexpectedly, not that of Lys-357. This indicated that labelling of Lys-357 was due to cross-linking of the remaining ligand, kirromycin. Details of the cross-linking characteristics of kirromycin will be published in a joint paper with Parmeggiani's group (Paris).

Two questions now arise: what is the location of the cross-linking sites on the three-dimensional EF-Tu model, and how can they be related to the tRNA binding sites I and II described above?

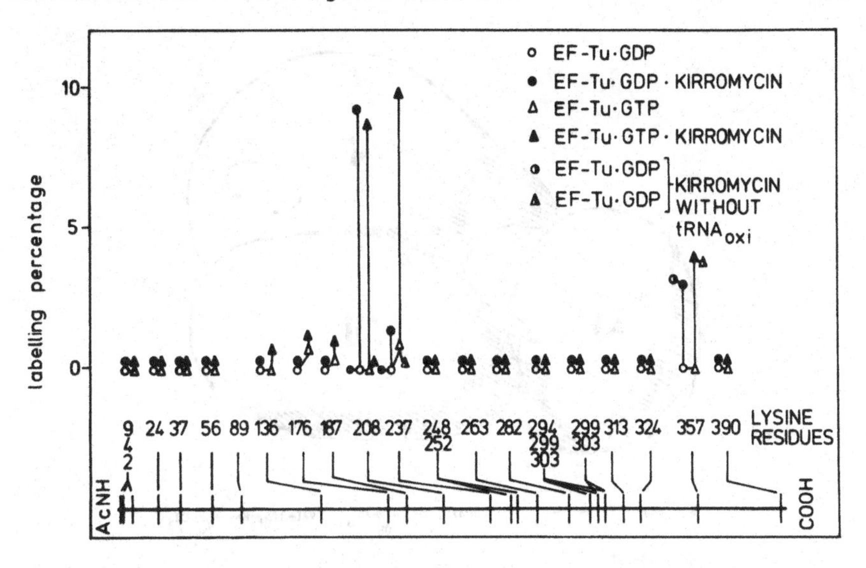

Fig. 7. *Efficiencies of* $tRNA_{oxi}$ *cross-linking to the lysines of EF-Tu in various complexes.* On the horizontal axis the lysine residue numbers are indicated. For further details see text.

X-ray diffraction of EF-Tu·GDP crystals (Clark *et al.*, 1982) has demonstrated that Lys-208 and Lys-237 are located on domain I of the protein model in Fig. 8, whereas Lys-357 is found on domain II at the inner edge of the cleft between domains I and II at the bottom of the molecule (joint manuscript in preparation). A more detailed picture of domain I (see Fig. 9) shows the position of Lys-208 on the loop between α-helix V and β-strand 6, and that of Lys-237 on α-helix VI. The distance between the two tRNA cross-linking sites may not be as large as Fig. 9 suggests, due to the fact that the loop with Lys-208 could not clearly be traced in the electron density map. On the other hand, the two sites are quite distinct, because Lys-208 of both EF-Tu·GDP and EF-Tu·GTP becomes labelled in the presence of kirromycin, whereas only Lys-237 is labelled in the latter complex.

From our previous modification studies we know that under the conditions of the cross-linking experiment with EF-Tu·GDP·kirromycin (linkage at Lys-208) mainly site II is filled (cf. Figs. 3 and 5). Also in the case of EF-Tu·GTP·kirromycin site II is filled preferentially but, as was measured for uncharged tRNA, now with a 30-fold lower affinity, which is comparable to the affinity of uncharged tRNA for site I in EF-Tu·GTP (Jonák *et al.*, 1980; Van Noort *et*

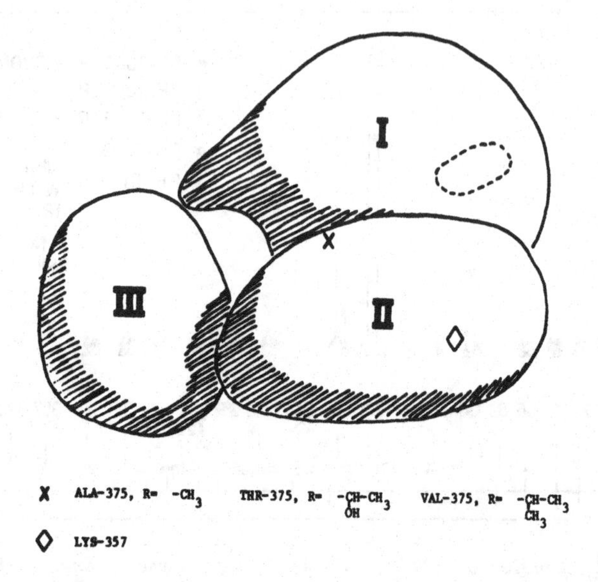

Fig. 8. *Three-dimensional representation of the overall shapes of the domains of EF-Tu.* The position of GDP is shown by the dotted region. Adopted from Clark *et al.* (1982).

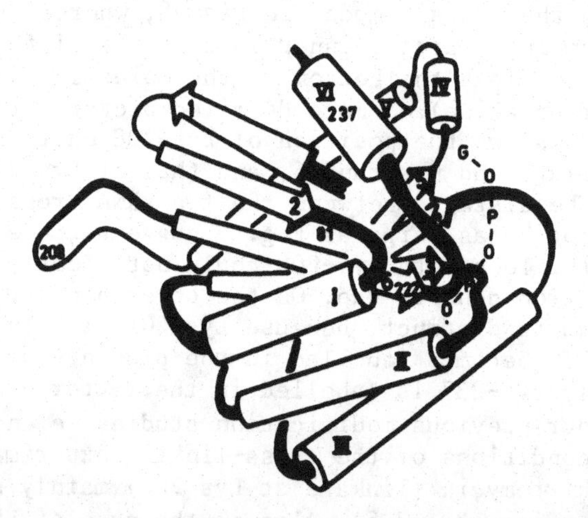

Fig. 9. *Structural cartoon of domain I (see Fig. 8) with arrows representing β-strands and cylinders representing α-helices.* All the C-terminal ends of the β-strands (arrow points) are found at one end of the central parallel β-sheet. Adopted from Clark *et al.* (1982).

al., 1982). This suggests that Lys-237 is related to site I. Obviously, no direct relation exists between the ratio of affinities and the relative efficiencies of cross-linking, as the latter values are greatly influenced by accidental steric factors. This may be illustrated by the fact that no cross-link was observed at site I in the experiment of EF-Tu·GTP without kirromycin.

Other support for the assignment of Lys-237 to site I comes from the fact that aminoacyl-tRNA and 3' terminal fragments thereof protect Cys-81 at the end of β-strand 2 against modification with N-tosylphenylalanine chloromethylketone (Jonák *et al.*, 1980). Finally, ε-N-bromoacetyllysyl-tRNA has been reported to cross-link to His-66 on β-strand 1 (Duffy *et al.*, 1981). All these data locate the 3' terminus of tRNA for site I in the same close region.

By consequence, we conclude that the identification of the two sites of tRNA cross-linking, Lys-208 and Lys-237, confirms in a more direct approach the previously obtained evidence from independent modification studies (Van Noort *et al.*, 1982) that there exists a second tRNA binding site on EF-Tu. A third line of supporting evidence for the two sites model comes from preliminary results with a defective mutant species, EF-TuB_0, which will be discussed at the end of the next section.

Finally, one might expect structural homology between the two sites capable of binding tRNA. Such a homology does exist between the region 203-208 (Glu-Arg-Ala-Ile-Asp-Lys) and 232-237 (Glu-Arg-Gly-Ile-Ile-Lys). Interestingly, both sequences are conserved in EF-1 from *Artemia salina* (Amons *et al.*, 1983). More details will be published in a paper together with Clark and coworkers (in preparation).

THE EFFECT OF SPECIFIC MUTATIONS ON THE EF-Tu FUNCTIONING

So far, two mutant EF-Tu species from kirromycin resistant *E. coli* cells with a single *tuf* gene have been described. One, designated EF-TuA_R, was isolated from strain LBE 2045 in which *tufB* was inactivated by bacteriophage Mu DNA insertion (Van der Meide *et al.*, 1980). Structural analysis showed that it contains threonine instead of alanine at position 375 (Duisterwinkel *et al.*, 1981b).

The other kirromycin resistant species was derived from strain D 2216. Both genetic and biochemical characterizations indicated that EF-Tu D 2216 is also a single gene product (Fischer *et al.*, 1977). Structural studies demonstrated (Duisterwinkel, 1981) that EF-Tu D 2216 is mutated at the same position as EF-TuA_R. In this case Ala-375 is replaced by Val. It was also found that EF-Tu D 2216 contains C-terminal glycine, like a *tufA* gene product, and that *tufB* (the gene product of which contains C-terminal serine) is not expressed in the cells.

Residue 375 is located in the cleft formed between EF-Tu domains I and II (see Fig. 8) and not far from and below the strand 44-58 (not drawn) that covalently connects the two domains. The two above-mentioned mutations cause a drop in the affinity for kirromycin of two orders of magnitude (see Table 1). In the preceding section it was already demonstrated that kirromycin cross-links at Lys-357 in a cleft at the opposite side of the central contact area between domains I and II (compare Fig. 8). It seems plausible, therefore, that alterations at position 375 exert long distance effects on the binding of kirromycin around position 357 by changing the mutual positions of domains I and II. Other effects of the mutations at position 375, as listed in Table 1, might also be explained in such an allosteric way. The GTPase centre, for example, seems to be not far away from the contact area between domains I and II. The binding of aminoacyl-tRNA provides protection against pro-

Table 1. *Kinetic features of various EF-Tu species*[a]

EF-Tu species	Wild type (Ala-375)	A_R (Thr-375)	D_{2216} (Val-375)
[Kirromycin] at 50% inhibition	0.1 μM	6 μM	20 μM
K_d' values at 5°C and 50 mM $[NH_4^+]$			
GDP binding	0.9 nM	0.5 nM	1.2 nM
GTP binding	770 nM	575 nM	91 nM
Relative GTPase activity at 37°C			
at 40 mM $[NH_4^+]$	1.0	1.1	4.5
at 400 mM $[NH_4^+]$	4.5	8.7	20.2
Relative affinity for Phe-tRNA	1.0	0.3	0.3[b]

[a] Values taken from Swart *et al.* (1982).
[b] According to A.P. Sam, A. Pingoud and L. Bosch (unpublished results).

teolysis of the strand 44-58 that connects domains I and II. The lowered affinity of EF-TuA_R for tRNA might be related to its much lower activity in the autogenous repression of *tufB* expression (Van der Meide *et al.*, 1983). The mutation in EF-TuA_R also appeared to have a negative effect on its functioning in the Qβ RNA polymerase complex (Blumenthal *et al.*, 1980). The complex seemed to be less stable than with wild type EF-Tu as the participating subunit, but the exact nature of the affected interaction has not yet been identified.

Another interesting mutant species is EF-TuB_O. It is the *tufB* product from the kirromycin-resistant strain LBE 2012 (Van der Meide *et al.*, 1980), that also contains the *tufA* product EF-TuA_R. Both species can be distinguished by isoelectric focusing, since the mutation in EF-TuB_O causes a lowering of the pI value with 0.1 unit. Because of the kirromycin-resistant phenotype, one could expect EF-TuB_O to be either resistant similar to EF-TuA_R, or non-functional, or sensitive in a recessive way. The latter possibility turned out to be the case (Duisterwinkel *et al.*, 1981a). Both EF-TuB_O·GDP and EF-TuB_O·GTP can bind aminoacyl-tRNA in the presence of kirromycin, but only the latter complex is able to associate with a ribosome·mRNA complex. This means, that after GTP hydrolysis on the ribosome, the latter ribosomal complex dissociates because the resulting EF-TuB_O·GDP has lost its affinity for the ribosome. We do not yet know whether the aa-tRNA remains in the ribosomal A-site, or accompanies EF-TuB_O·GDP on its release from the ribosome. The overall effect of the addition of kirromycin is, that EF-TuB_O·GDP does not become immobilized on the ribosome, which explains the recessive nature of the *tufB* mutation.

Preliminary experiments recently suggested that EF-TuB_O would be affected in the tRNA binding properties of site II. Chemical modification studies with N-ethylmaleimide at Cys-81 gave the wild type result for the complex EF-TuB_O·GTP·kirromycin. In the case of EF-TuB_O·GDP·kirromycin, however, the enhancement of Cys-81 reactivity characteristic for tRNA binding at site II was not observed and only protection of Cys-81 occurred by the tRNA binding at site I. By consequence, the inability of EF-TuB_O·GDP·aa-tRNA·kirromycin to bind to the ribosome could be explained by the defective affinity of site II for the tRNA at the ribosomal P-site. This would also mean that the immobilization of wild type EF-Tu on the ribosome by kirromycin primarily takes place via the tRNA at site II.

The amino acid mutation responsible for the altered EF-TuB_0 behaviour was recently located at position 222. DNA sequencing of a fragment of the cloned *tufB*$_0$ gene indicated that the original Gly has become an Asp. It can be seen in Fig. 9 that position 222 is at the C-terminal end of β-strand 6 running from the loop with Lys-208 towards the region of the GTPase centre and from there to α-helix VI. Gly-222 is part of a structural domain, characteristic for a variety of nucleotide binding enzymes, which has become known as a Rossman fold. All of the nucleotide binding enzymes studied so far contain Gly in a position corresponding to position 222 of EF-Tu. Investigations are in progress to study the effect of the Gly-222 → Asp replacement on GTP and GDP binding. More details will be published in a joint paper with Clark and Parmeggiani (in preparation).

NEW THOUGHTS ON THE MULTIFUNCTIONAL BEHAVIOUR OF EF-Tu

In the preceding section it has been shown that single amino acid mutations can evoke large effects on distant regions of the EF-Tu molecule. Many of these allosteric effects seem to be related to a change in the mutual positions of domains I and II. There is an antagonistic relation between position 357 (the kirromycin cross-link) and 375 (the kirromycin-resistant mutation). Both positions are in clefts between the domains I and II and opposite to each other in such a way that widening of one cleft might induce narrowing of the other one. Noteworthy is that kirromycin induces similar effects in EF-Tu as the elongation factor EF-Ts on one hand, and the ribosome on the other hand. Kirromycin, like EF-Ts, stimulates the GDP/GTP exchange and makes the connecting region 44-58 more accessible to proteolytic attack. In fact, kirromycin competes with EF-Ts for binding to EF-Tu. With regard to the ribosome, kirromycin similarly activates the GTPase centre on EF-Tu which is even further stimulated by aa-tRNA binding (Wolf *et al.*, 1974). The possibility may thus be envisaged that EF-Ts and the ribosome interact in turn at the cleft at the bottom of the EF-Tu molecule as depicted in Fig. 8.

The fact itself, that EF-Tu contains *two* tRNA binding sites, may have important implications for the current views on translational fidelity. From the literature it is known that the codon-anticodon interaction as such is not sufficiently sequence-specific to account for the accuracy of protein synthesis (Grosjean *et al.*, 1978). If such a limited selection step for a given tRNA would occur more than once, the overall accuracy for its selection might increase exponentially. In the past, a few authors such as Hopfield (1974), Gavrilova *et al.* (1981), and Thompson & Karim (1982), already suggested a direct involvement of EF-Tu in proof-reading, although they did not precisely mention a *structural* basis for such an interaction. It now looks very attractive to propose our two-site model of EF-Tu as such a structural basis, in the light of the general idea of conformational selection as previously discussed by Kurland (1979). The hypothesis (compare Fig. 10) is as follows.

Upon correct codon-anticodon interaction at the A-site the tRNA molecule undergoes a conformational rearrangement, by which it can pass the screening of sterical constraints imposed by the ribosomal A-site and EF-Tu site I together. This screening is therefore not restricted to the codon-anticodon area, but may also occur, for example, at the aminoacyl arm. In the case that a non-cognate tRNA would slip through and arrive at the P-site, a similar selection mechanism may work: only a correct codon-anticodon interaction induces the optimal fit for tRNA binding at the combined structure of the ribosomal P-site and EF-Tu site II. Evidence for such a second proof-reading step may be found in the work of Caplan and Menninger (1979) on mutants defec-

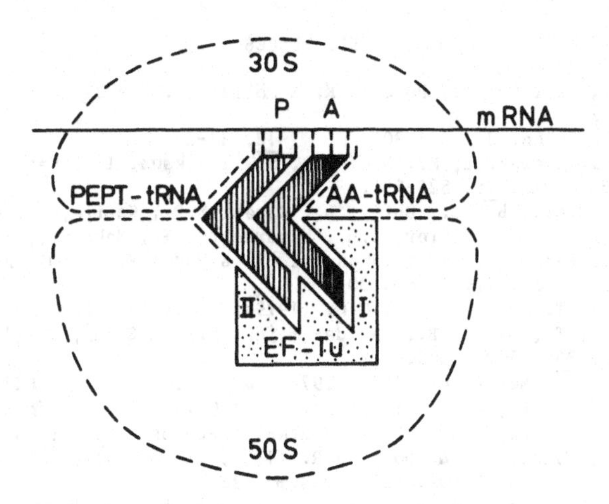

Fig. 10. *Hypothetical model of the EF-Tu mediated effect of conformational selection on the translational accuracy.*

tive in the hydrolysis of rejected peptidyl-tRNA.

The selectivity of the screening mechanism may be highly increased by the close packing of the two adjacent cognate tRNAs in their proper conformations. Structural elements of both the ribosome (such as its 16S rRNA; Thompson & Hearst, 1983) and EF-Tu may alleviate the constraints and prevent the cognate tRNAs from escaping. This could furthermore guarantee a tight joining of the incoming tRNA at site A-I to its neighbour at P-II (Fig. 10) and hence promote the maintaining of a correct reading frame. It perfectly explains why a filled P-site is obligatory for the EF-Tu mediated tRNA binding (Fig. 1). In this way, the biochemical mechanism which couples codon-anticodon matching with maintenance of the reading frame (Weiss & Gallant, 1983) might be explained, at least partially, by the allosteric behaviour of the elongation factor EF-Tu.

ACKNOWLEDGEMENTS

We thank Prof. B.F.C. Clark and coworkers for the communication of 3D details of their EF-Tu model and Prof. A. Parmeggiani for fruitful discussions.

This research was supported by the Netherlands Foundation for Chemical Research (SON), which is subsidized by the Netherlands Organization for the Advancement of Pure Research (Z.W.O.).

REFERENCES

Amons, R., Pluijms, W., Roobol, K. & Möller, W. (1983). *FEBS Letters*, 153, 37-42.

An, G. & Friesen, J.D. (1980). *Gene*, 12, 33-39.

Arai, K.-I., Kawakita, M., Nakamura, S., Ishikawa, I. & Kaziro, Y. (1974). *J. Biochem.*, 76, 523-534.

Arai, K., Clark, B.F.C., Duffy, L., Jones, M.D., Kaziro, Y., Laursen, R.A., L'Italien, J., Miller, D.L., Nagarkatti, S., Nakamura, S., Nielsen, K.M., Petersen, T.E., Takahashi, K. & Wade, M. (1980). *Proc. Nat. Acad. Sci., U.S.A.* 77, 1326-1330.

Blumenthal, T. & Carmichael, G.G. (1979). *Annu. Rev. Biochem.* 48, 525-548.

Blumenthal, T., Saari, B., Van der Meide, P.H. & Bosch, L. (1980). *J. Biol. Chem.* 255, 5300-5305.

Caplan, A.B. & Menninger, J.R. (1979). *J. Mol. Biol.* 134, 621-637.

Chinali, G., Wolf, H. & Parmeggiani, A. (1977). *Eur. J. Biochem.* 75, 55-65.

Clark, B.F.C., La Cour, T.F.M., Fontecilla-Camps, J., Morikawa, K., Nielsen, K.M., Nyborg, J. & Rubin, J.R. (1982). *FEBS Meeting on Cell Function and Differentiation*, Vol. 3, Symp. 12.

De Groot, N., Panet, A. & Lapidot, I. (1971). *Eur. J. Biochem.* 23, 523-527.

Duffy, L.K., Gerber, L., Johnson, A.E. & Miller, D.L. (1981). *Biochemistry*, 20, 4663-4666.

Duisterwinkel, F.J., De Graaf, J.M., Schretlen, P.J.M., Kraal, B. & Bosch, L. (1981a). *Eur. J. Biochem.* 117, 7-12.
Duisterwinkel, F.J., De Graaf, J.M., Kraal, B. & Bosch, L. (1981b). *FEBS Letters*, 131, 89-93.
Duisterwinkel, F.J. (1981). Ph.D. Thesis, University of Leiden, The Netherlands.
Fischer, E., Wolf, H., Hantke, K. & Parmeggiani, A. (1977). *Proc. Nat. Acad. Sci., U.S.A.* 74, 4341-4345.
Furano, A.V. (1975). *Proc. Nat. Acad. Sci., U.S.A.* 72, 4780-4784.
Gavrilova, L.P., Perminova, I.N. & Spirin, A.S. (1981). *J. Mol. Biol.* 149, 69-78.
Grosjean, H.J., De Henan, S. & Crothers, D. (1978). *Proc. Nat. Acad. Sci., U.S.A.* 75, 610-614.
Hansske, F. & Cramer, F. (1979). In *Methods in Enzymology* (Moldave, R. & Grossmann, L., eds.), Vol. 59, Academic Press, New York, pp. 172-181.
Hopfield, J.J. (1974). *Proc. Nat. Acad. Sci., U.S.A.* 71, 4135-4139.
Jonák, J., Smrt, J., Holý, A. & Rychlik, I. (1980). *Eur. J. Biochem.* 105, 315-320.
Jones, M.D., Petersen, T.E., Nielsen, K.M., Magnusson, S., Sottrup-Jensen, L., Gausing, K. & Clark, B.F.C. (1980). *Eur. J. Biochem.* 108, 507-526.
Jurnak, F., Rich, A. & Miller, D.L. (1977). *J. Mol. Biol.* 115, 103-110.
Kabsch, W., Gast, W.H., Schulz, G.E. & Leberman, R. (1977). *J. Mol. Biol.* 117, 999-1012.
Kaziro, Y. (1978). *Biochim. Biophys. Acta*, 505, 95-127.
Kurland, C.G. (1979). In *Ribosomes, structure, function and genetics* (Chambliss, G., Craven, G.B., Davies, J., Davis, K., Kahan, L. & Nomura, M., eds.), University Park Press, Baltimore, pp. 597-614.
Lührmann, L., Eckhardt, H. & Stöffler, G. (1979). *Nature*, 280, 423-425.
Miller, D.L. & Weissbach, H. (1977). In *Molecular Mechanisms of Protein Biosynthesis* (Weissbach, H. & Pestka, S., eds.), Academic Press, New York, pp. 323-373.
Miller, D.L., Nagarkatti, S., Laursen, R.A., Parker, J. & Friesen, J.D. (1978). *Mol. Gen. Genet.* 159, 57-62.
Parmeggiani, A. & Sander, G. (1980). In *Topics in Antibiotic Chemistry* (Sammes, P.G., ed.), Vol. 5, John Wiley and Sons, England, pp. 165-221.
Rubin, J.R., Morikawa, K., Nyborg, J., La Cour, T.F.M., Clark, B.F.C. & Miller, D.L. (1981). *FEBS Letters*, 129, 177-179.
Shibuya, M., Naskimoto, H. & Kaziro, I. (1979). *Mol. Gen. Genet.* 170, 231-234.
Thompson, R.C. & Karim, A.M. (1982). *Proc. Nat. Acad. Sci., U.S.A.* 79, 4922-4926.
Van der Meide, P.H., Borman, T.H., Van Kimmenade, A.M.A., Van de Putte, P. & Bosch, L. (1980). *Proc. Nat. Acad. Sci., U.S.A.* 77, 3922-3926.
Van der Meide, P.H., Vijgeboom, E., Talens, A. & Bosch, L. (1983). *Eur. J. Biochem.* 130, 397-407.
Van Noort, J.M., Duisterwinkel, F.J., Jonák, J., Sedláček, J., Kraal, B. & Bosch, L. (1982). *EMBO Journal*, 1, 1199-1205.
Weiss, R. & Gallant, J. (1983). *Nature*, 302, 389-393.
Wolf, H., Chinali, G. & Parmeggiani, A. (1974). *Proc. Nat. Acad. Sci., U.S.A.* 71, 4910-4914.
Wolf, H., Chinali, G. & Parmeggiani, A. (1977). *Eur. J. Biochem.* 75, 67-75.
Yokota, T., Sugisaki, H., Takanami, M. & Kaziro, I. (1980). *Gene*, 12, 25-31.

RIBOSOMAL CONTRIBUTIONS TO messenger RNA DIRECTED SELECTION OF transfer RNA

Måns Ehrenberg and C.G. Kurland

Department of Molecular Biology, The Biomedical Center, Box 590, 751 24 Uppsala, Sweden

Introduction

There exist mutated E.coli ribosomes which are more error prone than are wild type ribosomes (Gorini, 1971; Zimmerman et.al., 1971; Andersson et.al., 1982). There also exist ribosomal mutants which are more accurate than the wild type (Gorini & Kataja, 1964; Gorini, 1971). This pattern indicates that there exists an optimum choice of error levels in the bacterium. We wish to suggest that this optimum is a balance between the drawbacks of a too large fraction of faulty proteins on the one hand and the extra energetic costs of too low error frequencies on the other.

To evaluate such costs quantitatively we need to know some of the more fundamental parameters that characterize the translation process. One such parameter, which will be discussed in this report, is the intrinsic selectivity of codon-anticodon interactions on the ribosomes. This parameter corresponds to the maximum possible accuracy that can be obtained over a single step on the ribosome. However, more of the kinetic features of the translation apparatus must be known before a proper role can be ascribed to the intrinsic selectivity of the process. Hopfield (1974) and Ninio (1975) have described a mechanism for substrate proofreading in which the accuracy of a single selection step can be surpassed by a multiple step process in which at each step there is a preferential release of erroneous substrates. Since the amplification of the accuracy at each successive proofreading step requires the dissipation of free energy, proofreading schemes lead to an extra cost.We discuss some general features of kinetic proofreading and offer in an experimental section evidence for the existence of such a mechanism on the ribosome.

The intrinsic selectivity of aminoacylation and peptide elongation

Translation of a mRNA to a sequence of peptides involves several steps of high substrate selectivity. Charging of a particular isoacceptor tRNA with its proper aminoacid involves two selections, both of which are carried out by the tRNA-synthetase. One aminoacid has to be picked out from its twenty competitors and only those tRNAs which are cognate for that aminoacid should be charged.
The choice of aminoacyltRNAs on the programmed ribosome is template directed and depends on correct codon-anticodon interactions in the ribosomal A-site.

In order to separate correct substrates from incorrect ones the enzyme itself or the enzyme together with its template has to create one or several states where the cognate ligand has a favourable standard free energy in relation to the noncognate ones.

The maximum difference between correct and incorrect substrates in standard free energy on the pathway from substrate to product determines the intrinsic selectivity of the enzyme. It was long believed that this parameter constitutes an absolute upper limit for the accuracy of an enzymatic selection. Not until the basic concepts of kinetic proofreading, which will be discussed below, had been worked out by Hopfield (1974) and Ninio (1975) was it clear that a cleverly constructed enzymatic mechanism in principle could achieve accuracies far above any bounds set by the limited differences in standard free energies for different substrates. There are two aspects of the intrinsic selectivity in the codon reading on the ribosome and in the aminoacid-tRNA selections on tRNA-ligases that are of particular interest for the present discussion.

The first aspect is a theoretical one and it can be summarized by two questions: Does there exist for a given substrate pair, a theoretical upper limit to the maximal difference between them in standard free energy that can be created at the catalytic site of an enzyme? Furthermore, if such a limit exists what is its value? It appears that we can say yes to the first question on fairly general grounds but the second one is much more elusive. The second aspect is experimental and involves the problem of actually determining the intrinsic selectivity of tRNA-synthetases and of codon-anticodon interactions in the ribosomal A-site for different substrate pairs. Although this seems to be a well defined and straight forward task it has turned out to be almost as elusive as the theoretical aspects of the problem.

Pauling (1958) has discussed theoretical limits to the intrinsic selectivity of tRNA-ligases for different pairs of aminoacids. He predicted that aminoacids with similar side groups would be very poorly separated. This poor separation would be a particularly severe problem when the competing, incorrect aminoacid has a smaller side group than the correct one or one which is isosteric so that the enzyme cannot use a small site to exclude the erroneous substrate. So for instance, valine might replace isoleucine with an error frequency not smaller than 5% and possibly considerably higher.

Pauling's predictions were rapidly refuted by experiments. It could be shown by pyrophosphate exchange that the proper ligase for isoleucine and isoleucyl-tRNA activates isoleucine about 200-fold more efficiently than it activates the competitor valine (Fersht, 1977). In this case Pauling's theoretical estimate of the maximum conceivable accuracy was off by at least one order of magnitude and his predictions came out even more poorly in other cases when they were eventually checked experimentally. Fersht (1977) has compiled data about the selectivity of the pyrophosphate exchange reaction for several pairs of aminoacids. He identifies the accuracies of these reactions with the intrinsic selectivities of the synthetases that are involved. Unfortunately it is not possible to make such identifications without one further qualification: One has to know what fraction of the intrinsic selectivity is expressed in the accuracy of the reaction. Since this fraction may range between zero and one in the experiments referred to by Fersht, these results can only be used to obtain lower limits for the intrinsic selectivity of the pathway. On the one hand, therefore, the results imply a rejection of Pauling's too small theoretical estimates but on the other, a major ambiguity persists which is that the experimental data themselves may substantially under estimate the intrinsic selectivity of these enzymes.

The most intense efforts to determine the intrinsic selectivity of different codon-anticodon combinations in the translation of the genetic code have been directed toward a model system originally discovered by Eisinger (1971). Eisinger observed that the equilibrium constant between two tRNAs with matching anticodon triplets is about six orders of magnitude larger than the affinity between the corresponding two triplets when they are free in solution. A large part of this difference can be attributed to an entropic effect (Eisinger, & Gross, 1974) but also other parameters such as stacking free energies seem to be involved (Grosjean et al., 1976). Grosjean et al. (1978) have taken advantage of this model system to make an extensive investigation of how well matching anticodon-anticodon interactions are separated from nearly matching ones. They have used temperature jump techniques and have chosen the relative differences in average lifetimes of the tRNA-tRNA complexes as a criterium for the intrinsic selectivity of these interactions. They observe a strong correlation between long lifetimes for matching anticodon triplets on the one hand and short lifetimes for nearly matching anticodon triplets on the other. However, their data set does not allow this strong correlation to be identified with a strict rule. In several cases they cannot observe differences in lifetimes between "correct" and "incorrect" complexes. This directly implies that their model system at least in some cases does not properly reflect the intrinsic selectivity of real translation systems.

We may extend our criticism of the relevance of this type of experiment one step further. One subset of the data clearly shows that the ribosome must play an important role for the accuracy of codon-anticodon interactions (Gorini, 1971). It is therefore not unreasonable to expect considerable upward adjustments of their selectivity parameters for the whole data set in the future. Accordingly, we are left with strong support for the notion of some sort of ribosome contribution such as conformational selection as described by Kurland et al. (1975), but with very little confidence in the estimates of the intrinsic selectivity of real translation systems. However, the above model system has one property which probably also characterizes codon-anticodon interactions on the ribosome and which, in a beautiful way, illustrates one aspect of the conformational selection model (Kurland et al. 1975). It is well known that G-C base pairs are considerably more stable than A-U base pairs in ordinary RNA-double helices (Borer et al., 1974). Thus, a G-C pair has three hydrogen bonds while an A-U base pair only has two, so the former should be considerably more stable than the latter. The straightforward prediction from these basepairing rules is that the affinity between tRNAs which can form G-C rich anticodon-anticodon complexes is several orders of magnitude larger than the affinity between tRNAs which form A-U rich complexes. In contrast to these predictions Grosjean et al. (1978) observe almost no lifetime differences between G-C rich and A-U rich complexes. The particular steric configuration of the tRNA anticodon loop appears therefore to change radically the binding properties of its anticodon triplet compared to the situation in which the same three nucleotides are allowed to adapt a different conformation. It is probably safe to predict more surprises of this kind when more is known experimentally about tRNA-messenger interactions on the ribosome itself.

Kinetic proofreading. Its concepts.

One problem that kinetic proofreading (Hopfield, 1974; Ninio, 1975) solves is the following. Let us assume that in an enzymatic selection the intrinsic selectivity as defined above is limited by an upper bound. Assume further that the error level of the selection is required to be much lower than the reciprocal of the intrinsic selectivity of the system. This greater selectivity can be achieved provided that the pathway is constructed so that is fulfills some general criteria which characterize all kinetic proofreading mechanisms.

A first requirement is that the preferential selection of the correct substrate is partitioned into several steps. An initial selection step must be followed by one or several subsequent ones in such a way that a particular step can be reached to a good approximation only from the preceeding one. Secondly, the enzyme-substrate complex must be able to dissociate along a direct pathway from each of its intermediate states so that these states can contribute to the accuracy.

A third requirement is that the frequency of substrate association directly to these intermediate states is very small. If such a low association rate is to be combined with a high probability of dissociation for the incorrect substrates then a second driving force becomes necessary. Apart from the thermodynamic driving force which gives a net flow from substrate to product, there must be another displacement which favours the dissociation of substrate from the intermediate states. Such a second displacement over the proofreading branches to drive the discard of substrate can be obtained when the dissociation of a substrate molecule is coupled to the hydrolysis of a nucleoside triphosphate as pointed out by Hopfield, (1974) and Ninio (1975). These nucleosidetriphosphates are in vivo displaced by eight orders of magnitude or more from equilibrium with their hydrolytic product (Kurland, 1978). Thus the thermodynamic driving force for the proofreading branch(es) appears to be sufficient for the demands of aminoacylation and peptide elongation.

This necessary coupling between discard events and nucleoside triphosphate hydrolysis leads to an experimental issue of practical significance. The number of hydrolyzed nucleoside triphosphates per product formed for the incorrect substrate reaction divided by the corresponding number for the correct substrate is a ratio which defines how much the proofreading part of the enzymatic selection contributes to its accuracy. This ratio can be used to identify kinetic proofreading. If the ratio is near one then the selection is conventional and if the ratio is much larger than one the existence of proofreading is strongly implicated.

The accuracy of an enzymatic selection with kinetic proofreading step(s) is related to the dissipative losses of the reaction in a way that is new and goes beyond what characterizes a conventional selection scheme (Kurland, 1978). The relationship between dissipation and accuracy, which in some form is present in all selections, becomes particularly obvious in kinetic proofreading because of the coupling between the flows of the different substrates e.g. aminoacyl tRNA and GTP. The enhancement of accuracy over a particular proofreading step can be directly related to the number of nucleoside triphosphates that is hydrolyzed in excess over the number of substrates that has survived this particular step and has been effectively transferred to the next. When the actual enhancement of accuracy in a proofreading step is near the intrinsic selectivity of that step then the number of nucleoside triphosphates hydrolyzed per correct product formed is very large (Blomberg et al. 1980). When several proofreading steps are coupled after each other in a sequence then a given error level can be achieved at a smaller dissipative cost than with a single proofreading step (Blomberg *et al*. 1980). Within certain limits a multistep proofreading enzyme can achieve error levels corresponding to the reciprocal of the product of the intrinsic selectivity in all of its steps. When this lower bound to the error level becomes small enough it must be replaced by another bound which is set by the magnitude of the driving force of the discard branches. It is not possible to magnify the accuracy by proofreading, irrespective of how many steps that are coupled in sequence, above a limit set by the displacement of the participating nucleoside triphosphate from equibibrium with its hydrolytic products (Blomberg, et al., 1980).

Proofreading in translation

Elongation of polypeptides on the ribosomes is driven by the hydrolysis of two guanine nucleotides (Lucas- Lenard & Lipmann, 1971). One or both of these guanine nucleotides could in principle drive the discard branches of a proofreading mechanism in the translation of mRNA. tRNA enters the ribosome in complex with EF-Tu and GTP in a ternary complex (Lucas-Lenard & Lipmann, 1971). When the mRNA codon matches the anticodon of the tRNA on EF-Tu the guanine nucleotide becomes hydrolyzed. If there is proofreading on the ribosome then the tentatively accepted tRNA must have a possibility of being discarded after this GTP-molecule has been hydrolyzed. In other words, EF-Tu must necessarily go through more cycles during proofreading than there are peptide bonds formed. One approach to the identification of proofreading in this system is to relate the number of hydrolyzed GTP:s associated with a correct peptide bond to the number of nucleotides hydrolized per erroneously formed peptide bond as suggested by Hopfield (1974). If the ratio between the latter number and the former is much larger than one proofreading is indicated. Furthermore the accuracy enhancement of this step can be determined as described above.

Two different experimental approaches to this problem have been taken. In the first a poly(U)-programmed in vitro system which lacks elongation factor G (EF-G) is used to make single cycle measurements on the number of GTP:s that are required to form correct or incorrect dipeptides (Thompson & Stone, 1977; Thompson & Dix, 1982). In this system the general error in elongation is undefined, the temperature is kept far below the physiological one so that the rate of peptide bond formation is almost two orders of magnitude smaller than in vivo and the buffer conditions are known to be suboptimal for translation (Jelenc & Kurland, 1979). These drawbacks of the approach make it difficult to evaluate the positive outcome of the search for proofreading with this technique. First, EF-G may change the kinetics of the elongation cycle in a way that is impossible to foresee in a system that lacks this factor. Secondly, kinetic proofreading relies on a proper balancing of a forward rate constant and a discard rate constant for the correct and incorrect substrates on their passage to product formation. One possible candidate for the forward rate constant in a proofreadig mechanism on the ribosome is the rate of peptide bond formation. In vivo this rate must be in the order of $20s^{-1}$ or considerably larger to be compatible with the rate of elongation (Gausing, 1974). In contrast, Thompson and Dix (1982) report a value of $0.4s^{-1}$ for this rate constant. It is not easy to know whether or not the balancing between the forward rate and the discard rate will stay the same when the forward rate constant by proper adjustments of temperature and of buffer composition is brought near its in vivo value.

The other approach(Ruusala et al., 1982b) involves the design of an in vitro translation system in such a way that its most relevant parameters, its speed and its accuracy, mimic those characteristics of the in vivo situation. It involves furthermore a complete system operating in the steady state with all of its factors and in particular EF-G present.

A first step in this direction was taken by Jelenc & Kurland (1979). Here it was shown that in a buffer, with ions and other constituents like polyamines, which resembles the solvent composition of bacteria, a poly(U)-translating in vitro system could elongate peptides at an accuracy close to that in vivo and yet at a considerable speed. Although, the elongation rate in this system was compatible with the maximum rate of all known in vitro translation systems at the time it was still more than two orders of magnitude slower than the elongation rate in vivo of 16-17 aminoacids per second per ribosome (Gausing, 1974).

A next step in the development of this in vitro system was taken by Wagner et al. (1982). They showed that the slow rates of poly(U)-translating in vitro systems did not in general result from a slow elongation cycle per se. Instead most systems were inhibited by a very slow activation of the ribosomes. They preincubated the ribosomes together with poly(U) and N-acetyl-Phe-$tRNA^{Phe}$ and could in this way bring the system to a preinitiated state. By a subsequent addition of all other factors necessary for peptide elongation and by observing the incorporation of aminoacids during a short time they were able to translate poly(U) with a rate of between 8 and 12 cognate amino-acids per second per ribosome. This rapid burst of polyphenylalanine synthesis lasts for about 20 seconds and subsequently fades when the activated ribosomes reach the end of their messengers. Only between 10% and 20% of the ribosomes could be coordinately preactivated in this way.

One difficulty in assessing the number of GTP-molecules that are necessary to forward a peptide bond is the existence of other and unrelated GTP:ase activities in the system. It is known for instance that EF-G has a large GTP:ase activity uncoupled from peptide elongation (Lucas-Lenard & Lipmann, 1971). It was therefore necessary to develop a more selective method of determining the number of cycles of EF-Tu necessary to forward a correct peptide bond (Phe) in relation to the number of cycles associated with incorrect peptide bond formation (Leu). The solution to this problem could be found by using a remarkable property of the cycle of EF-Tu. In the absence of elongation factor Ts (EF-Ts) the dissociation rate constant of the binary complex EF-Tu.GDP is very small and under our conditions the average life time of this complex is about 100 seconds. (Ruusala et al. 1982A). When EF-Ts is added in saturating amounts the dissociation rate is accelerated by more than three orders of magnitude (Ruusala et al. 1982A). Accordingly, only a moderate rate of hydrolysis of ternary complex on the programed ribosome would in the absence of EF-Ts be necessary to saturate the cycle of EF-Tu by trapping the factor in its EF-Tu.GDP state. When the cycle of EF-Tu is saturated in this way by the correct $tRNA^{Phe}$ then the rate of peptide bond formation is determined by three variables: The input concentration of active EF-Tu (Tu_o), the dissociation rate constant of GDP from the elongation factor (k_d) and by the average number of cycles of EF-Tu necessary to forward a correct peptide bond (f_c).

$$\dot{j}_c = \frac{Tu_o \cdot k_d}{f_c}$$

When the correct substrate $tRNA^{Phe}$ is replaced by an incorrect tRNA which almost matches the poly(U)-triplet by its anticodon sequence then a saturated cycle of EF-Tu will lead to a flow rate determined by a similar expression but where f_c is now replaced by f_w which is the average number of cycles of EF-Tu associated with an incorrect peptide bond.

$$\dot{j}_w = \frac{Tu_o \cdot k_d}{f_w}$$

The ratio between f_w and f_c (the F-factor), is the degree to which proofreading contributes to the accuracy of peptide bond formation as explained above. Experimentally the realization of these two extreme cases was accomplished in a single plot by varying the charging level of the correct tRNA and by simultaneously observing the rate of Leu incorporation as a function of the rate of phe-incorporation. (Ruusala *et al*. 1982B).

The results of this study indicate first that proofreading indeed appears to be associated with the translation process on E.coli ribosomes. Furthermore, the accuracy is normally partitioned between the initial selection step and the subsequent proofreading in such a way that each part contributes roughly the same factor. So, for instance the typical error frequency associated with the isoacceptor Leu-$tRNA_4^{Leu}$ is 6.10^{-4} in competition with Phe-$tRNA^{Phe}$ and the proofreading factor is about 40. For another isoacceptor, Leu-$tRNA_2^{Leu}$, the error is 10^{-4} and the proofreading factor approximately 100. This equipartitioning between the two steps can however be perturbed by variations in the composition of ions in the system: The selectivity of the first step is only weakly dependent on the concentration of magnesium ions while the proofreading branch shows a much stronger response to such changes (Ruusala, unpublished results, 1983). At low magnesium ion concentrations the error is decreased considerably and the major part of this decrease can be attributed to changes in the proofreading factor F.

This result implies that the proofreading part of the selection of aminoacyl tRNAs in translation normally operates below, and maybe far below, its maximum attainable selectivity.

This conclusion is further supported by estimates of f_c, the number of guanine nucleotides that are dissipated over the EF-Tu cycle per correct peptide bond.

If k_d is estimated from a nucleotide exchange assay and the concentration of active EF-Tu is determined from filter binding of radioactive GDP in complex with the factor the rate of polyphenylalanine synthesis in the absence of EF-Ts can be used to determine f_c according to the formula above. We have found that f_c probably lies between 1.0 and 1.2 but exactly how close to one is impossible to say. If the proofreading part involves only a single step then f_c can be used to obtain a lower limit to the intrinsic selectivity of that step. If for instance f_c=1.2 then the intrinsic selectivity is at least five times larger than the F-factor. If on the other hand there exist multiple steps such a simple relationship does not hold any more and the intrinsic selectivity might in fact be much smaller. What appears to be clear, however, is that the system is designed in such a way that accuracy has been sacrificed to some extent so that in wild type ribosomes the excess losses of guanine nucleotides are kept at small values.

How strong is the evidence for proofreading?

Although all our experimental results seem consistent with a proofreading mechanism in translation one property of our,(Wagner et al., 1982) as well as of other translation systems (Thompson & Dix, 1982), motivates a more thorough discussion of a conceivable artifact. As mentioned above it has not been possible to preactivate more than between 10% and 20% of the ribosomes. Therefore all the results that we report come from a minor fraction of active ribosomes in a background of ribosomes which do not produce polypeptides. Although these inactive ribosomes cannot elongate they might still be able to catalyse the hydrolysis of GTP in ternary complexes. In this way the excess turnover of EF-Tu that we observe for leucine but not for phenylalanine and which we interpret as proofreading might arise from a side reaction unrelated to peptide bond formation.

The number of excess cycles of EF-Tu associated with poly(Phe) synthesis (f_c) is very near one while the excess cycles associated with leucine incorporation (f_w) is much larger than one. This shows that the inactive ribosomes cannot recognize ternary complexes with the same precision as the elongating ones since then f_c would be similar to f_w and both would be much larger than one if the background activity were high. We can also immediately exclude another extreme case of the artifact, namely that the inactive, "killer" ribosomes have no codon specificity at all. If that were the case the different error frequencies observed for different tRNA-isoacceptors would have an exact counterpart in the F-factors for these tRNAs. Accordingly, $tRNA^{Leu}$ (P_E=6.10^{-4}) would have an F-factor which is six times smaller than the F-factor for $tRNA_2^{Leu}$ (P_E=1.1.10^{-4}) and this is clearly not the case as discussed above and by Ruusala et al. (1982B). In order to explain that the accuracy is roughly equipartitioned between initial selection and proofreading for different tRNAs one must ascribe to the killer ribosome an intermediate codon specificity : It is not as selective as the elongater but not indifferent to codon-anticodon interactions either. These observations make the killer ribosome so special that it becomes a rather unlikely alternative to proofreading. Further indirect evidence in the same direction can be found from the effects of streptomycin on the accuracy of translation (Ruusala, manuscript in preparation, 1983). In the presence of streptomycin the two tRNA isoacceptors $tRNA_4^{Leu}$ and $tRNA_2^{Leu}$ incorporate leucine with substantially elevated error frequencies in comparison with translation in the absence of the drug. The point we want to make here is that even in the presence of streptomycin poly(U)-programmed ribosomes accept $tRNA_2^{Leu}$ less efficiently than $tRNA_4^{Leu}$ and both these tRNAs are accepted much more poorly than $tRNA^{Phe}$. The excess turnover of EF-Tu is for all three tRNAs very near one in spite of their different activity in peptide formation. This is compatible with a mechanism where the drug streptomycin preferentially deactivates the discard branch of a proofreading scheme and effects the initial selection only three to five fold. If instead this result is caused by a background activity this would imply the following. Although streptomycin appears to bind to all ribosomes indiscriminately and not only to the active ones (Ruusala, manuscript in preparation, 1983) the effects of the drug must be very selective. In the absence of streptomycin the background turnover of EF-Tu would dominate over peptide elongation for $tRNA^{Leu}$-isoacceptors. In the presence of streptomycin,however,the background must be significantly smaller than the turnover of EF-Tu associated with peptide formation. This is possible but does not seem very likely since the drug binds to all ribosomes and one would expect a considerable residual background turnover of EF-Tu also in its presence.

A nominally streptomycin resistant ribosome mutant (Sm3) which has been described by Zengel et.al. (1977) has in vitro translation properties which are relevant for the present discussion (Ruusala, unpublished results, 1983). In the absence of streptomycin the mutant has an f_c almost twofold that of the wild type and a significantly reduced error frequency. When streptomycin is added to these ribosomes the f_c-value is reduced to near one and becomes indistinguishable from the wild type.

Concomitant with this change in f_c is a twofold increase in k_{cat}/K_m for cognate ternary complexes.
All these features are consistent with a proofreading mechanism: The reduced error in the mutant has its counterpart in an increased f_c. When streptomycin is added the error goes up and at the same time f_c drops to its wild type value so that the losses of cognate tRNA over the discard step (s) become very small. This reduction of excess

losses would in turn explain the observed increase in k_{cat}/K_m.

In the alternative interpretation we would have to argue that in the absence of streptomycin the turnover of EF-Tu induced by inactive ribosomes is the same as the turnover associated with the elongating ones (f_c=2). Addition of streptomycin would subsequently eliminate the effect of the inactive ribosomes on the cycling of EF-Tu (f_c=1). It is difficult to explain this reduction of the killer activity below the detection limit of our system solely by the observed twofold increase in k_{cat}/K_M of the active ribosomes when streptomycin is added. More selective effects of the drug on the two ribosome types have to be invoked.

We have, in conclusion, found no evidence against the interpretation that a substantial part of the accuracy in translation is due to proofreading. An alternative view where the inactive ribosomes act as scavengers for ternary complexes, appears to be a remote possibility. However, all evidence against this latter interpretation is indirect and it cannot be definitively excluded.

Concluding remarks.

We are optimistic that our in vitro system for translation will help to answer a number of questions about proteinsynthesis. One such question is the as yet unresolved problem of how large is the intrinsic selectivity of codon-anticodon interactions under optimal conditions. Another important issue is to determine f_c with enough precision so that the excess losses of GTP-molecules in protein-synthesis can be assessed properly. Related to this question is how many steps that are involved in the proofreading part of the ribosomal selection of tRNA. Knowledge of these and other mechanistic details of translation are necessary when we want to know how the bacterium optimizes the three parameters accuracy, rate and energy dissipation so as to obtain its high rate of growth.

References

Andersson,D.J., Bohman,K., Isaksson,L. & Kurland,C.G. (1982). Mol.Gen.Genet. 187, 467-472.

Blomberg,C., Ehrenberg,M. & Kurland, C.G. (1980). Quart.Rev.Biophys. 13, 231-254.

Borer,P.N., Dengler,R., Tinoco,I. & Uhlenbeck,O. (1974). J.Mol.Biol. 86, 843-853.

Bouadloun,F. & Kurland,C.G. (1983). EMBO J. In press.

Eisinger,J. (1971). Bioch.Bioph.Res.Comm. 43, 854-861.

Eisinger,J. & Gross,N. (1974). J.Mol.Biol. 88, 165-174.

Fersht,A. (1977). Enzyme Structure and Mechanism. W.H. Freeman and Company. Reading and San Francisco.

Gorini,L. & Kataja,E. (1964). Proc.Natl.Acad.Sci. U.S.A 51, 487-493.

Gorini,L. (1971). Nature New Biology, 234, 261-264.

Grosjean,H., Söll,D. & Crothers,D.M. (1976). J.Mol.Biol. 103, 499-519.

Grosjean,H., de Henau,S. & Crothers,D.M. (1978). Proc.Natl.Acad.Sci.U.S.A 75, 610-614.

Hopfield,J.J. (1974).Proc.Natl.Acad.Sci.U.S.A. 71, 4135-4139.

Jelenc,P.C. & Kurland,C.G. (1979). Proc.Natl.Acad.Sci. U.S.A. 76, 3174-3178.

Kurland,C.G., Rigler,R., Ehrenberg,M. & Blomberg,C. (1975). Proc.Natl.Acad.Sci.U.S.A. 72, 4248-4251.

Kurland,C.G. (1978). Biophys.J. 22, 373-388.

Loftfield,R.B. (1963). Biochem.J. 89, 82-92.

Loftfield,R.B. & Vanderjagt,D. (1972). Biochem.J. 128, 1353-1356.

Ninio,J. (1975). Biochimie, 57, 587-595.

Norris,A. & Berg,P. (1964). Biochemistry, 52, 330-337.

Norris Baldwin,A. & Berg,P. (1965). J.Biol.Chem. 242, 839-845.

Pauling,L. (1958). In Festschrift Arthur Stoll. Birkhauser, A.G. Basel, 597-602.

Rosenberger, R.F. & Foskett,G. (1981). Mol.Gen.Genet., 185, 561-563.

Ruusala,T., Ehrenberg,M. & Kurland,C.G. (1982A). EMBO J. 1, 75-78.

Ruusala,T., Ehrenberg,M. & Kurland,C.G. (1982B). EMBO J. 1, 741-745.

Springgate,C.F. & Loeb,L.A. (1975). J.Mol.Biol.97, 577-591.

Thompson,R.C. & Stone,P.J. (1977). Proc.Natl.Acad.Sci. U.S.A. 74, 198-202.

Thompson,R.C. & Dix,D.B. (1982). J.Biol.Chem. 257, 6677-6682.

Wagner,E.G.H., Jelenc,P.C., Ehrenberg,M. & Kurland,C.G. (1982). Eur.J.Biochem. 122, 193-197.

Zengel,J.M., Young,R., Dennis,P.P. & Nomura, M. (1977). J.Bacteriol. 129, 1320-1329.

Zimmerman,R.A., Garvin,T. & Gorini,L. (1971). Proc.Natl. Acad.Sci. U.S.A. 68, 2263-2267.

EVOLUTION OF RIBOSOMAL STRUCTURE AND FUNCTION

James A. Lake

Molecular Biology Institute & Department of Biology

University of California, Los Angeles, Calif. 90024.

INTRODUCTION

Ribosomes from all organisms have similar three-dimensional structures and functional domains. They are generally subdivided into two domains, each associated with a set of specific functions (Bernabeu and Lake 1982). Ribosomes contain a Translational Domain consisting of the apparatus for reading the message and an Exit Domain that is the site of emergence of the nascent protein chain and of membrane binding. The locations of these sites within the general scheme of ribosomal organization is shown in Figure 1.

THE TRANSLATIONAL DOMAIN

The most detailed information regarding the translational domain comes from immune mapping studies on ribosomes from the eubacterium, E. coli. Some of the regions in this domain are shown in Figure 1. On the small subunit (at left) they include the messenger RNA binding site, on the 30S "platform" just visible in the space between subunits; the initial site of EF-Tu binding (i.e. the R site), presumed to be on the external surface of the small subunit, near 50S proteins, L7/L12; and the binding site for elongation factor EF-G, between both subunits near the attachment site of L7/L12 and near the central protuberance of the large subunit (at the top). On the large subunit (at the right), these latter proteins, the only ribosomal proteins present in multiple copies (four per ribosome), form the finger-like projection, the "L7/L12 stalk" that extends from the large subunit. The peptidyl transferase, found on the large subunit on the reverse side of the central protuberance, is not visible in this figure. (These locations are reviewed in detail in Lake (1981)). Without exception, the components that are directly involved in the translation mechanism are in the top region of the ribosome.

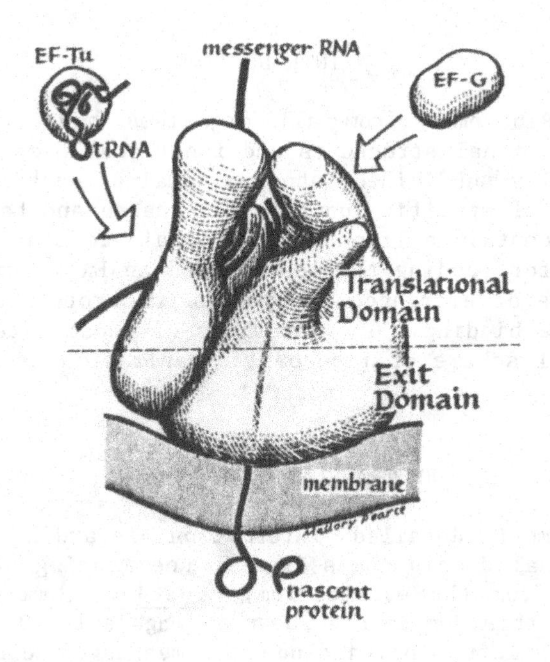

Figure 1. An illustration of the general organization of ribosome structure. Adapted from Bernabeu and Lake, 1982.

THE EXIT DOMAIN

Until recently, the function of the exit domain, i.e. the lower half of the ribosome, was unknown. An early clue to the existence of specialized ribosomal structures involved in protein secretion was provided by the work of Unwin (1979). In the lizard, Lacerta sicula, ordered arrays of membrane-bound ribosomes form in the ovarian follicles during its winter hibernation. Density maps calculated by three-dimensional reconstruction techniques from these arrays showed that a structure (RNA and/or protein) extended from the large ribosomal subunit into the rough endoplasmic reticulum (RER). This structure appears to form an attachment site linking the ribosome and membrane. Although direct evidence is lacking, because of its size and location, this ribosomal structure is a candidate to be the signal recognition particle (Walter and Blobel, 1982).

A role for the bottom half of the ribosome emerged when nascent polypeptide chains were localized there. Using antibodies specific for the beta-galactosidase molecule we (Bernabeu and Lake, 1982) showed that the nascent protein chain exits from the bottom of the ribosome. As shown in the Figure 1, this exit site is directly opposite the translational domain and the peptidyl transferase. The exit site of the nascent chain subsequently has been mapped at this same position on the cytoplasmic eukaryotic ribosome (Bernabeu et al., 1983). These, and other data, suggest that the large-scale organization of eukaryotic and procaryotic ribosomes are similar.

With these results, the organization of the exit domain soon became clear. When the structures of the lizard and E.coli. ribosomes were compared, the membrane attachment site was found to be near the nascent chain exit site (Bernabeu and Lake, 1982). The adjacent locations of the nascent chain and of the membrane-binding site probably correspond to the two types of interactions binding ribosomes to the RER. The attachment that can be released by treatment with the antibiotic puromycin is thought to occur through an anchoring of the nascent chain, and the interaction that is sensitive to high concentrations of monovalent salts possibly corresponds to the membrane binding complex.

The emerging view of ribosomal organization is that the translational domain is located as far as possible from the membranes of the RER, so that tRNAs, factors, and other ligands can gain access to the ribosome from the cytoplasm. In contrast, the functions involved in protein secretion are located in the exit domain in order to face the membranes of the RER.

EVOLUTION WITHIN DOMAINS

Within domains, however, archaebacteria, eubacteria and eucaryotes have evolved different substructures for performing basic functions. These differences occur in both RNA and protein. The discovery of archaebacteria (Woese and Fox, 1977), in particular, was based on oligonucleotide catalogs of small subunit rRNAs. These catalogs implied a pattern of primary sequence in archaebacterial rRNA unlike those of either eubacterial or eukaryotes. Evidence for different protein arrangements within ribosomes was suggested when Matheson _et al_. (1980), observed that the amino acid sequence of a L7/L12-like ribosomal protein in an extreme halophile, H. cutirubrum an archaebacterium had a pattern more like that of eukaryotic L7/L12 than that of eubacterial L7/L12.

A survey of the structures of the ribosomes from these three classes of organisms is shown in Figure 2. Electron micrographs of representable eubacteria are shown in the first column. Archaebacterial and eukaryotic ribosomes are shown in the middle and left columns respectively.

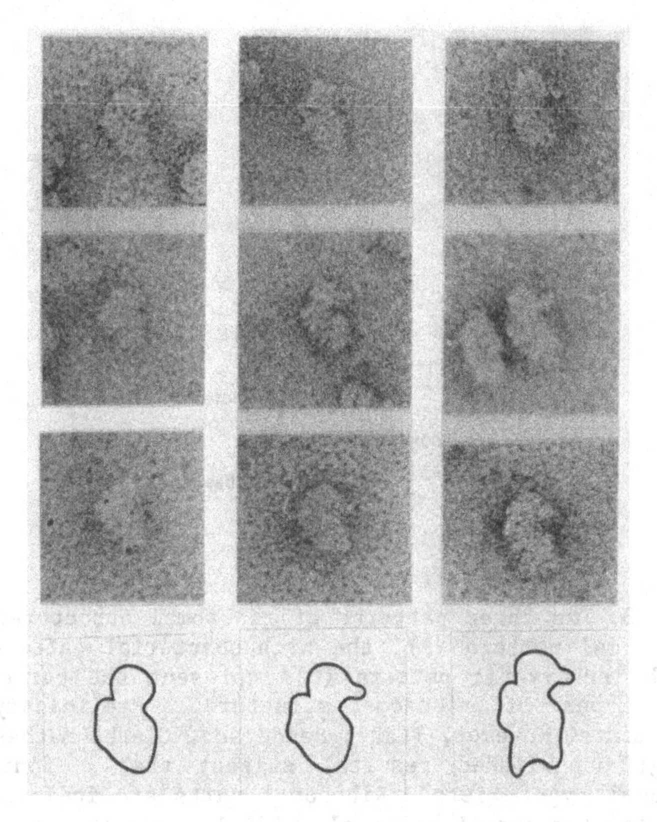

Figure 2. Electron micrographs of small ribosomal subunits representative eubacteria (left column), archaebacteria (middle column), and eukaryotes (right column). The eubacterial organisms are (from top to bottom) Bacillus stearothermophilus, a gram positive thermophilic bacterium; Synechocystis 6701, a cyanobacterium, and Spinacia oleracea chloroplast. The archaebacterial organisms are Methanobacterium autotrophicum, a methanogen, H. cutirubrum, an extreme halophile, and Sulfolobus acidocaldarius, an extreme thermophile. The eukaryotic organisms are Saccharomyces cerevisias, a yeast; Triticum aestivum, a wheat; and Rattus rattus, a rat. Adapted from Lake et al. (1982).

These ribosomes are divided into three separate patterns (Lake et al., 1982) and correspond to the eubacterial, eucaryotic and archaebacterial lineages. Within each of these lineages ribosome structure is highly conserved. The three general patterns of small subunit structure are shown in Figure 3.

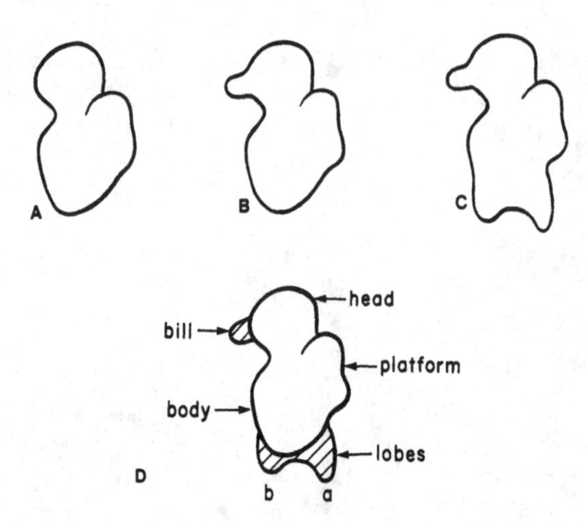

Figure 3. The three patterns of ribosomal structure. The eubacterial pattern (A), the archaebacterial pattern (B) and the eukaryotic pattern (C) represent the currently known types of ribosome structure. Preliminary experiments, however, (Lake, Henderson, Clark, Matheson and Zillig; unpublished results) suggest that a fourth ribosomal type exists. Ribosomal parts are indicated in (D). Adapted from Lake et al. (1982).

Small subunits show a common overall structure except that in the translational domain, small subunits from archaebacteria and from the cytoplasmic component of eukaryotes both contain a feature on the head of the subunit, the archaebacterial bill, that is absent in eubacteria, and in the exit domain, eukaryotic small subunits contain regions of density of the base at the subunit, the eukaryotic lobes, that are absent in archaebacteria and eubacteria.

Within the translational domain, archaebacterial and eukaryotic ribosomal small subunits have an archaebacterial bill, whereas eubacterial subunits do not. The bill is most likely protein and has a volume corresponding to a molecular weight of about 55,000. The correlation of biochemical and structural evidence, although indirect, suggests the bill may function in the factor related steps of protein synthesis. The position of the bill in the archaebacterial 70S ribosome is shown in Figure 4. A similar arrangement also applies for the eukaryotic ribosome (not shown). The bill is adjacent to the L7/L12 stalk and to the binding site of EF-G (as determined for the E.coli ribosome by Girschovich et al. (1981)). The proximity of the bill and the stalk suggests that the L7/L12 amino acid sequence similarities found in archaebacteria and eukaryotes may simply reflect compensatory modifications of the stalk that are necessary to accommodate an adjacent bill. Consistent with the proximity of the bill and EF-G, Kessel and Klink (1980), find that the diptheria toxin specificity of EF-2 the eukaryotic equivalent of EF-G, is present in the corresponding archaebacterial factor, although as has long been known, it is lacking in the eubacterial factor. Thus one anticipates that these correlated changes in the translation domain could be related to alternative mechanisms for performing the factor dependent steps of protein synthesis.

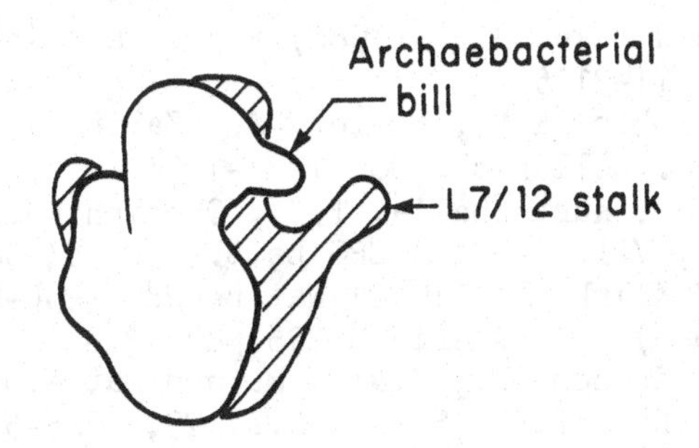

Figure 4. Diagram of the archaebacterial ribosome. Adapted from Bernabeu and Lake (1982).

Within the exit domain, the major change among lineages involves the eukaryotic lobes. These lobes are present in eukaryotic ribosomes but not in the other two ribosomal types. They are sufficiently large to contain about 300 nucleotides if their primary component is RNA.

The approximately 200 nucleotide long insert that is present in the 18S rRNA sequence, and lacking in the 16S sequence (starting at nucleotide 620 in the E. coli numbering system, see Noller and Woese (1981) for a review), is the prime candidate for comprising much of the lobes. The function of the eukaryotic lobes is, however, unknown. If they have a role in protein secretion, as their location within the exit domain suggests, then one can anticipate that their properties will attract much research interest.

Rapid progress is being made in understanding the domain organization of ribosomes. Indeed, recognition of two types of domains and the recognition of a third form of ribosomal structure are providing baselines for relating the functional differences of ribosomes to their structural differences.

REFERENCES

Bernabeu, C. and Lake, J.A. (1982) Proc. Nat. Acad. Sci. U.S.A. 79, 3111-3115.
Bernabeu, C., Tobin, E.M., Fowler, A., Zabin, I. and Lake, J.A. (1983) J. Cell Biol. 96, 1971-1974.
Girshovich, A.S., Kurtskhalia, T.V., Orchinnikov, Yu A., and Vasiliev, V.D. (1981) FEBS Lett. 130, 54-58.
Kessel, M. and Klink, F. (1980) Nature 287, 250-251.
Lake, J.A. (1981) Sci. American 295, 87-97.
Lake, J.A., Hinderson, E., Clark, M. and Matheson, A.T. (1982) Proc. Nat. Acad. Sci. U.S.A. 79, 5948-5952.
Matheson, A.T., Nazar, R.N., Willick, G.E. and Yaguchi, M. (1980) In : Genetics and Evolution of RNA Polymerases, tRNA and Ribosomes (Osawa, et al, eds.) pp 625-645, University of Tokyo Press.
Noller, H. and Woese, C. (1981) Science 212, 403-411.
Unwin, P.N.T. (1979) J. Mol. Biol. 132, 69-84.
Walter, P. and Blobel, G. (1982) Nature 299, 691-695.
Woese, C. and Fox, G. (1977) Proc. Natl. Acad. Sci. U.S.A. 74, 5088-5090.

Effect of Elongation Inhibitors on the Accuracy of Poly(U) Translation

ABRAHAM K. ABRAHAM AND ALEXANDER PIHL*

Department of Biochemistry, University of Bergen
Årstadveien 19, N-5000 Bergen, Norway

Norwegian Cancer Research Institute,
Montebello, Oslo 3, Norway

INTRODUCTION

In view of the fact that numerous steps are involved in protein synthesis, the number of translational errors occurring *in vivo* is surprisingly low. It now seems clear that translational errors involve either a mismatching at the codon-anticodon level (Edelman and Gallant, 1977) or a mischarging of tRNA (Loftfield and Vandervagt, 1972). In the poly(Phe) synthesizing system which is extensively used in translational fidelity studies, the latter possibility is minimal since the two amino acids Phe and Leu are widely different in structure. Recent work has revealed several important factors governing the fidelity of *in vitro* polypeptide synthesis. The nature and concentration of the polycations present for example, are important.

Influence of Polyamines on Translational Accuracy

Several years ago, we first demonstrated (Abraham et al., 1979) that polyamines such as spermidine and spermine have a pronounced effect on the accuracy of poly(U) translation in wheat germ cell-free system (Fig. 1).

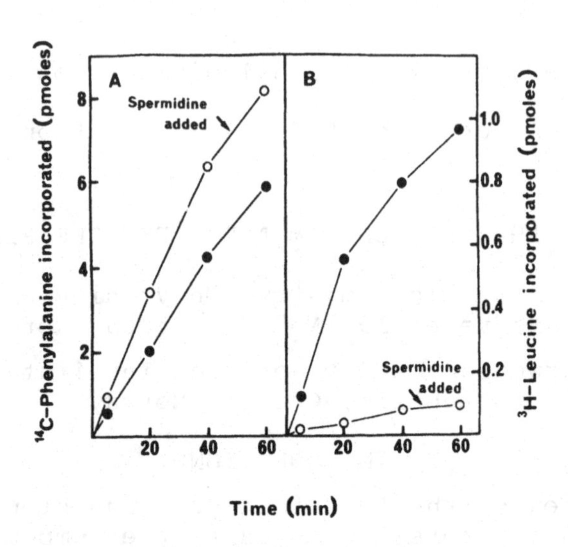

Fig. 1. Poly(U) Directed Incorporation of Phenylalanine and Leucine in a Wheat germ system. For experimental details see Abraham et al. 1979. (●) 10 mM Mg^{2+}. (o) 2 mM Mg^{2+} and 1.6 mM spermidine.

This error reducing effect of polyamines is not so predominant in all eukaryotic cell-free systems. For example, polyamines do not influence the accuracy of poly(U) translation in a nuclease treated rabbit reticulocyte lysate, a cell-free system most widely used for the translation of eukaryotic messengers, (unpublished observation). We therefore looked for conditions that produce polyamine requirement for accurate poly(U) translation in this system. Figure 2 shows the effect of gel filtration of lysates at low (0.5 mM) and high (5 mM) Mg^{2+} containing buffers. Nuclease-treated lysate, gel filtrated in a 5 mM Mg^{2+} containing buffer, requires added polyamines for accurate translation of poly(U). This effect is less significant in a lysate gel-filtrated in buffer containing low Mg^{2+}.

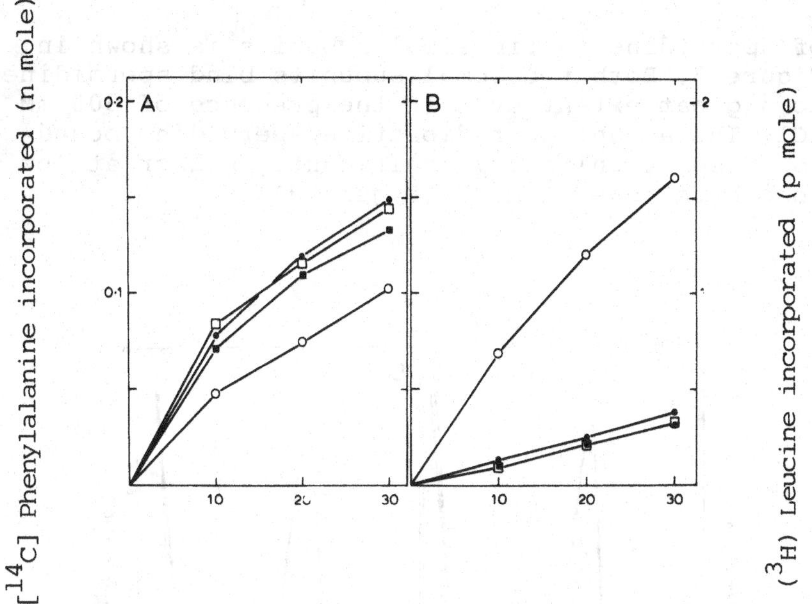

Fig. 2. Effect of Gel-Filtration on the Fidelity of poly(U) Translation in Rabbit Reticulocyte Lysate. For experimental details see Abraham and Flatmark 1982. Nuclease-treated lysate gel filtered in 0.5 mM Mg^{2+} (□, ■) and 5 mM Mg^{2+} (o,●). Open circles and squares spermidine (0.66 mM) added.

These results indicate that the inability of crude lysate to respond to added polyamines is due to the presence of sufficient amounts of these amines tightly bound to the components of the cell-free system. Removal of these tightly bound polyamines is most effective if gel-filtration is carried out in a buffer containing fairly high concentrations of Mg^{2+}.

Effect of Mg^{2+} on the binding of Spermidine to Ribosomes.

Polyamines are known to interact strongly with nucleic acids and nucleic acid containing structures (Sakai & Cohen, 1976). The two main components of the protein synthesising system that interact strongly with polyamines are tRNAs and ribosomes. The influence of Mg^{2+} on the binding of spermidine to ribosomal subunits is shown in Figure 3. Both ribosomal subunits bind spermidine to a great extent even in the presence of 500 mM KCl. The amount of radioactive spermidine bound to these subunits is significantly higher at low concentrations of Mg^{2+} (Figure 3A).

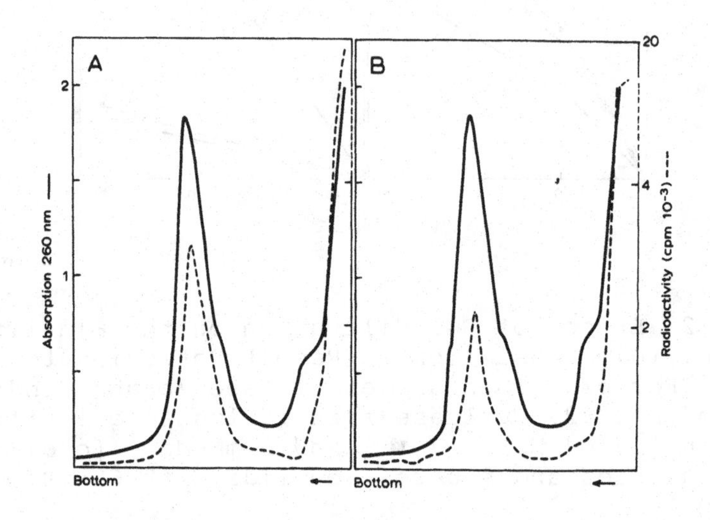

Fig. 3. Spermidine Binding to Ribosomal Subunits. For experimental details see Abraham and Flatmark (1982). A. 0.5 mM Mg^{2+}, B. 5 mM Mg^{2+}.

Another important factor that influences translational accuracy at the ribosomal level is the ratio between GTP and its hydrolysis products GMP and GDP. The effect was clearly demonstrated in the bacterial system by Kurland and coworkers (Jelene & Kurland 1979). A high ratio of GTP to its hydrolysis products is important for maximal fidelity also in the eukaryotic systems (unpublished observations), and we have maintained efficient energy regeneration in our cell-free systems.

Effect of Substrate Concentration on Translational Accuracy

Errors at the ribosomal level occur due to competition between the correct substrate phenylalanyl $tRNA_{Phe}$ and incorrect substrate Leucyl $tRNA_{Leu}$. The degree of accuracy is therefore dependent on the concentrations of these two competing species. In the poly(U) translating system, the correct substrate is consumed faster than the incorrect one, producing a situation where the concentration of the incorrect substrate is maintained at a high level and the correct one at a low level. A similar situation does not occur during the translation of natural messengers.

Addition of preformed phenylalanyl $tRNA_{Phe}$ reduced misreading significantly in our system. Also the concentration of poly(U) influences translational fidelity in this system (Table 1). The reason for the increase in accuracy with increasing poly(U) concentration is not clearly

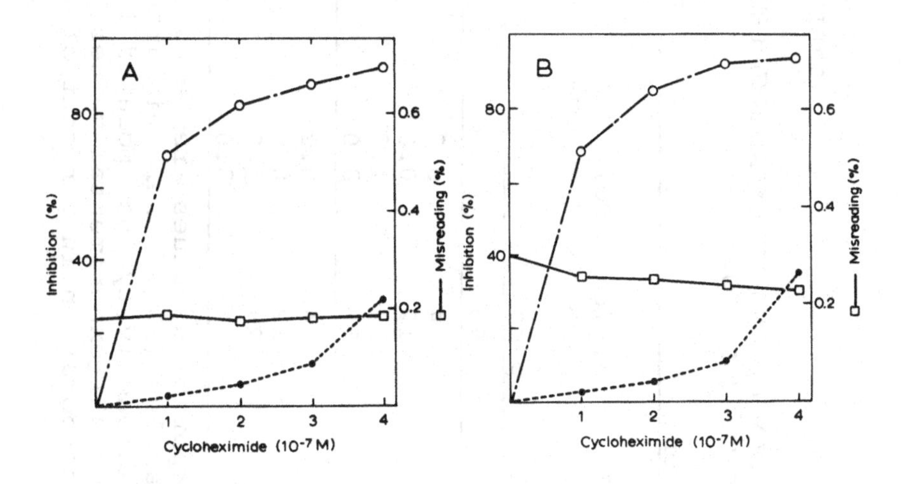

Fig. 4. Effect of Cycloheximide on Translational Fidelity. A. Wheat germ system. B.Reticulocyte Lysate. (o) Globin synthesis (o) Poly(Phe)synthesis.

TABLE I

EFFECT OF POLY(U) AND PHENYLALANYL tRNA CONCENTRATIONS ON RATE AND FIDELITY OF TRANSLATION

Conditions			Radioactivity incorporated		Misreading
			A	B	
^{14}C Phenyl-alanyl -tRNA (n mole)	Poly(U) (p mole)	Poly(U)/ ribosome	^{14}C Phenyl-alanine (p mole/ ribosome)[x]	^{3}H Leucine (p mole/ ribosome	$\frac{B}{A}$ x 100
0.3	60		39.0	0.38	0.97
1.3	4	0.2	3.5	0.010	0.30
	12	0.6	11.0	0.026	0.24
	24	1.2	22.7	0.055	0.24
	60	3.0	61.0	0.105	0.17
5.3	4	0.2	3.3	0.010	0.30
	12	0.6	10.2	0.025	0.25
	24	1.2	23.0	0.051	0.23
	60	3.0	64.0	0.066	0.10

[x] The incorporation values are based on the total number of ribosomes present. Samples (10 µl) were withdrawn in triplicate after 10 min. at 28°. Radioactivity incorporated in the absence of poly(U) was subtracted from all experimental values. Misreading is expressed as the leucine incorporation in per cent of the phenylalanine incorporation.

understood at the moment. It appears from these observations that there is a greater possibility of imperfect base pairing between codon and anticodon when several ribosomes are bound to the same poly(U) molecule, which unlike natural messengers contain 3 different reading frames.

There are several conflicting reports on the relationship between elongation rate and translational accuracy (Abraham 1983). Earlier publications from several laboratories have implied that these two parameters are inversely related (Gavrilova et al 1976). Experimental support for this view comes from studies using factor deficient, slowly elongating systems and elongation inhibitors. The opposite view is that speed and accuracy are directly related to each other. There are several data which support this view (Laughrea, 1981). Using the optimised system described above, we have studied the effect of several translational inhibitors on the accuracy of poly(U) translation in eukaryotic systems.

Effect of Protein Synthesis Inhibitors

In experiments shown in Figure 4 we have reexamined the effect of cycloheximide on translational fidelity. It is seen that in both systems studied cycloheximide in concentrations which drastically inhibited the translation of globin mRNA, had only a slightly inhibiting effect on the synthesis of polyphenylalanine. In the wheat germ system cycloheximide had no effect on the fidelity of translation whereas in the reticulocyte system the misreading was slightly decreased at low concentrations of cycloheximide which barely affected the poly(Phe) synthesis, but no further reduction in error rate was observed at higher, inhibiting cycloheximide concentrations. These results stand in striking contrast to those of Kurkinen (1981) who reported that addition of cycloheximide to a reticulocyte system strongly reduced the error rate. It should be noted that in his experiments the error rate was about twenty fold higher than in our experiments.

TABLE II

EFFECT OF ABRIN, RICIN AND CYCLOHEXIMIDE ON TRANSLATIONAL FIDELITY UNDER SUBOPTIMAL CONDITIONS

Addition	Inhibition of Phe incorporation (%)	Misreading (%)
None	0	0.93
Cycloheximide ($4x10^{-7}M$)	42	0.63
" ($1x10^{7}M$)	65	0.52
Abrin (2 ng)	27	0.65
" (5 ng)	69	0.41
Ricin (2 ng)	38	0.58
" (5 ng)	73	0.40

Translational assay was carried out in the rabbit reticulocyte system not supplemented with exogenous phenylalanyl-tRNA.
Misreading is expressed as the incorporation of leucine in per cent of the phenylalanine incorporation.

In view of the findings in bacterial systems (Wagner et al 1982) the possibility was considered that the error-reducing effect of cycloheximide may be expressed only under suboptimal conditions. The results in Table II indicate that this is indeed the case. Thus it is seen that when the system was not supplemented with phenylalanyl-tRNA, addition of cycloheximide definitely decreased the error rate, in general agreement with the earlier observations (Kurkinen 1981).

Abrin and ricin inhibit polypeptide chain elongation by a mechanism entirely different from that of cycloheximide, viz. by inactivating the large ribosomal subunit (Olsnes and Pihl 1972). The effect of these inhibitors on the translational accuracy is demonstrated in Tables II and III. When the poly(U) system was not supplemented with phenylalanyl tRNA, abrin and ricin, like cycloheximide, clearly reduced the relatively high error rate observed under these conditions (Table II). However, when the system was supplemented with 100 µM phenylalanyl-tRNA, the toxins in concentrations which inhibited poly(Phe) synthesis by approximately 70%, had no effect on the misreading (Table III). It should be noted that addition of abrin and ricin to the suboptimal system did not reduce the error rates below that observed in the optimized system.

TABLE III

EFFECT OF ABRIN AND RICIN ON FIDELITY ON TRANSLATION

Additions	Inhibition of Phe incorporation (%)	Misreading (%)
None	0	0.24
Abrin (5 ng)	32	0.23
" (20 ng)	70	0.23
Ricin (4 ng)	40	0.24
" (20 ng)	68	0.23

Translational assay was carried out in the rabbit reticulocyte system, supplemented with 100 µM phenylalanyl-tRNA. Misreading is expressed as the incorporation of leucine in per cent of the phenylalanine incorporation.

Taken together the above results show that the translational accuracy is not directly linked to the elongation rate as such. The finding that the effect of the inhibitors on the error rate depends on the concentration of the cognate substrate, shows that the inhibitors have no direct effect on the mechanism of aminoacyl-tRNA selection by the ribosomes, but affect fidelity by influencing the kinetics of the system, in agreement with the conclusions drawn from bacterial systems (Kurland 1982).

Effect of Temperature

Further evidence that the rate of elongation per se does not determine the error rate, was obtained in experiments where the temperature was varied. The results in Fig. 5A show that although in the wheat germ system the increase in poly(Phe) synthesis with increasing temperature was associated with a decrease in the missense frequency, there was a discrepancy between the temperature optima for elongation and fidelity. Thus the optimal temperature for the rate of poly(Phe) synthesis was higher than that for accuracy of translation. In the reticulocyte system the temperature optima for synthetic rate and accuracy almost coincided (Fig. 5B).

Measurements showed that the levels of aminoacyl-tRNAs were constant in the system (data not shown). Hence, the difference in temperature optima cannot be attributed to an effect on the levels of substrates.

In the reticulocyte system the effect of temperature on misreading was studied in the presence of a ricin concentration that decreased the rate of poly(Phe) synthesis by about 70%. It is seen (Fig. 6) that both in the presence and absence of ricin the error rate decreased with increasing temperature up to about 36°C. It is evident that the two curves coincided completely, showing that the influence of the incubation temperature on the fidelity was the same whether or not the rate of elongation was inhibited by ricin.

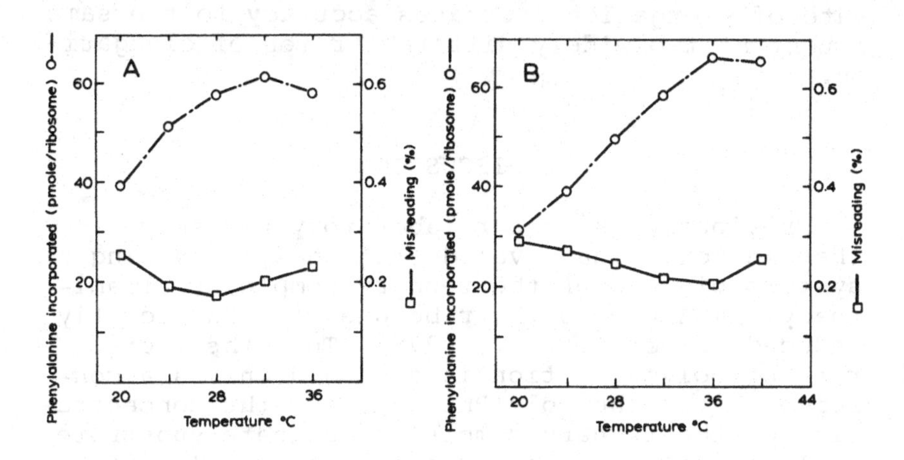

Fig. 5. Effect of Temperature on Translational Accuracy in Phenylalanyl tRNA Supplemented Systems. A. Wheat germ system. B. Reticulocyte lysate.

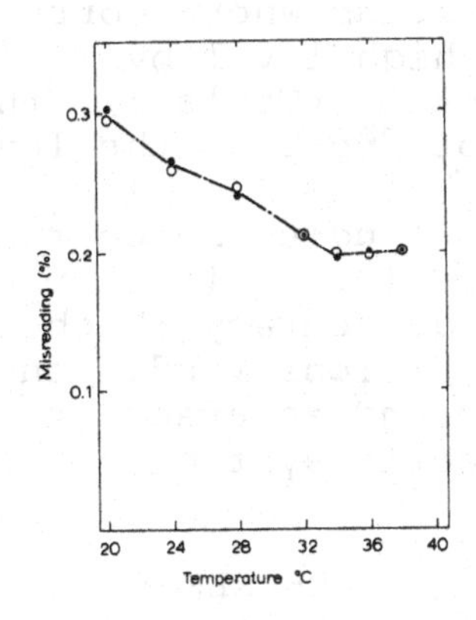

Fig. 6. Combined Effect on Temperature and Ricin on Translational Accuracy in the Reticulocyte system.
(o) Control (•) 20 ng recin present.

Thus, an increase in temperature and consequent rate of elongation increases accuracy to the same extent at two widely different rates of elongation.

DISCUSSION

Previous work in our laboratory has shown that in toxin-inactivated protein synthesizing systems binding of the ternary complex EF-1:aminoacyl-tRNA:GTP to the ribosome is significantly reduced (Benson et al., 1975). Thus the overall rate of polymerization is slowed down. As a consequence, in the poly(Phe) system, the concentration of the ternary complex of cognate substrate will increase, while the competing complex of non-cognate substrate will remain at the same fixed level. Therefore, the apparent increase of translational accuracy of non-optimisted systems observed earlier (Kurkinen 1981) and here, can be attributed to an effect on the kinetics of the system in the presence of elongation inhibitors. Thus in systems where correct substrate is maintained at a high level by adding preformed phenylalanyl $tRNA_{Phe}$ (Table 3), these elongation inhibitors do not increase the level of translational accuracy.

Our inhibitor studies support the view that there is no general relationship between rates of elongation and accuracy which hold under all experimental conditions (Anderson et al 1982). Apparently speed and accuracy are two independent outputs of a kinetic system.

REFERENCES

Abraham, A.K. , Olsnes, S. and Pihl, A. (1979) FEBS Lett. 101, 93-96.
Abraham, A.K. (1982) Prog. Nucleic Acid Res.Mol. Biol. 28, 81-100.
Abraham, A.K. and Flatmark, T. (1982) Advances in Polyamine Research vol. 4. 255-265. (Bachrach, U., Kaye, A. and Chayen, R. Eds.)Raven Press, New York.

Andersson, D.J., Bohman, K., Isaksson, L.A. and Kurland, C.G. (1983) Mol.Gen.Genet. 187, 467-472.
Benson, S., Olsnes, S., Pihl, A., Skorve, L. and Abraham, A.K. (1975) Eur.J.Biochem. 59,573-580.
Edelman, P. and Gallant, J. (1977) Cell, 10, 131-137.
Gavrilova, L.P., Kostiashkina, O.E., Koteliansky, V.E., Rutkevitch, N.M. and Spirin, A.S. (1976) J.Mol.Biol., 101, 537-552.
Jelenc, P.C. and Kurland, C.G. (1979) Proc.Natl. Acad.Sci. USA, 76, 3174-3178.
Kurkinen, M. (1981) FEBS Lett., 124, 79-83.
Kurland, C.G. (1982) Cell, 28, 201-202.
Loftfield, R.B. and Vanderjagt, D. (1972) Biochem.J., 128, 1353-1356.
Laughrea, M. (1981) Biochemie, 63, 145-168.
Olsnes, S. and Pihl, A. (1972) Nature, 238, 459-464.
Sakai, T.T. and Cohen, S. (1976) Prog.Nucleic Acid Res. and Mol.Biol., 17, 15-41.
Wagner, E.G.H., Ehrenberg, M. and Kurland, C.G. (1982) Mol.Gen.Genet., 185, 269-274.

ACKNOWLEDGEMENT

This work was supported by the Norwegian Cancer Society.

TRANSLATIONAL AMBIGUITY AND CELL DIFFERENTIATION IN A LOWER EUCARYOTE

M. Picard-Bennoun, E. Coppin-Raynal,
M. Dequard-Chablat.

Laboratoire de Génétique - Université de
Paris XI - Bâtiment 400
91405 ORSAY Cedex - France

INTRODUCTION

Control of translational accuracy has been well documented in bacteria, especially E. coli. The key role played by the ribosome has been emphazised through the use of aminoglycoside antibiotics (such as streptomycin) and the study of mutations which enhance or decrease the rate of errors in translation (see Piepersberg et al, 1980 for a review).

Our present knowledge of the control of translational fidelity in eucaryotes is very poor and scattered. Mutations which modify (or might modify) the error level in translation have been described in a few lower eucaryotes, especially yeasts (see discussion for the references). However, a systematic search for such mutations remained to be performed to answer two main questions : How many genes are involved in the control of translational fidelity ? What is the response of an eucaryotic cell when the error rate is increased or lowered ?

This paper summarizes our efforts to develop a genetic system in which these two questions might be answered. We present our rationale and strategies for a systematic search of mutants altered in translational fidelity. Moreover, the fungus we use, Podospora anserina, develops in a more sophisticated way than yeasts. This allowed us to observe that sexual reproduction does not occur when translational fidelity is too high . This and other data suggest that inaccuracy might be used as a translational control of gene expression during cellular differentiation.

MATERIALS and METHODS

Podospora anserina is a filamentous fungus, related to *Neurospora crassa*. The main properties of and culture techniques for *Podospora* were first described by Rizet and Engelmann (1949) and later reviewed by Esser (1974). The origin and phenotypic properties of the strains used in this paper have been previously described (see the text for references).

RESULTS

(A) - Rationale in the Search of Mutations which Modify the Rate of Errors in Translation

1 - *Mutations which increase translational ambiguity act as omnipotent informational suppressors*. Informational suppressors are mutations in genes coding for macromolecules involved in protein synthesis, either tRNAs or ribosomal proteins. The suppressor mutation gives new decoding properties to the translational apparatus. Therefore, tRNA and ribosomal suppressors share the property to relieve the phenotypic defects caused by mutations lying in several unrelated genes. Two criteria allow to distinguish these two kinds of suppressors. First, tRNA suppressors are codon-specific while ribosomal suppressors act as omnipotent suppressors capable of correcting nonsense, missense and frameshift mutations. Second, tRNA suppressors are mainly dominant while ribosomal suppressors are mostly recessive (see Gorini, 1970 for a review). This background was our guide in searching for and analysing mutations which increase translational ambiguity. Two screening procedures were used : restoration of pigmentation in spore-color mutants and restoration of prototrophy for an auxotrophic (*leu1-1*) mutant strain. Data are reported in table 1. Omnipotent suppressors lie in two major (*su1*-*su2*) and three minor genes (*su3* - *su5* - *su7*) (Picard, 1973 and unpublished data). In vitro analysis of *su1* and *su2* ribosomes confirmed the hypothesis suggested by genetic analysis : the mutations affect the miscoding capacity of the ribosomes (Coppin-Raynal, 1982 and unpublished data).

2 - *Mutations which decrease translational ambiguity act as antisuppressors*. In *E. coli*, mutations expressing resistance to certain aminoglycosides have been shown to decrease misreading. These so-called "restrictive" (or antisuppressor) mutations diminish the efficiency of tRNA and

Suppressors	Screening procedure		
	spore pigmentation (a)	prototrophy (b)	restoration of sporulation (c)
Omnipotent	71	20	2
locus sul	51	13	
su2	13	7	
su3	4		
su5	1		
su7	2		
sull			1
sul2			1
tRNA like	1	23	
locus su4	1	6	
su8		2	
sul0		15	

Table 1 : Isolation of informational suppressors
Strains used (a) three mutants altered in spore pigmentation, mainly 193
(b) leul-1
(c) AS7-2

ribosomal suppressors (see Piepersberg *et al*, 1980 , for a review).

We used these two properties to look for such mutations in *Podospora*. Ten mutations conferring resistance to the aminoglycoside paromomycin were obtained. They lie in two genes (*Pml* and *Pm2*). Although biochemical analysis suggests that they are ribosomal mutations, they do not display an antisuppressor effect and they do not seem to modify the misreading level (Dequard *et al*, 1980 and unpublished data). Direct selection for antisuppressor mutations was performed using the *193 sul-1* strain. The *sul-1* mutation is an omnipotent suppressor which relieves the pigmentation defect caused by the *193* mutation. Mutations which decreased spore pigmentation in this strain were isolated. Besides trivial mutations, 6 antisuppressors were obtained. They lie in two genes, *AS1* and *AS2* (Picard-Bennoun, 1976 and table 2).

Biochemical analysis of the AS1-1 ribosomes showed that this mutation can restrict translational ambiguity, at least in the presence of paromomycine (Picard-Bennoun, 1981).

Screening procedure	Antisuppressor loci							antagonistic suppressors
	AS1	AS2	AS3	AS4	AS5	AS6	AS7	
Decrease of suppressor efficiency[(a)]	5	1						
Restoration of female fertility		1[(b)]						9[(c)]
Restoration of growth in the presence of Pm						1	2	1[(c)]
Cold-sensitivity			1			1		
Restoration of sporulation in a strain carrying the su4-1 mutation			4	45	1			

Table 2 : Isolation of mutations displaying an antisuppressor effect
Strains used (a) 193 su1-1
(b) su1-31
(c) su2-5. The antagonistic mutations are su1 suppressors

3 - Phenotypic properties of ribosomal suppressors and antisuppressors open new screening procedures for mutations modifying translational misreading

. Strains carrying a high efficient suppressor are unable to differentiate female organs. The level of pigmentation of a spore-color mutant (193) in the presence of a suppressor gives an estimation of its efficiency. The most efficient suppressors specifically alter differentiation of female organs and spore germination (Picard, 1973 and unpublished data). We looked for mutations capable of restoring female fertility in these strains. All the mutations decreased the efficiency of the suppressor. One of them was an AS2 allele. Others showed that two omnipotent suppressors (su1 and su2) could interact in an antagonistic way (Picard-Bennoun, 1976 and table 2).

. Susceptibility to paromomycin is increased by most suppressors. In E. coli, the ram mutations which increase misreading provide hypersensitivity to the aminoglycoside streptomycin (see Piepersberg et al, 1980, for a review). A similar situation has been observed in Podospora (Coppin-Raynal, 1981). Therefore, we looked for mutations relieving the susceptibility to paromomycin caused by ribosomal suppressors. Two kinds of mutations were observed. Some of them did not seem to be involved in the control of translational fidelity although they conferred resistance to paromomycin. They were localized at the previously identified Pm1 and Pm2 loci (see above) and a new Pm3 locus. The other mutations decreased the efficiency of the suppressor. Again, this decrease can be achieved either by an antisuppressor mutation or by an antagonistic suppressor (Coppin-Raynal and Le Coze, 1982 and table 2). In vitro analysis confirms that AS7 mutations are involved in the ribosomal control of translational fidelity (Coppin-Raynal, 1982). Resistance of AS7 ribosomes to the paromomycin-induced misreading is a property of the 40S subunit (Coppin-Raynal, unpublished data).

. Strains homozygous for the AS7 mutations are unable to sporulate. Several antisuppressors mutations disturb either female organ development or sporulation. Mutations capable of restory sporulation in AS7 strains have been obtained. The most efficient increase the level of translational errors both in vivo and in vitro. They map in two suppressor loci not previously identified (Dequard and Coppin-Raynal, manuscript in preparation and table 2).

. Indirect selection for antisuppressor mutations : cold-sensitivity. Cold-sensitivity has been thought to be a good screen to obtain ribosomal mutations. Twenty-four mutant strains unable to grow at low temperature were isolated in Podospora. They lie in 24 different genes. None of them display a suppressor effect while two are antisuppressors (Picard-Bennoun and Le Coze, 1980 and table 2). The AS6-1 ribosomes are resistant to the ambiguity effect of paromomycin in vitro (Picard-Bennoun, 1981).

(B) - Translational Misreading and Cellular Differentiation

The most efficient suppressors and several antisuppressors cause phenotypic alterations, mainly defects in spore germination, female organ development and sporulation. Several arguments demonstrate that these defects are directly related to the level of translational errors.

1 - Female organ differentiation is disturbed when the error level is too high. The striking facts which support this conclusion are as follows.

. When strains carry two (*su1* and *su2*) suppressors, three kinds of interaction are observed at the level of suppression : additivity, no effect, or antagonistic action. The kind of interaction observed depends on the combination of the *su1* and *su2* alleles considered (Picard-Bennoun, 1976). While suppressors of low efficiency do not cause female sterility, double-mutant strains carrying two such suppressors, which are subject to an additive effect, do not develop female organs. The converse is observed : antagonistic action between two suppressors restore fertility (table 2).

. Female sterility caused by a suppressor is reversed only by mutations which decrease suppressor efficiency (table 2).

. When paromomycin is added to the culture medium at low concentrations that do not slow the growth rate, production of female organs is reduced. Those female organs which do develop are unable to promote the whole process of sporulation. Cycloheximide, used as a control, produced no effect on female organ differentiation even at doses that slow the growth rate by a factor of two (Coppin-Raynal, 1981).

2 - Sporulation does not occur when the error level is too low. Autofertilization of a strain homozygous for the *AS7* mutations does not result in the production of spores although fructifications develop normally. Cytological analysis shows that the defect occurs after meïosis, at the end of the sporulation process. The direct relationship between this phenotypic alteration and the low level of errors in these strains was ascertained in two ways (Dequard-Chablat and Coppin-Raynal, manuscript in preparation). First, paromomycin completely relieves the sporulation defect. Second, mutations which restore a wild-type level of spore production increase the level of translational errors (table 2).

3 - Strains homozygous for the tRNA suppressor su4 do not sporulate. As reported in table 1, several tRNA-like suppressors have been obtained in Podospora. They are dominant and act only upon putative nonsense mutations. While su8 and su10 strains develop normally, strains homozygous for the su4 mutations are defective in the process of sporulation. In order to understand this phenomenon, we looked for mutations capable of restoring sporulation in these strains. This procedure gave two kinds of mutations : second-site mutations which abolish or lower the suppressor effect and antisuppressor mutations (table 2). In fact, all the mutations reducing the efficiency of this suppressor restore sporulation. On the other hand, although strains heterozygous for the mutation ($su4/su4^{+}$) are able to sporulate, the presence of a ribosomal suppressor abolishes sporulation (in strains $su4.su1/su4^{+}.su1$). These data show that the presence of the wild-type $su4^{+}$ gene is neither necessary nor sufficient to promote sporulation. The lack of sporulation appears to be related to suppression. In the same way, su8.su2 strains are unable to sporulate although $su8.su2^{+}$ strains develop normally.

DISCUSSION

(A) - How many Genes are Involved in the Control of Translational Fidelity ?

. In E. coli, ram mutations, which increase translational misreading, have so far been localized in two structural genes for ribosomal proteins (see Isono, 1980 for a review). In Saccharomyces cerevisiae, omnipotent suppressors lie mainly in two genes (Hawthorne and Leupold, 1974 ; Surguchov et al, 1980 ; Surguchov et al, 1981) although a new one has been recently characterized (Ono et al, 1981 ; Masurekar et al, 1981). In Schizosaccharomyces pombe, two loci for omnipotent suppressors have also been identified (Hawthorne and Leupold, 1974). In Podospora, most of the omnipotent suppressors (84 out of 93) have been localized in two loci su1 and su2 (table 1). If we had only used the classical method to select suppressors (i.e., auxotrophic mutants), only these two loci would have been characterized. Spore pigmentation appears to be a broader screen for omnipotent suppressors, especially those which display a weak efficiency. Su11 and su12 have been obtained in a very different way : they have been screened against an antisuppressor. Therefore, their isolation is reminiscent of the way used to obtain ram mutations in E. coli. In fact, ram mutations have been selected because they antagonize restrictive Sm^R or Sm^D mutations (see Piepersberg et al, 1980 for a review).

The striking fact is that in bacteria as well as in eucaryotes, mutations capable of increasing misreading lie mainly in two genes. However, our data suggest that unidentified genes for omnipotent suppression remain to be found in yeasts and possibly in bacteria.

. In E. coli, restrictive (antisuppressor) mutations lie in three genes coding for ribosomal proteins (see Isono, 1980 for a review). Two additional genes, not so well characterized, have also been described (Elseviers and Gorini, 1975 ; Garvin and Gorini, 1975). Antisuppressors acting exclusively on omnipotent suppressors and localized in six genes have been isolated in S. pombe (Thuriaux et al, 1975). One of them was shown to be a ribosomal gene (Coddington and Fluri, 1977). In S. cerevisiae, antisuppressors that reduce the efficiency of nonsense tRNA suppressors have been isolated (Mc Cready and Cox, 1973). It was not determined whether they act on omnipotent suppressors. However, many antisuppressors have been recently selected against omnipotent suppressors by means of different procedures (Liebman and Cavenagh, 1980 ; Liebman et al, 1980 ; Ishiguro, 1981). They lie in three genes. In one case, the gene has been shown to be ribosomal (Ishiguro, 1981).

With the use of five different screening procedures, we have been able to identify 7 antisuppressor loci in Podospora (table 2). Three of them, at least, are ribosomal genes. The number of mutations identified in each locus suggests that new loci remain to be found.

. Omnipotent suppressors and antisuppressors are scattered on the whole genetic map of Podospora. However, it is noteworthy that there are several cases where suppressor and antisuppressor mutations lie in the same locus. At this time, there is no genetic procedure available to establish whether the two kinds of mutations affect two closely linked genes or the same gene. The situation happens for the following couples of mutations : su2-AS2, su3-AS7, and su11-AS4. It would be interesting to learn whether mutations increasing and mutations decreasing misreading may affect the same gene. Further biochemical experiments are needed to solve this problem.

(B) - What is the Response of an Eucaryotic Cell when the Error Rate is Increased of Lowered ?

The results we have obtained in Podospora with paromomycin and mutations which modify the error rate in translation lead us to the following conclusion. Cell differentiation involved in sexual reproduction occurs only within a certain range of misreading frequencies. These processes are impaired both when the error level is too high and when it is too low, even though vegetative growth continues under these extreme conditions. Two striking facts support this conclusion. First, female sterility caused by a highly efficient suppressor is relieved only when the efficiency of the suppressor is lowered. Second, the defect in sporulation associated with the AS7 mutations is relieved by paromomycin and mutations which increase misreading. These data show that a precise level of translational errors is needed at these key points of the life cycle. Although indirect and scattered, a few other facts lead us to assume that the error rate might be increased during these processes. In yeast (Rothstein et al, 1977) as well as in Podospora, some nonsense suppressor mutations disturb sporulation when the cells are homozygous for the suppressor. Physiological study of this phenomenon led the authors to assume that this defect is caused by an increased efficiency of suppression. In Podospora, we were able to show that this (putative) increase of nonsense suppression efficiency during sporulation is controlled at the ribosomal level (Picard-Bennoun, 1976 and this paper). In the alga Chlamydomonas reinhardii, P. Bennoun (pers. com.) observed that several nuclear mutants defective in photosynthesis are leaky when the cells have differentiated into gametes but not during vegetative growth. A functional leakiness induced by gametogenesis *per se* appears unlikely.

Defects in sporulation caused by some tRNA nonsense suppressors in S. cerevisiae and in Podospora can be explained if these tRNAs carry out some termination codon readthrough which is not deleterious during growth but which is lethal to the process of sporulation. In the same way, nonsense readthrough or ribosomal frameshifting induced by ribosomal suppressors might inhibit cell differentiation without affecting growth. However, the other data cannot be explained in such a trivial way. First, leakiness associated with gametogenesis in Chlamydomonassuggests new decoding properties of the translational apparatus when compared to vegetative cells. Second, the physiological defects caused by the high fidelity mutations (antisuppressors) suggest that termination codon readthrough or frameshift of some messages is necessary for differentiation at least in *Podospora*.

Geller and Rich (1980) were the first to assume that control of gene expression might be achieved at the level of termination of translation in eucaryotic cells. This control might be used at key points in cellular development for the synthesis of regulatory proteins. At least three parameters could be involved in this putative modification of the translational apparatus.

. *Synthesis of new species of tRNAs or specific modifications of pre-existing ones*. Geller and Rich (1980) reported a UGA suppressor activity which produces a -haemoglobin readthrough protein in rabbit reticulocytes. Diamond *et al*, (1981) characterized an opal serine tRNA from bovine liver. A tyrosine tRNA from *Drosophila* is either able or unable to read the termination codon UAG depending on the modification of a Q base in its anticodon (Bienz and Kubli, 1981).

. *Passive increase of translational errors caused by starvation through depletion of some aminoacyl-tRNA.* It may be relevant to recall that the cellular processes needed for sexual reproduction in the lower eucaryotes mentioned above are induced by starvation, in particular by nitrogen starvation. It is well established that bacteria have developed a mechanism, the stringent response, as a defence against miscoding generated by amino acid starvation (see Gallant, 1979 for a review). We do not know, at this time, whether eucaryotic cells have analogous mechanisms.

. *Changes at the ribosomal level causing an enhanced misreading*. In higher eucaryotes, extensive structural modifications of the ribosomes during the processes of differentiation and development have not been reported except for *Drosophila* (Lambertson, 1975). However, phosphorylation of the ribosomal protein S6 has been well documented (see Leader, 1980 for a review). Several stimuli have been demonstrated to increase S6 phosphorylation. However, it has not yet been possible to explain the biological role of these phosphorylations. In lower eucaryotes, the situation could be quite different. In fact, in *Dictyostelium* (Ramagopal and Ennis, 1981) and in *Tetrahymena* (Hallberg *et al*, 1981) cell differentiation is accompanied by changes in ribosome structure and function. Gametogenesis in *Chlamydomonas* (Martin *et al*, 1976) and sporulation in *S. cerevisiae* (Pearson and Haber, 1977) are both characterized by extensive turn-over of vegetative ribosomes and synthesis of new ribosomes. Chlamydomonas ribosomes isolated from gametes differ strikingly from those of vegetative cells, especially in their susceptibility to aminoglycosides (Picard-Bennoun and Bennoun, in preparation).

In summary, our data suggest that readthrough or frame-shifting is needed at key points of the life cycle in lower eucaryotes. Further genetic and biochemical experiments will help to learn more about the control of translational fidelity and the putative role of inaccuracy in eucaryotes.

Acknowledgments - We thank D. Le Coze and S. Largeault for their technical assistance. This work was supported by grants from the C.N.R.S. (ATP Microbiologie), from the D.G.R.S.T. (Interactions dynamiques entre macromolecules biologiques) and the Fondation pour la Recherche Médicale.

REFERENCES

Bienz, M. and Kubli, E. (1981) Nature. 294, 188-190
Coddington, A. and Fluri, R. (1977) Molec. gen. Genet. 158, 93-100
Coppin-Raynal, E. (1981) Biochem. Genet. 19, 729-740
Coppin-Raynal, E. (1982) Current Genetics. 5, 57-63
Coppin-Raynal, E. and Le Coze, D. (1982) Genet. Res. 40, 149-164.
Dequard, M., Couderc, J-L., Legrain, P., Belcour, L. and Picard-Bennoun, (1980) Biochem. Genet. 18, 263-280
Diamond, A., Dudock, B. and Hatfield, D. (1981) Cell. 25, 497-506
Elseviers, D. and Gorini, L. (1975) Molec. gen. Genet. 137, 277-287
Esser, K. (1974) in Handbook of Genetics (R.C. King, ed., Plenum Press.) 1, 531-551.
Gallant, J.A. (1979) Ann. Rev. Genet. 13, 393-415
Garvin, R.T. and Gorini, L. (1975) Molec. gen. Genet. 137, 73-78.
Geller, A.I. and Rich. A. (1980) Nature 283, 41-46
Gorini, L. (1970) Ann. Rev. Genet. 4, 107-134
Hallberg, R.L., Wilson, P.G. and Sutton, C. (1981) Cell. 26, 47-56.
Hawthorne, D.C. and Leupold, U. (1974) Curr. Top. Microbiol. Immunol. 64, 1-47

Ishiguro, J. (1981) Curr. Genet. 4, 197-204
Isono, K. (1980) in Ribosomes : Structure, Function and Genetics (G. Chambliss, G.R. Craven, J. Davies, K. Davis, L. Kahan and M. Nomura, eds. University Park Press) pp. 641-669
Lambertson, A.G. (1975) Molec. gen. Genet. 139, 133-144
Leader, D.P. (1980) in Protein phosphorylation in regulation (P. Cohen, ed. Elsevier) pp. 203-233.
Liebman, S.W. and Cavenagh, M.M. (1980) Genetics. 95, 49-61
Liebman, S.W., Cavenagh, M.M. and Bennett, L.N. (1980) J. Bacteriol. 143, 1527-1529.
Mc Cready, S.J. and Cox, B.S. (1973) Molec. gen. Genet. 124, 305-320
Martin, N.C., Chiang, K.S. and Goodenough, V.W. (1976) Dev. Biol. 51, 190-201
Masurekar, M., Palmer, E., Ono, B.I., Wilhelm, J. and Sherman, F. (1981) J. Mol. Biol. 147, 381-390
Ono, B.I., Stewart, J.W. and Sherman, F. (1981) J. Mol. Biol. 147, 373-379
Pearson, N.J. and Haber, J.E. (1977) Mol. gen. Genet. 158, 81-91
Picard, M. (1973) Genet. Res. 21, 1-15
Picard-Bennoun, M. (1976) Molec. gen. Genet. 147, 299-306
Picard-Bennoun, M. (1981) Molec. gen. Genet. 183, 175-180
Picard-Bennoun, M. and Le Coze, D. (1980) Genet. Res. 36, 289-297
Piepersberg, W., Geyl, D., Hummel, H. and Böck, A. (1980) In : Genetics and evolution of transcriptional and translational apparatus (Kodansha Scientific, Tokyo) pp 359-377
Ramagopal, S. and Ennis, H.L. (1981) Proc. Nat. Acad. Sci. USA. 78, 3083-3087
Rizet, G. and Engelmann, C. (1949) Rev. Cytol. Biol. Veg. 11, 201-304
Rothstein, R.J., Esposito, R.E. and Esposito, M.S. (1977) Genetics. 85, 35-54
Surguchov, A.P., Berestetskaya, Yu.V., Forminykch, E.S., Pospelova, E.M., Smirnov, V.N., Ter-Avanesyan, M.D. and Inge-Vechtomov, S.G. (1980) Febs Letters. 111, 175-178
Surguchov, A.P., Berestetskaya, Yu.V., Smirnov, V.N., Ter-Avanesyan, M.D. and Inge-Vechtomov, S.G. (1981) FEMS Letters. 12, 381-384
Thuriaux, P., Minet, M., Hofer, F. and Leupold, U (1975) Molec. gen. Genet. 142, 251-261

This paper is dedicated to the memory of Luigi Gorini

Strategies of Translation

C.G. Kurland

Department of Molecular Biology
Biomedical Center
Box 590, 751 24 Uppsala, Sweden

I wish to remark on possible differences between, among other things, prokaryotic and eukaryotic ribosomes. In order to do so, I will introduce you to what I call a molecular strategy. This is a simple notion which can be made accessible with the help of a riddle. The riddle is: If errors of translation are to be avoided, why are there common mutant ribosomes that are more accurate than wild type ribosomes? One class of relevant bacterial mutants are those that are streptomycin resistant (rpsL). These have been shown to lower the missense frequencies in translation due to suppressor tRNA's as well as those arising spontaneously. Why, then aren't all bacterial ribosomes as accurate as some of the streptomycin resistant ones?

There are at least two possible answers to this conundrum. One of these will be illustrated by Marguerite Bennoun when she describes the evidence for the proposition that the proper expression of the cellular program in certain fungi requires a minimum error rate. In other words she will show that all errors are not necessarily destructive. Some may even be required.

The second sort of answer will be forthcoming in the talks of R. Loftfield, Måns Ehrenberg, and B. Kraal, who discuss some of the partial reactions in protein synthesis that are coupled to the consumption of nucleoside triphosphates. Thus, there now is data suggesting that both the ATP-dependent selection of amino acids by the synthetases as well as the GTP-dependent selection of aminoacyl-tRNA by ribosomes involve proofreading functions. This means that the accuracy of both synthetase as well as ribosome function are directly related to their energy consumption. If there is an extra energetic cost associated with higher accuracies of translation, the evolutionary optimization of the translation process may settle at a modest accuracy bought at the expense of a modest dissipation of energy.

Now, the optimization of the performance characteristics of the translation apparatus with respect to accuracy and dissipative losses is a particularly simple example of what I mean by a molecular strategy. A more complicated example is the stringent response in bacteria. The point is that there are virtually infinitely many ways to arrange the individual rate constants as well as the concentration terms affecting the partial reactions that make up the translation process. Under competitive conditions some of these arrangements are more "effective" than others; by effective is simply meant that the bacterium or organism characterized by that particular kinetic arrangement survives under some particular set of external constraints in competition with bacteria having other kinetic arrangements. I choose the term molecular strategy to describe these kinetic arrangements in order to stress the fact that these arrangements are not unique, but are selected by competition out of many conceivable arrangements, and that the succesful arrangement will tend to optimize more than one performance characteristic of the system. Above all, while I wish to avoid any implication of volition for the systems I discuss here, the effective strategies selected by evolution will appear to be solving "problems" in the sense that the boundary conditions of the selective environment will impose constraints on what can be an effective strategy. In order to avoid the necessity of writing paragraphs such as this one whenever discussing the constraints on the kinetic arrangements of translation I will use the terms "strategies" and "problems" in the sense defined above.

The point then is that the problem confronting E.coli when it translates a messenger is not simply to get an accurate protein copy. Among other things, it must do so at an energetic cost that does not set too high a burden on its metabolism. Furthermore, the metabolic state of an organism such as E.coli will be exquisitely responsive to the growth medium. Hence, the metabolic burden of synthesizing a protein will in such an organism change with the growth medium. We might therefore expect that the design strategies of the translation apparatus are optimized with respect to the repertoire of metabolic adaptations that E.coli commands. Here, for example, the stringent response is of clear relevance.

Contrast this metabolic adaptability with the relatively constant environment of most of the cells of a complex metazoan. If the strategies of the translation system are selected by the problems confronting the cell, they should be very different in these two extreme examples. Accordingly, when A. Abraham and J. Lake discuss their work with eukaryotic, archebacterial and eubacterial ribosomes it would be useful for us to attend to the structural and kinetic differences between these different systems; they might provide clues to important differences in the biological problems confronting these systems.

It seems more than a little relevant here that metazoan cells seem not to have the elements to support a stringent response of the sort found in bacteria. A related fact is that the eukaryotic cytosol does not seem to have a catalytic factor resembling the EF-Ts of bacteria. Clearly the guanine nucleotide fluxes that drive and control translation in bacteria, are arranged differently in the eukaryotic cytosol. Why?

By far the most bizarre molecular strategies associated with gene expression are those found in plastids. Typically a mitochondrion will synthesize a small fraction of its own proteins. It does so with the aid of a translation system that resembles that of bacteria. Nevertheless, all but one of the ribosomal proteins in mitochondria are coded by nuclear genes, synthesized in the cytosol and then transported into the plastid for assembly together with RNA of mitochondrical origin into a ribosome. What problem is solved by this elaborate and as yet totally inexplicable arrangement?

A formidable amount of detailed work has been carried out on the bacterial ribosome during the last two decades. In the light of this effort, it is natural to ask what sort of things are worthwhile doing with ribosomes from other sources. I wish to suggest that one possible answer is that attention should be focused on uncovering the different molecular strategies employed by these different systems, rather than merely to duplicate blindly the earlier work done with bacteria.

4. INTRACELLULAR PROTEIN TRANSPORT

DEFECTIVE ENDOSOME ACIDIFICATION IN MAMMALIAN CELL MUTANTS "CROSS-RESISTANT" TO CERTAIN TOXINS AND VIRUSES

William S. Sly*, Michael Merion*, Paul Schlesinger¶, Joan M. Moehring† and Thomas J. Moehring†

*Departments of Pediatrics and ¶Physiology and Biophisics, Washington University School of Medicine, St. Louis, MO 63110, and †Department of Medical Microbiology, University of Vermont, Burlington, VT 05405

ABSTRACT

Many physiological ligands, toxins, and viruses enter mammalian cells through receptor-mediated endocytosis. After entry in coated vesicles, they appear in uncoated vesicles called endosomes, which serve as intermediates in the transfer of ligands from the cell surface to lysosomes. Evidence suggests that once internalized, the toxic subunit of diphtheria toxin and the nucleic acid of several animal viruses gain access to the cytosol of host cells through an acidic intracellular compartment(s). A mutation conferring resistance to diphtheria toxin that also confers resistance to these viruses could indicate a defect in the acidification of this compartment(s). In this report, we discuss our recent evidence that one class of "cross-resistant" mutants from CHO-K1 cells have such a defect.

Control cell lines and "cross-resistant" mutants were allowed to internalize FITC-Dextran, and endosomes and lysosomes were isolated by subcellular fractionation. Fluorescence measurements of subcellular fractions permitted the measurement of the internal pH of the isolated vesicles. Our results demonstrated that: (i) endosomes and lysosomes isolated from parental Chinese hamster ovary cells (CHO-K1) maintain an acidic pH; (ii) acidification of both endosomes and lysosomes is mediated by a Mg^{++} and ATP-dependent process; (iii) GTP can satisfy the ATP requirement for lysosome acidification, but not for endosome acidification; (iv) at least one class of mutant CHO-K1 cells that are "cross-resistant" to toxins and animal viruses have a defect in the ATP-dependent acidification of their endosomes.

These studies provided biochemical and genetic evidence that the mechanisms of acidification of endosomes and lysosomes are distinct, and that a defect in acidification of endosomes is one biochemical basis for "cross-resistance" to toxins and viruses.

INTRODUCTION

Receptor-mediated endocytosis is a specific pathway by which extracellular substances can enter the intracellular vesicular system (1-5). Steps in the entry pathway include: i) binding to cell surface receptors; ii) internalization within coated vesicles; iii) distribution of ligand to appropriate intracellular compartments; and iv) in many cases, return of the receptor to the cell surface. For some pathogenic viruses and toxins, step (iii) includes transfer of a portion of the ligand to the cytosol (6-16). Recently, it has become apparent that this transfer is critical for the biologic activity of these viruses and toxins. The discovery that lysosomotropic amines, which raise the pH of normally acidic intracellular vesicles (17-19), block viral infection without impairing either viral adsorption or internalization (9,10,20,24), suggested that endocytosis and acidification of endocytic vesicles are essential for viral infection (20-25). Helenius and co-workers have provided direct evidence implicating an acidic environment in viral penetration (22-24). Although lysosomes are known to be acidic, the recent reports that pre-lysosomal, endocytic vesicles are rapidly acidified (19) make these smooth membrane bound structures suitable sites for the transfer of the viral genome to the cytoplasm.

Diphtheria toxin (DT) binds receptors on the cell surface of mammalian cells, and enters cells by receptor-mediated endocytosis (11-13). Two pieces of evidence indicate that DT also enters the host cell cytosol from an acidic intracellular environment. First, lysosomotropic amines block the action of DT (13-15). Second, one can bypass the resistance conferred by amines by exposing cells to which toxin has been bound, to a low pH shock (13,15,16). The recent evidence that the endosome is the site at which internalized ligands are first exposed to an acidic environment (19) makes the endosome a possible site for toxin entry also.

Amines are thought to prevent the toxicity of some viruses and DT by raising the internal pH of the compartment(s) through which the viral nucleic acid and the toxic component of DT gain access to the cytosol. If these agents enter the same acidic compartment, or if they enter different compartments that are acidified by the same mechanism, cells selected for resistance to one agent, which simultaneously acquire resistance to the others, are good candidates for mutants which have lost the ability to acidify the compartment(s) through which these agents gain access to the cytosol. Such mutants were isolated from KB cells by Moehring and Moehring (26). These mutants which were selected for resistance to DT, but proved also to be resistant to a number of RNA viruses were called "cross-resistant" mutants. More recently, the same laboratory has isolated "cross-resistant" mutants of CHO-K1 cells (27). Among the "cross-resistant" mutants of CHO-K1 cells is a class, designated DPV^r, which was selected on the basis of resistance to Pseudomonas exotoxin A (PT), and which proved 10-100 times more resistant to DT than CHO-K1 cells, and to infection by Sindbis, Semliki Forest, and vesicular stomatitis viruses (VSV). The resistance of this class of mutants to DT was overcome by exposing the mutant cells to low pH (27), as would be expected for mutants with a defect in acidification.

Merion and Sly recently reported (28) experiments using a two-step Percoll density gradient fractionation which resolved two pre-lysosomal endocytic compartments through which several physiological ligands pass enroute to lysosomes. These two compartments were less buoyant than plasma membrane markers, but more buoyant than secondary lysosomes. We recently (28A) extended these studies to CHO-K1 cells, and to DPV^r mutants derived from them. There we demonstrated: i) resolution of endosomes and lysosomes in CHO-K1 cells by this technique; ii) that loading of the endocytic compartments with fluorescein labeled dextran (FITC-Dextran) prior to subcellular fractionation, permits measurement of the internal pH of the isolated compartment by the method of Ohkuma and Poole (17); iii) that both endosomes and lysosomes of normal CHO-K1 cells possess a Mg^{++} and ATP-dependent acidification system; iv) that GTP can replace ATP in acidification of isolated lysosomes, but not of isolated endosomes; v) that DPV^r mutants of CHO-K1 cells have a defect in the ATP-dependent acidification of endosomes. In addition, we presented studies of acidification of endocytic vesicles in whole cells which demonstrated impaired acidification in DPV^r mutants. These studies which we described below suggested that this class of "cross-resistance" of CHO-K1 cells to PT, DT, and animal viruses can be explained by a mutational loss in the ability to acidify the endosome.

MATERIALS AND METHODS

Four different cell lines were examined in our recent report. These included, as the normal control line, CHO-K1, the parental cells from which the mutant cells were isolated. Two independently isolated "cross-resistant" lines were used, RPE.28 and RPE.44 (27). Both cell lines were selected for resistance to Pseudomonas exotoxin A (PT) and found to be resistant to Diphtheria toxin (DT) and to several viruses. A control for isolated resistance to DT was provided by RE.31 (29), a cell line selected for DT resistance which retained normal sensitivity to viruses. Culture conditions were described elsewhere (28). The method for isolation of endosomes and lysosomes was described previously (28). It involved a two-step Percoll density gradient fractionation. The first step resolved typical secondary lysosomes (which localized at a peak modal density of 1.07 g/ml) from more buoyant structures. In order to resolve endosomes from plasma membrane, fractions from the buoyant region of the first gradient (modal density 1.040 g/ml) were pooled and applied to a second Percoll density gradient. This gradient resolved two (so far physiologically indistinguishable) peaks of endocytic vesicles from plasma membrane. These peaks have been referred to as heavy endosome (density 1.043 g/ml) and light endosomes (density 1.040). Endocytic vesicles were loaded with FITC-Dextran by allowing cells to internalize this compound for varying periods (5 minutes or longer) prior to fractionation. Internalization for 5 minutes loads only the endosome peaks. Internalization for 15 minutes or longer permits accumulation of FITC-Dextran in secondary lysosomes as well. The internal pH of endocytic vesicles (isolated endosomes and lysosomes) was determined by measuring the emission intensity at 550 nm after excitation at both 495 and 450 on subcellular fractions, and using the standard curve relating the pH to the ratio of fluorescence emitted at 550 when excited at 495 to that emitted when excitation is at 450 nm, as described by Ohkuma and Poole (17,18).

RESULTS

Acidification of Isolated Lysosomes and Endosomes

In order to examine the internal pH of isolated endosomes and lysosomes (28A), CHO-K1 cells were incubated with EMEM containing FITC-Dextran (15 mg/ml) for 15 minutes at 37°, prior to fractionation. Lysosomes were isolated from the first gradient, and the two endosome peaks from the second gradient, as described in "methods." Fractions were then buffered with Hepes/KOH (50 mM, pH 7.0), and the intravesicular pH determined using the 495/450 ratio as described in Fig. 1.

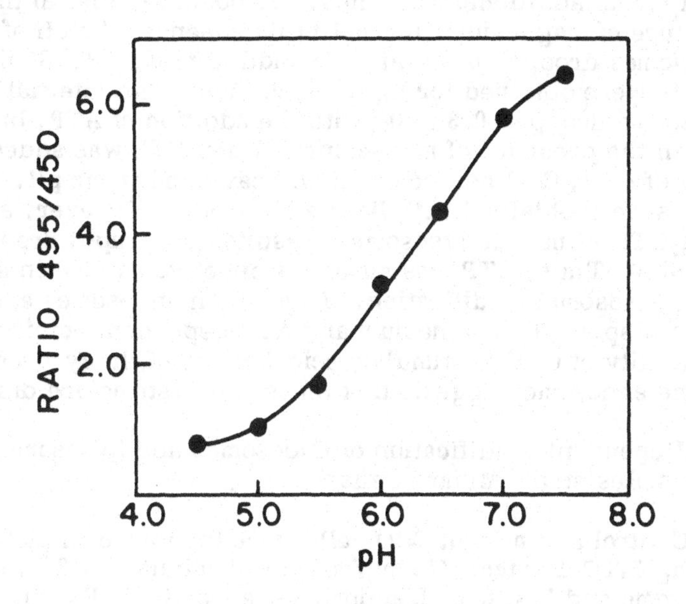

Fig. 1. Standard Curve for FITC-Dextran Fluorescence. The fluorescence emission at 550 nm after excitation at 495 and 450 nm were measured for FITC-Dextran (5 μg/ml) in buffers at various pHs. The buffer used at, and above pH 6 was sodium phosphate (100 mM), and the buffer used below pH 6 was sodium acetate (100 mM). The ratio of emission intensities at 550 nm, after excitation at 495 and 450 nm (495/450) was then expressed as a function of pH. Values from similar fluorescence measurements on subcellular fractions were used with this standard curve to measure the internal pH of FITC-Dextran loaded endosomes and lysosomes.

The internal pH of light and heavy endosomes were 5.9±0.2, and that of lysosomes was 5.5±0.2. To examine the effect of ATP on the internal pH of isolated endosomes and lysosomes, fractions were buffered with Hepes/KOH (50 mM, pH 7.0) in the presence or absence of $MgCl_2$ (5 mM) and the pH monitored for 10 minutes. K_2ATP (1 mM) was then added, and the pH monitored for an additional 5 minutes. We observed that in the presence of magnesium, but not in its absence, the pH of heavy endosomes dropped 0.5 units after addition of ATP. Similar results were obtained for light endosomes. The internal pH of lysosomes dropped 0.3 units with the addition of ATP, but again, only in the presence of magnesium. When GTP was added instead of ATP, GTP had no effect on heavy endosome pH. Similar results were obtained with light endosomes. However, addition of Na_2GTP (1 mM) to lysosomes, resulted in a rapid drop of 0.2 pH units. Thus, GTP was almost as effective as ATP in stimulating lysosomal acidification. Thus, both endosomes and lysosomes displayed a magnesium and ATP dependent acidification. The ability of GTP to stimulate acidification of the lysosome but not the endosome suggests that these two systems are distinct (28A)

ATP Dependent Acidification of Endosomes and Lysosomes from "Cross-Resistant" Mutant Cells

Control and mutant CHO cells were incubated in EMEM containing FITC-Dextran (15 mg/ml) for 15 minutes at 37°, and endosome and lysosome fractions were isolated. Fractions were buffered with Hepes/KOH (50 mM, pH 7.0) containing $MgCl_2$ (5 mM). After the pH was monitored for 5 minutes, K_2ATP (1 mM) was added, and the pH monitored for an additional 10 minutes. As reported elsewhere (28A), the addition of ATP to endosomes and lysosomes isolated from normal CHO-K1 cells (Fig. 2A) resulted in a rapid acidification of both compartments. The same was true for RE.31 cells, a cell line resistant only to DT (Fig. 2D). However, the addition of ATP to endosomes isolated from "cross-resistant" RPE.28 and RPE.44 cells (Fig. 2B, C), did not affect the pH. Lysosomes isolated from these two cell types (Fig. 2B,C) like the lysosomes isolated from normal CHO-K1 and RE.31 cells (Fig. 2A,D) acidified in response to ATP. However, the extent of lysosome acidification in the two "cross-resistant" mutants was different. While the acidification of RPE.28 lysosomes was almost as great as that of CHO-K1 lysosomes, the extent of acidification of RPE.44 lysosomes was reduced. Both RPE.28 and RPE.44 cells are "cross-resistant", and have in common a defect in the acidification of the endosome.

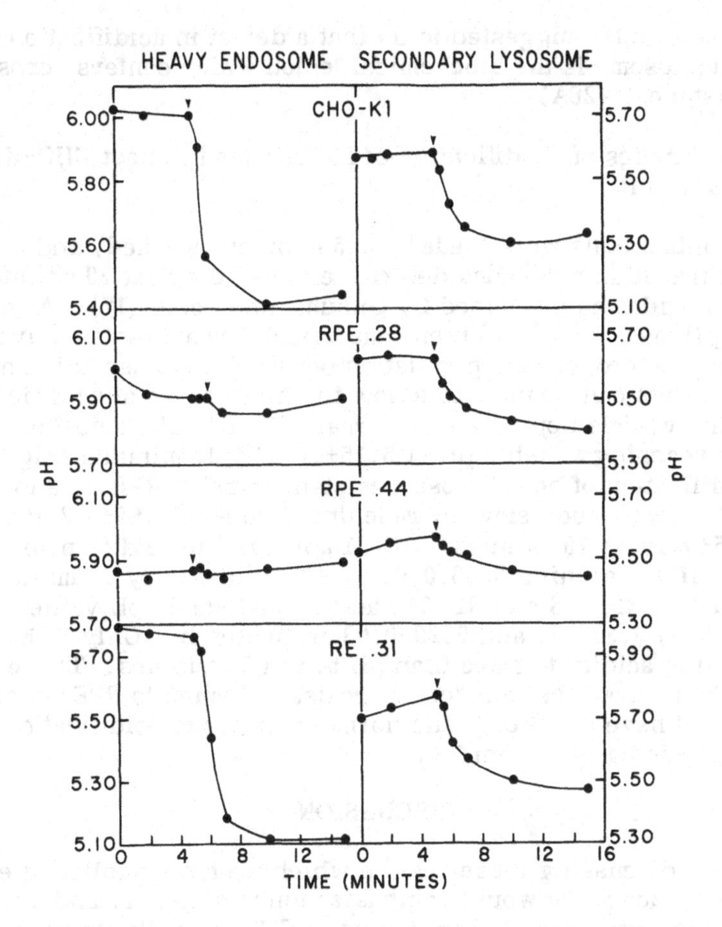

Fig. 2. Acidification of Isolated Endosomes and Lysosomes. Cells were incubated with FITC-Dextran (15 mg/ml) for 15 min. at 37°, and then subjected to the two-step Percoll density gradient fractionation. Heavy endosome from the second gradient and secondary lysosome from the first gradient were buffered in Hepes/KOH (50 mM, pH 7.0) containing 5 mM $MgCl_2$. After the pH was monitored for 5 min, 1 mM K_2ATP was added (▼), and the pH monitored for an additional 10 min. Cell types were: A) Parental CHO-K1 cells; B) "cross-resistant" cell line RPE.28; C) "cross-resistant" cell line RPE.44; D) RE.31 cells (resistant only to DT). Modified from Merion et al. (28A).

These results suggested to us that a defect in acidification of the endosome is the biochemical lesion which confers "cross-resistance" (28A).

The Kinetics of Acidification of Endosomes in Intact CHO-K1 and Mutant Cells

Intact cells were loaded for 5 minutes, washed, and the pH of intracellular vesicles determined over the next 20 minutes on glass slides as described by Ohkuma and Poole (17). Although the pH measured at a given time actually represents an average from a heterogeneous population of vesicles, consecutive measurements can be used to follow the time course of acidification of this vesicle population. Normal CHO-K1 cells acidified rapidly reaching a stable pH of 5.15±0.05 by 10 minutes (Fig. 3). Acidification of both "cross-resistant" strains RPE.28 and RPE.44 was much slower, reaching values of 5.65±0.2 and 5.75±0.15 by 10 minutes. RE.31 appeared to acidify nearly normally, reaching 5.45±0.02 by 10 minutes. By 20 minutes CHO-K1, RPE.28 and RE.31 had reached stable pH values of 5.1±0.1, 5.3±0.2, and 5.29±0.03 respectively. Only RPE.44 failed to acidify to more than pH 5.5 in 20 minutes. These studies show that both "cross-resistant" mutants RPE.28 and RPE.44 have a defect in the normal process of acidification of endocytic compartment(s).

DISCUSSION

In discussing these results which we have published elsewhere (28A), we would emphasize four points: i) Endosomes and lysosomes isolated from normal CHO-K1 cells maintain an acidic pH; ii) Acidification of both endosomes and lysosomes is mediated, at least in part, by a magnesium and ATP dependent process; iii) GTP can replace the ATP requirement for lysosome acidification, but not for endosome acidification; iv) The DPV^r mutant CHO-K1 cells that are "cross-resistant" to toxins and animal viruses display a defect in the normal acidification process of their endosomes.

A variety of physiological ligands have been shown to enter a pre-lysosomal, or endosomal compartment, prior to their transfer to lysosomes (28). We have shown (28A) that the fluid phase marker FITC-Dextran also proceeds through compartments of similar density prior to its transfer to lysosomes. The data summarized here indicate that the internal pH of isolated endosomes and lysosomes is acidic, and that the addition of ATP

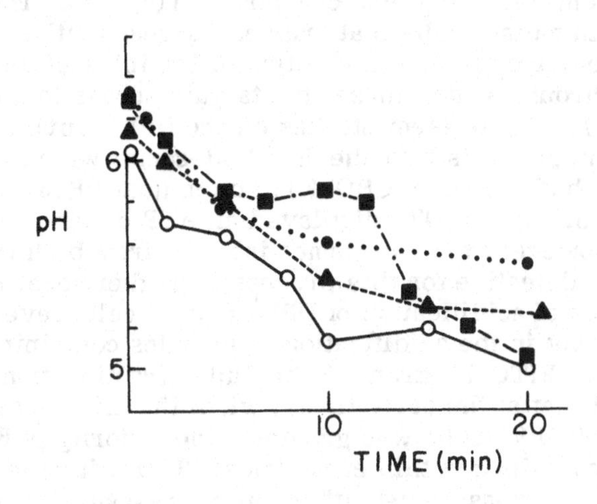

Fig. 3. Kinetics of Acidification of the Endocytic Compartment(s) in Intact Cells. Cells on glass plates were loaded with FITC-Dextran, washed, and then placed in a thermostated cuvette containing HBSS/Hepes (pH=7.4), after which the pH was determined at the indicated times using the emission at 516 nm when excitation was at either 495 nm or 450 nm. The media (HBSS/Hepes, pH=7.4) was replaced 15 seconds before each determination to remove extracellular FITC-Dextran. Cells were CHO-K1 (○), RE.31 (▲), RPE.28 (■) and RPE.44 (●) cells. Modified from Merion et al. (28A).

---

in the presence of magnesium, results in further acidification of both compartments. Thus, at least in part, both endosomes and lysosomes are acidified through a magnesium and ATP dependent process. Ohkuma _et al_. (18) have shown that both ATP and GTP stimulated acidification of rat liver lysosomes, and we confirmed this observation in CHO cells. However, we have shown that, in contrast to lysosomes, endosome acidification is not stimulated by GTP (28A). This suggests that these two mechanisms are distinct.

It has been proposed that DT (13-15), and several enveloped animal viruses (20,22,23,25,34) gain access to the cytosol through acidic intracellular compartment(s). We proposed that certain mutant cells that are "cross-resistant" to these agents are defective in acidification of the intracellular compartment(s) through which these agents gain access to the cytosol (27,35,36). The present studies on the DPV^r mutants of CHO-K1 cells support this hypothesis. Endosomes which were isolated from both parental CHO-K1, and mutant RE.31 cells which are resistant only to DT, displayed an ATP dependent acidification. However, the endosomes isolated from both DPV^r mutants were defective for this process. Furthermore, studies of the kinetics of acidification of DPV^r mutant cells revealed a substantial delay in the acidification of vesicles containing newly internalized FITC-Dextran. Subcellular fractionation studies suggested that at the early times, when the difference between control and DPV^r cells was greatest, the majority of FITC-Dextran was contained within endosomes. Thus, it appears that this class of "cross-resistant" mutants possess a defect in the acidification of endosomes. This might represent a defect in the proton pump itself, ie. the proton generating ATPase, or a defect in another component of the membrane essential for maintaining a pH gradient.

By 25 minutes following internalization, the vesicles containing FITC-Dextran in intact RPE.28 cells were as acidic as those in CHO-K1 and RE.31 cells. Subcellular fractionation studies showed that by this time, a large portion of the internalized marker was localized within lysosomes. These observations suggest that the lysosomes in these cells acidify normally, or at least they reach the same final pH. This conclusion is supported by studies of lysosomes isolated from these cells that show normal ATP-dependent acidification. Studies with intact RPE.44 cells indicated that vesicles containing internalized dextran had not reached the pH of lysosomes in CHO-K1 cells even after 25 minutes. Studies of lysosomes isolated from these cells also showed a diminished response to the addition of ATP. Thus, the "cross-resistance" in cell strain RPE.28 appears to be associated with a defect that affects acidification of the endosome primarily, if not exclusively. The defect in acidification in RPE.44 cells clearly includes the endosome, but may also affect the lysosome. However, lysosomes isolated from RPE.44 after an overnight load of FITC-Dextran were as acidic as those isolated from

CHO-K1 cells under similar conditions. The relationship between the systems responsible for acidification of these two compartments is not yet clear. The fact that the defect in RPE.28 cells clearly affects only one of them, together with the difference in response of the two systems to GTP, suggests that endosomes and lysosomes in CHO-K1 cells are acidified through two similar, but distinct mechanisms. We suggest that a defect in one of them, ie. in endosome acidification, is responsible for the "cross-resistance" of these cells to toxins and viruses.

Although the DPV^r mutants have a defect in the ATP dependent acidification of endosomes, the internal pH of their endosomes is nearly 1 pH unit below that of the surrounding buffer when isolated. One factor contributing to this acidification might be a Donnan equilibrium created by impermeant anions (sialic acid residues of membrane glycoproteins and polar head groups of phospholipids on one side of the vesicle membrane coupled with a selective permeability for cations). Another system that might contribute to acidification is the Na^+/H^+ exchange system known to be present in plasma membrane (37). However, this degree of acidification is not sufficient for DT toxicity (38) and successful infection by several animal viruses (22-24). The studies reported here suggest that the additional acidification required for virus and toxin action is produced by the ATP dependent acidification system for which the DPV^r mutants are defective.

Robbins et al. (39) recently reported the isolation and characterization of mutant CHO Cells with some similarities to those studied here. These mutants were selected for resistance to DT and proved to have a pleiotropic defect affecting receptor-mediated endocytosis of acid hydrolases. While these mutants resembled those studied here in their "cross-resistance" to DT and animal viruses, they differed from those reported herein their sensitivity to Pseudomonas toxin.

REFERENCES

1. Goldstein, J.L., Anderson, R.G.W. & Brown, M.S. (1979) Nature (Lond.) 279: 679-688.
2. Willingham, M.C., Pastan, I.H., Sahagian, G., Jourdian, G.W. & Neufeld, E.F. (1981) Proc. Natl. Acad. Sci. USA 78: 6967-6976.
3. Neufeld, E.F. & Ashwell, G. (1980) in The Biochemistry of Glycoproteins and Proteoglycans, ed. Lennarz, W.J. pp. 241-266.
4. Stahl, P.D., Schlesinger, P.H. (1980) Trends Biochem. Sci. 5: 194-196.

5. Gonzalez-Noriega, A., Grubb, J.H., Talkad, V. & Sly, W. S. (1980) J. Cell Biol. 85: 839-852.
6. Dourmashkin, R.R. & Tyrell, D.A.J. (1974) J. Gen. Virol. 24: 129-141.
7. Helenius, A., Kartenbeck, J., Simons, K. & Fries, E. (1980) J. Cell Biol. 84: 404-420.
8. Matlin, K.S., Reggio, H., Helenius, A. & Simons, K. (1981) J. Cell Biol. 91: 601-613.
9. Matlin, K.S., Reggio, H., Helenius, A. & Simons, K. (1982) J. Mol. Biol. 156: 609-631.
10. Marsh, M. & Helenius, A. (1980) J. Mol. Biol. 142: 439-454.
11. Ucida, T., Pappenheimer, A.M., Jr. & Harper, A.A. (1972) Science 175: 901-903.
12. Dorland, R.B., Middlebrook, J.L. & Leppla, S.H. (1979) J. Biol. Chem. 254: 11337-11342.
13. Draper, R.K. & Simon, M.I. (1980) J. Cell Biol. 87: 828-832.
14. Dorland, R.B., Middlebrook, J.L. & Leppla, S.H. (1981) Exp. Cell Res. 134: 319-327.
15. Sandvig, K. & Olsnes, S. (1982) J. Biol. Chem. 257: 7504-7513.
16. Marnell, M.H., Stookey, M. & Draper, R.K. (1982) J. Cell Biol. 93: 57-62.
17. Ohkuma, S. & Poole, B. (1978) Proc. Natl. Acad. Sci. USA 75: 3327-3331.
18. Ohkuma, S., Moriyama, Y. & Takano, T. (1982) Proc. Natl. Acad. Sci. USA 79: 2758-2762.
19. Tyko, B. & Maxfield, F.R. (1982) Cell 28: 643-651.
20. Helenius, A., Marsh, M. & White, J. (1982) J. Gen. Virol. 58: 47-61.
21. Miller, D.K. & Lenard, J. (1981) Proc. Natl. Acad. Sci. USA 78: 3605-3609.
22. White, J. & Helenius, A. (1980) Proc. Natl. Acad. Sci. USA 77: 3273-3277.
23. White, J., Matlin, K. & Helenius, A. (1981) J. Cell Biol. 89: 674-679.
24. White, J., Kartenbeck, K. & Helenous, A. (1980) J. Cell Biol. 87: 264-272.
25. Lenard, J. & Miller, D.K. (1981) Virology 110: 479-482.
26. Moehring, T.J. & Moehring, J.M. (1972) Infect. Immun. 6: 487-492.
27. Didsbury, J.R., Moehring, J.M. & Moehring, T.J. Mol. Cell Biol. In press.
28. Merion, M. & Sly, W.S. (1983) J. Cell Biol. 96: 644-650.

28A. Merion, M., Schlessinger, P., Brooks, R.M., Moehring, J.M, & Moehring, T.J. (1983) Proc. Natl. Acad. Sci. USA In press.
29. Moehring, J.M. & Moehring, T.J. (1979) Somat. Cell Genet. 5: 453-468.
30. Glaser, J.H. & Sly, W.S. (1973) J. Lab. Clin. Med. 82: 969-977.
31. Merion, M. & Poretz, R.D. (1981) J. Supra. Struct. Cell Biochem. 17: 337-346.
32. De Belder, A.N. & Granath, K. (1973) Carbohydrate Res. 30: 375-378.
33. Rome, L.H., Garvin, A.J., Allietta, M.M. & Neufeld, E.F. (1979) Cell 17: 143-153.
34. Marsh, M., Wellsteed, J., Bolzau, E. & Helenius, A. (1982) J. Cell Biol. 95: 418a.
35. Sly, W.S., Merion, M., Grubb, J., Moehring, T. & Moehring, J. (1982) J. Cell Biol. 95: 426a.
36. Moehring, J.M. & Moehring, T.J. (1983) Infect. and Immun. In press.
37. Rothenberg, P., Glaser, L., Schlesinger, P. & Cassel, D. (1983) J. Biol. Chem. 258: 4883-4889.
38. Sandvig, K. & Olsnes, S. (1980) J. Cell Biol. 87: 828-832.
39. Robbins, A.R., Peng, S.S. & Marshall, J.L. (1983) J. Cell Biol. 96: 1064-1071.

ACKNOWLEDGEMENTS

The authors gratefully acknowledge Drs. K. Creek and V. Talkad for helpful discussions, Joanne Nielsen for cell culture, and Sabra Lovejoy and Margaret Kane for help in manuscript preparation.

This work was supported by U. S. Public Health Service grants No. GM 21096 and GM 31988 to W. S. S., HL 26300 to P. S., and A1-09100 to T. J. M. and J. M. M., and by a gift from the Ranken Jordan Trust.

Entry of Toxic Proteins into Cells

KIRSTEN SANDVIG, ANDERS SUNDAN AND SJUR OLSNES

Norsk Hydro's Institute for Cancer
Research and The Norwegian Cancer Society,
Montebello, Oslo 3, Norway

1. INTRODUCTION

A number of toxic proteins from bacteria and plants exert their effect by entering into the cytoplasm of cells where they carry out enzymatic activity. In recent years considerable efforts have been made to elucidate how these proteins cross the membrane and enter the cytosol. To some extent this entry represents the reverse of the transport of secretory proteins. Possibly certain physiological proteins enter by the same mechanisms. Furthermore, because of the great interest at the present time in constructing hybrid toxins selective for malignant cells, more information about the way toxins enter into the cells is needed.

The toxins that will be discussed here all appear to consist of two functionally different parts. An enzymatically active moiety which enters into the cytosol is connected by a disulfide bond to a moiety which binds the toxin to cell surface receptors. These toxins comprise the plant toxins abrin, ricin, modeccin and viscumin as well as the bacterial toxins diphtheria toxin, *Pseudomonas aeruginosa* exotoxin A, Shigella toxin, cholera toxin and *Escherichia coli* heat labile enterotoxin. Most of these toxins (with exception of the last two) exert their action by inhibiting protein synthesis.

2. STRUCTURE AND FUNCTION OF TOXINS

The functional division in two moieties with different functions is most obvious in the plant

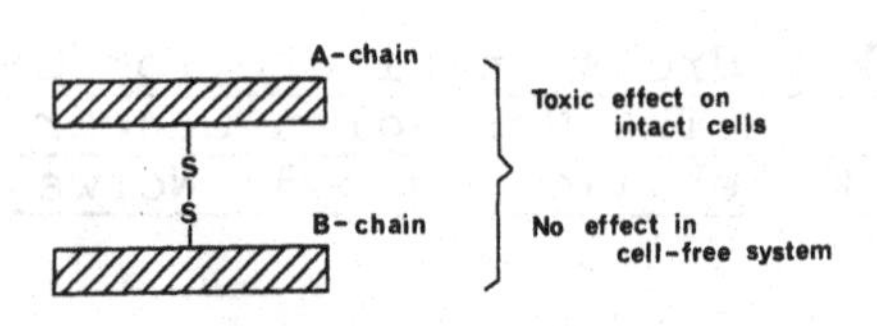

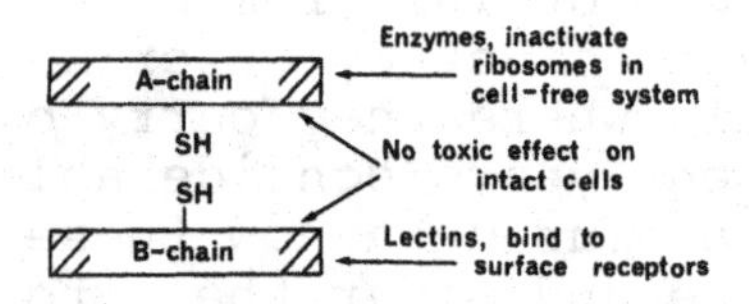

Abrin, Modeccin, Ricin, Viscumin

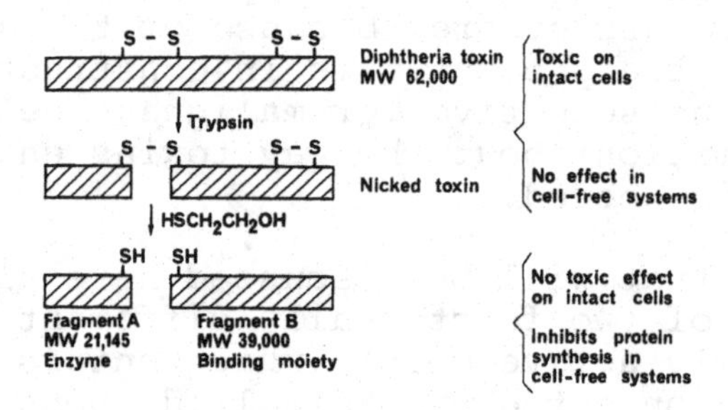

Diphteria toxin

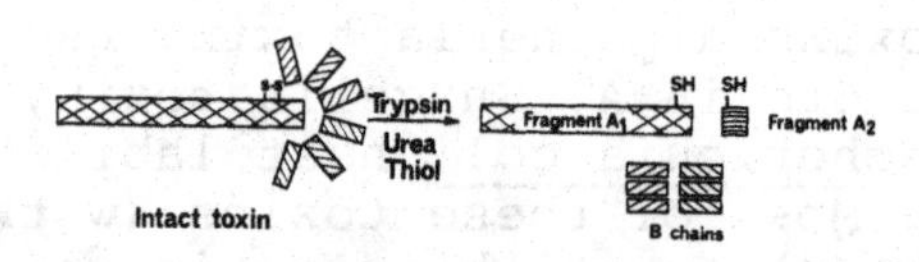

Shigella toxin

FIG. 1. Schematic structure of toxins and their constituent chains and fragments.

toxins abrin, ricin, modeccin and viscumin which all consist of two polypeptide chains linked by a disulfide bridge (Fig. 1). The toxins have molecular weights of approximately 60,000. After reduction of the disulfide bridge they can be dissociated into A- and B-chains which in most cases can be easily separated (for review, see Olsnes and Pihl, 1982). The A-chain carries enzymatic activity which consists in inactivating the 60 S ribosomal subunits whereas the B-chain binds to galactose-containing receptors at the cell surface. It is not known what modification of the 60 S subunit the A-chains introduce, but the modification appears to be irreversible and related to the binding site of elongation factor 2. Toxin treated ribosomes are unable to elongate already initiated polypeptide chains.

Diphtheria toxin is synthesized as one polypeptide chain which contains a trypsin-sensitive region as indicated in Fig. 1. After trypsin cleavage the two fragments are linked by a disulfide bridge and can be separated after reduction (for review, see Pappenheimer, 1977). The A-fragment enters the cytosol and inactivates elongation factor 2 by ADP-ribosylating an unusual amino-acid, diphthamide, present in this protein (Van Ness *et al.* 1980). The B-fragment is involved in binding of the toxin to cell surface receptors and apparently assists in the entry of the A-fragment. *Pseudomonas aeruginosa* exotoxin A has the same intracellular action as diphtheria toxin, but it binds to other receptors and appears to enter by a somewhat different mechanism (Iglewski & Sadoff, 1979).

Shigella toxin consists of an A-chain linked by weak interactions to 6-7 B-chains (Olsnes *et al.* 1981). The A-chain contains a trypsin sensitive region. After cleavage by trypsin and treatment with reducing agents, the disulfide bridge linking the two chains (Fig. 1) is cleaved and the different components can be separated. The A_1 fragment thus obtained contains the enzymatic activity. Like the plant toxins it inactivates the

60 S ribosomal subunit and inhibits elongation (Reisbig et al. 1981). It is not clear if Shigella toxin exerts the same effect on the ribosomes as the plant toxins.

3. ENTRY OF DIPHTHERIA TOXIN AT LOW pH

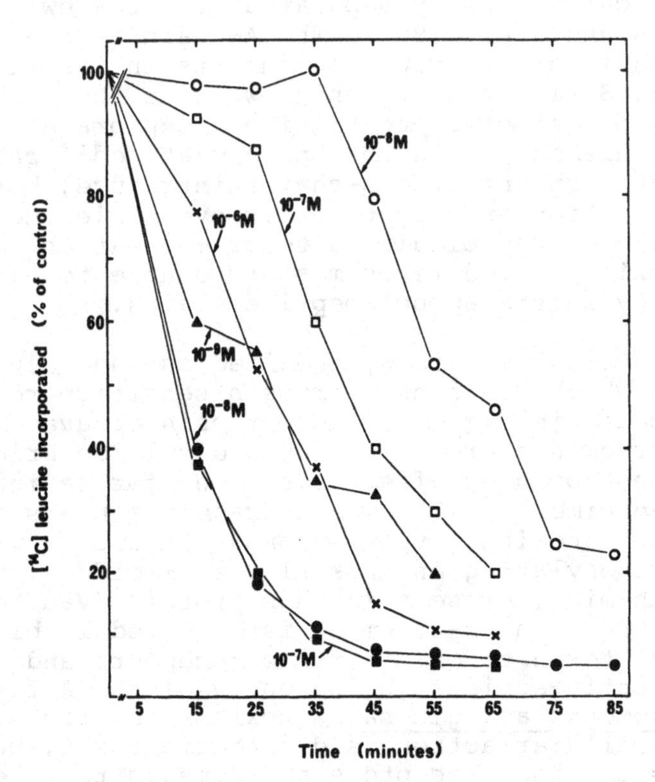

FIG. 2. Effect of pH on the rate of protein synthesis inhibition after exposure of cells to diphtheria toxin. Diphtheria toxin was added to cells in 24-well disposable trays. After binding for 1 h at 0°C the medium (and unbound toxin) was removed and buffer, pH 4.5 (filled symbols), or pH 7.5 (open symbols) was added (time zero). After 10 min at 37°C the buffer was removed and medium prewarmed to 37°C was added. After various periods of time, as indicated on the abscissa, the ability of the cells to incorporate ^{14}C leucine into acid-precipitable material during a 10 min period was measured.

When diphtheria toxin is added to cells and the rate of protein synthesis is measured after different periods of time, inhibition increases with the toxin concentrations (Fig. 2). If the cells are exposed to pH 4.5 for a brief period of time, the rate of protein synthesis decreases much more rapidly than in the absence of the low pH-treatment (for review, see Olsnes & Sandvig 1983). As shown in Fig. 2, 10^{-9} M diphtheria toxin inhibited protein synthesis as rapidly after a low pH pulse as 10^{-6} M toxin in the absence of the low pH pulse. This 1000-fold increase in the rate of intoxication by the low pH indicates that low pH is necessary for entry of diphtheria toxin. As will be discussed below, the toxin is normally exposed to low pH in intracellular acidic vesicles.

The low pH has a direct effect on the structure of the toxin. Thus a hydrophobic region present in

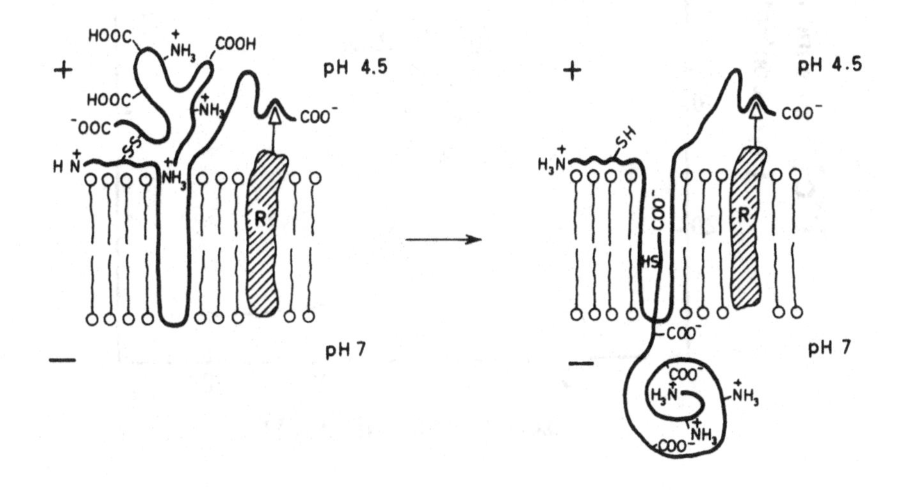

Fig.3. Hypothetical model of the transfer of diphtheria toxin fragment A across the membrane. The B-fragment inserts itself into the membrane at low pH and forms a hydrophilic channel. The A-fragment is partly unfolded by the low pH and guided through the channel by the membrane potential.

the B-fragment of the toxin is normally hidden, but when the pH is lowered to approximately 4.5 this region is exposed. Electrophysiological studies have shown that the hydrophobic region can insert itself into lipid bilayers and form ion-permeable channels. The requirement for this to occur is a membrane potential which is

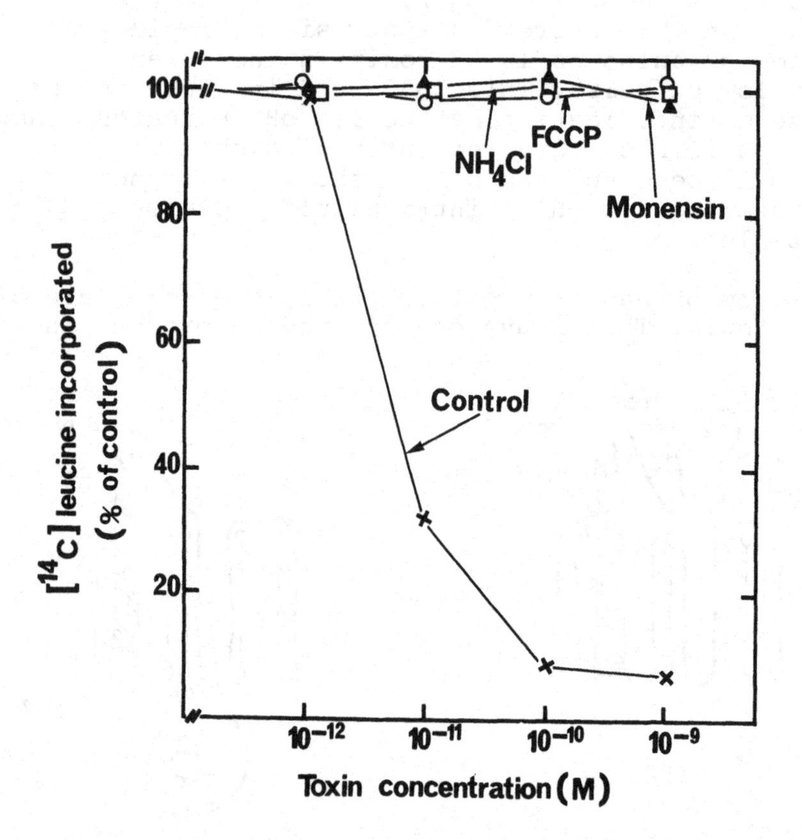

FIG. 4 Ability of NH_4Cl and ionophores to protect cells against diptheria toxin. Hepes medium with the indicated compounds was added to Vero cells in disposable trays and 15 min later increasing amounts of toxins were added. After incubation at 37°C for 1 hour the rate of ^3H leucine incorporation was measured.

positive on the same side of the membrane as the toxin (Kagan & Finkelstein, 1981; Donovan _et al_.1981). This is actually the same polarity as exists across the plasma membrane. In Fig. 3 is shown a hypothetical model of how the pH gradient across the membrane as well as the membrane potential might assist in transferring the A-fragment into the cytosol.

4. ROLE OF ENDOCYTOSIS

There is now evidence that all the toxins here discussed enter the cytosol from endocytic vesicles. The best evidence for this has been obtained with diphtheria toxin incubated with cells at neutral pH (for review, see Olsnes and Sandvig, 1983). Under these conditions the only place the toxin can be exposed to low pH is inside acidic vesicles. A variety of compounds which have been shown to increase the pH in lysosomes and other acidic vesicles were found to protect efficiently against diphtheria toxin under these conditions. It is an old observation that NH_4Cl protects against diphtheria toxin. The effect is illustrated in Fig. 4. At neutral pH some NH_3

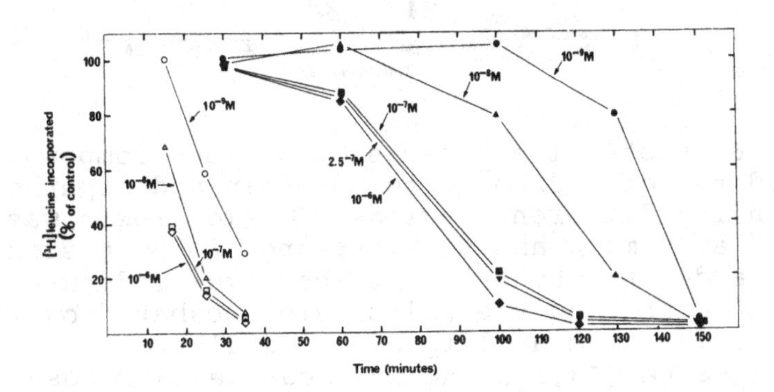

FIG. 5. Lag time after addition of diphtheria toxin and modeccin to cells. Cells were incubated with the indicated concentrations of diphtheria toxin (open symbols) or modeccin (closed symbols) and, after different periods of time, the rate of protein synthesis was measured.

diffuses through the membranes of the cell. In acidic vesicles it is protonized and thus increases the pH of the vesicle. Ionophores like monensin which are able to dissipate proton gradients by electroneutral exchange of protons for monovalent cations as K^+ and Na^+, as well as protonophores like FCCP and CCCP also protect very efficiently against diphtheria toxin (Fig. 4 A). The same agents also strongly protect against modeccin It is, however, not possible to induce rapid entry of modeccin from the cell surface with low pH and it is so far not clear which role low pH plays in the entry of this toxin.

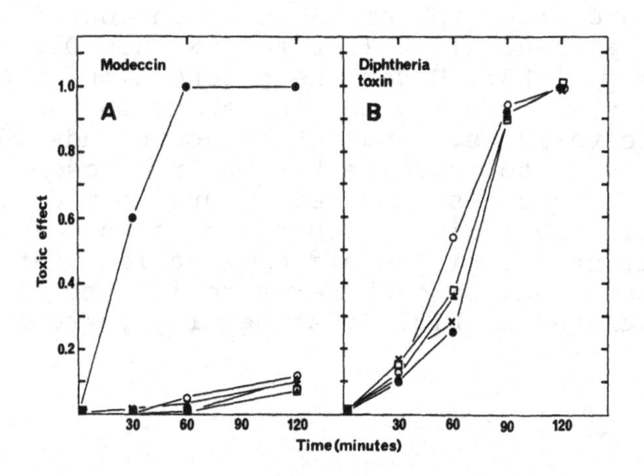

FIG. 6. Protective effects of various compounds added to cells exposed to modeccin and diphtheria toxin for different periods of time. Toxin was added at time 0 and the compounds to be tested were added to the cells at the times indicated on the abscissa and the cells were incubated over night. Then the ability of the cells to incorporate [^{3}H]leucine was measured and dose-response curves were constructed. The toxic effect expressed as $1/ID_{50}$ was related to that in control cells treated with toxin alone. The additions were: (●), 1 µl/ml antiserum; (o), 10 mM NH_4Cl; (▲), 10 uM FCCP; (x), 10 uM monensin; (□), 10 mM procaine; (△), 15 µM FCCP.

When added to cells diphtheria toxin inhibits protein synthesis more rapidly than modeccin (Fig. 5). This indicates that
after endocytosis diphtheria toxin is rapidly transferred across the membrane of the vesicle whereas modeccin is transferred only after a certain period of time (Sandvig _et al._ 1983). That this is so is further supported by the finding that agents which increase the pH of intracellular vesicles are able to reduce the toxic effect of modeccin even when added hours after the toxin is endocytosed. This is not the case with diphtheria toxin (Fig. 6). Entry of modeccin may require that the endocytic vesicles must fuse with some other intracellular vesicular compartment, _e.g._ the lysosomes, before the toxin can enter into the cytosol. Thus at low temperatures (20°C and below) and in the presence of high KCl-concentrations (Fig. 7) which has been reported to inhibit vesicle fusion

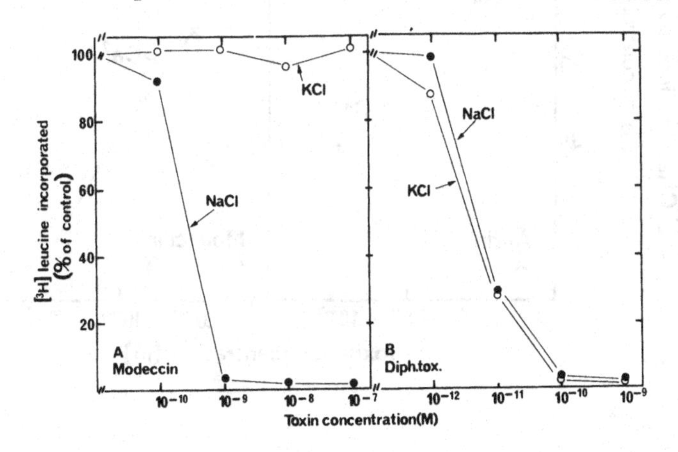

FIG. 7. Sensitivity of cells to modeccin and diphtheria toxin in buffers containing either 0.2 M NaCl or 0.2 M KCl. Cells were preincubated with buffer containing 0.2 M NaCl (●) or 0.2 M KCl (o) for 1 hour. Then modeccin (A) or diphtheria toxin (B) was added and the incubation was continued for 3 hours more. Finally the rate of protein synthesis was measured.

(Baenziger and Fiete. 1982), modeccin is unable to enter the cytosol whereas diphtheria toxin can enter under such conditions.

Also abrin and ricin appear to enter from endocytic vesicles. Thus when cells were exposed to the toxins under conditions where endocytosis occur, but entry into the cytoplasma is inhibited, and the cells were then washed and treated with antitoxin to inactivate any toxin present at the cell surface, the cells were intoxicated when returned to normal medium which allows toxin to enter (Sandvig & Olsnes, 1982 b).

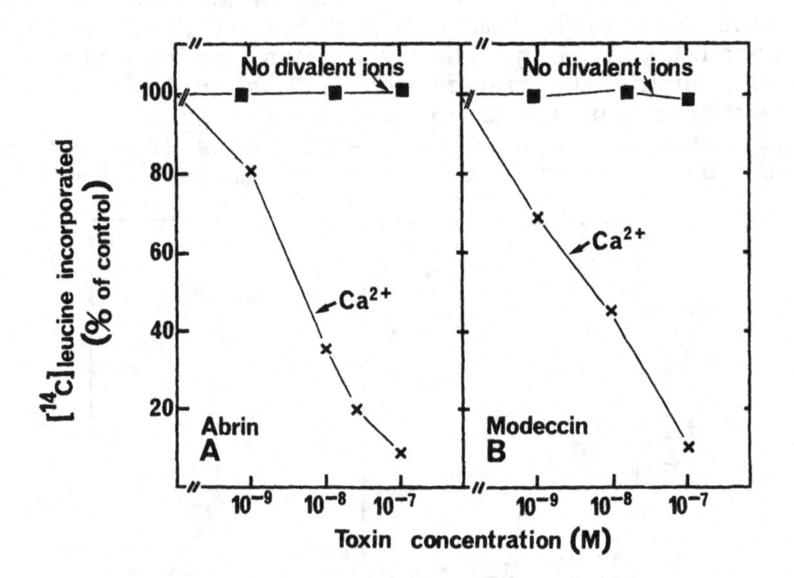

FIG. 8. Requirement for Ca^{2+} for entry of abrin and modeccin. Vero cells in disposable trays were rinsed with EGTA and incubated with isotonic NaCl-Hepes buffer, pH 7.3, with and without 2 mM $CaCl_2$. After 15 minutes at 37°C increasing amounts of abrin (A) or modeccin (B) were added and the incubation was continued for 50 minutes (A) or 80 minutes (B). Finally the ability of the cells to incorporate [^{14}C] leucine was measured.

5. ION-REQUIREMENTS FOR ENTRY

Two kinds of ions play an important role in entry of toxins, viz. Ca^{2+} and Cl^-. As shown in Fig.8 cells were completely protected against abrin and modeccin in the absence of Ca^{2+} and their sensitivity to ricin was strongly reduced. Apparently an influx of Ca^2 into the cells is required since inhibitors like verapamil and Co^{2+}, which inhibit Ca^{2+} entry into the cells, were able to protect in the presence of Ca^{2+}. Interestingly, also the Ca^{2+}-ionophore A 23187 protected very strongly against abrin and modeccin (Sandvig & Olsnes, 1982 a). Possibly this is due to dissipation of Ca^{2+}-gradients across membranes of intracellular vesicles. Another possibility is that increased Ca^{2+}-concentration in the cytosol may inhibit a required Ca^{2+}-influx.

Diphtheria toxin and modeccin do not appear to enter into the cytosol in Cl^--free medium made isotonic with mannitol. As shown in Fig. 9, the toxic effect of diphtheria toxin and modeccin

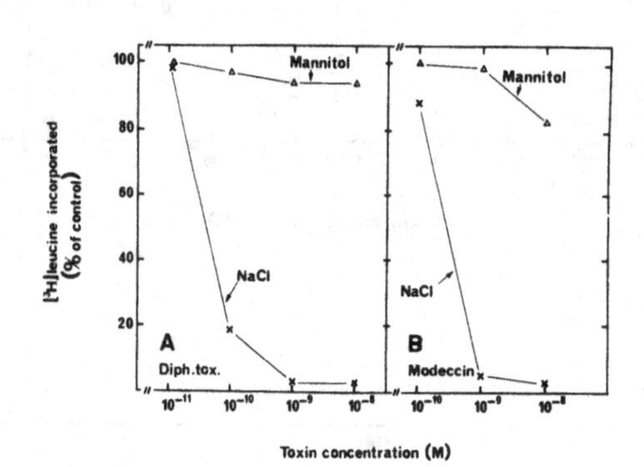

FIG. 9. Requirement of Cl^- for sensitivity of cells to diphtheria toxin (A) and modeccin (B). Cells were incubated with buffer containing 0.14 M NaCl or 0.26 M mannitol for 1 hour, then toxins were added and the incubation was continued for 50 minutes (A) or 3 hours (B) more. Finally the rate of protein synthesis was measured.

was strongly reduced in NaCl-free buffer. When toxin was prebound to cells before the NaCl-concentration was reduced, as little as 2 mM NaCl fully sensitized the cells and 0.5 mM was almost as efficient. Abrin and ricin did not require NaCl for entry (data not shown). Only Br^- could replace Cl in the case of diphtheria toxin and modeccin. Apparently, Cl^- entry into cells is necessary and not only its presence in the medium. Thus a variety of anions and compounds (like SITS) which inhibits Cl^- entry, protected against the toxin in the presence of Cl^-.

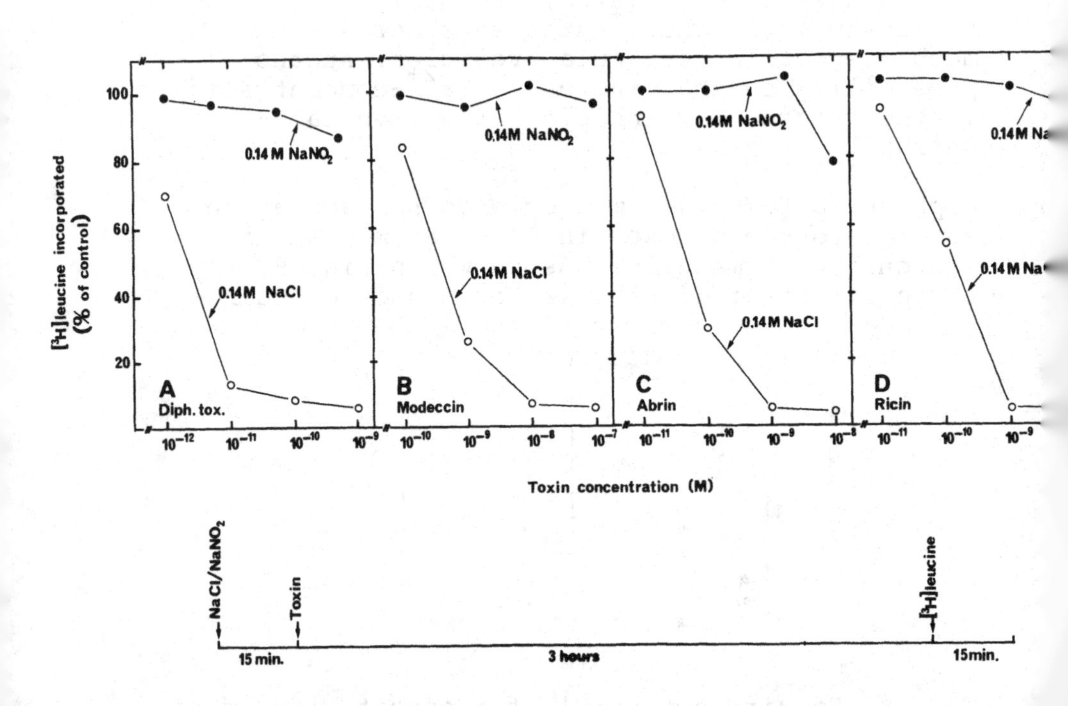

FIG. 10. Ability of $NaNO_2$ to protect cells against toxins. Cells were incubated for 15 minutes with Hepes-buffer containing 0.14 M NaCl or $NaNO_2$ and then toxins were added. After further incubation for 3 hours the rate of protein synthesis was measured. (o), NaCl; (•), $NaNO_2$;

6. EFFECT OF WEAK ACIDS

Salts of several weak acids (acetate, nitrite, salicylate, benzoate etc.) were found to protect against all toxins tested (Fig. 10). These acids are able to pass membranes in their protonized form. Possibly they protect by acidifying the cytoplasm or certain intracellular vesicles with alkaline or neutral pH.

7. CONCLUSIONS

Different toxic proteins with intracellular sites of action enter the cytosol in similar, but not identical manners. In all cases endocytic uptake seems to be required. Low pH is required for the entry of diphtheria toxin, modeccin and Pseudomonas toxin, whereas low pH does not appear to be required for the entry of abrin, ricin and viscumin. Ca^{2+} is required for the entry of abrin, modeccin, viscumin and Pseudomonas toxin, whereas Cl^- is required in the case of diphtheria toxin and modeccin. Intracellular vesicle fusion may be required for the entry of modeccin and Shigella toxin, but not for diphtheria toxin. We cannot exclude the possibility that the differences only represent small deviations from a common entry route.

REFERENCES

Baenziger, J.U. & Fiete, D. (1982) J.Biol.Chem. 257, 6007-6009.

Donovan, J.J., Simon, M.I., Draper, R.K. & Montal, M. (1981) Proc.Natl.Acad.Sci.USA. 78, 172-176.

Iglewski, B.H. & Sadoff, J.C. (1979) Methods Enzymol. 60, 780-793.

Kagan, B.L., Finkelstein, A. & Colombini, M. (1981) Proc.Natl.Acad.Sci. USA. 78, 4950-4954.

Olsnes, S. & Pihl, A. (1982) In "The molecular action of toxins and viruses". (Eds. P. Cohen and S. van Heyningen) Elsevier/North Holland, Amsterdam, pp. 51-105.

Olsnes, S., Reisbig, R. & Eiklid, K. (1981) J.Biol.Chem. 256, 8732-8738.

Olsnes, S. & Sandvig, K. (1983) In "Receptor-mediated endocytosis", Receptors and Recognition, Series B Vol. 15 (Eds. P. Cuatrecasas & T.F. Roth) Chapman and Hall, London, pp. 187-236.

Pappenheimer, A.M., Jr. (1977) Ann.Rev.Biochem 46, 69-94.

Reisbig, R., Olsnes, S. & Eiklid, K. (1981) J.Biol.Chem. 256, 8739-9744.

Sandvig, K. & Olsnes, S. (1980) J.Cell Biol. 87, 828-832.

Sandvig, K. & Olsnes, S. (1982a) J.Biol.Chem. 257, 7495-7503.

Sandvig, K. & Olsnes, S. (1982 b) J.Biol.Chem. 257, 7504-7513.

Sandvig, K., Sundan, A. & Olsnes, S. (1983) J.Cell Biol. Submitted.

Van Ness, B.G., Howard, J.B. & Bodley, J.W. (1980) J.Biol.Chem. 255, 10710-10716

INTRACELLULAR TRANSPORT OF ENDOCYTOSED GLYCO-PROTEINS IN RAT HEPATOCYTES

H. TOLLESHAUG, R. BLOMHOFF, and T. BERG

Institute for Nutrition Research

University of Oslo, Blindern, Oslo

INTRODUCTION

The mammalian liver consists of four main cell types: the parenchymal cells, the macrophages (Kupffer cells), the endothelial cells, and the stellate cells (1). Apart from the stellate cells all of these cell types are able to remove macro-molecules from the blood by endocytosis. This function contributes to the homeostatic system which keeps the concentration of plasma proteins and lipids relatively constant. The various ligands endocytosed by the liver are to some extent divided between the cells. For instance, only hepatocytes take up chylomicron remnants or asialoglycoproteins (2). Endothelial cells are particularly active in the uptake of proteins with increased negative charge such as acetylated LDL (3) or formaldehyde-treated albumin (4), while Kupffer cells are specialized phagocytic cells. The main function of the stellate cells seems to be storage of retinol received from the hepatocytes (5,6).

Endocytosis of various ligands can be studied in isolated, purified rat liver cells. Such investigations have been performed in parenchymal cells, Kupffer cells, and endothelial cells . After isolation, these cells maintain their ability to bind and take up the specific ligands. We have used rat hepatocytes as a model system to study binding, intracellular transport, and degradation of various glycoproteins. Isolated rat hepatocytes lend themselves particularly well to studies of endocytosis. The cells are obtained in high yield and can be kept in suspension or in monolayers. They take up various ligands very efficiently, which makes determinations of binding and uptake easy. Contrary to many other cell types, hepatocytes are relatively easy to homogenize without much disruption of organelles; the cell-free preparation can be fractionated in gradients of sucrose, Nycodenz or Percoll, and this offers one possibility to follow the intracellular movement of the ligands. The cells take up several ligands by receptor-mediated endocytosis: glycoproteins with terminal galactose (asialoglycoproteins) (7), chylomicron remnants (8), and hemoglobin in association with haptoglobin (9). Uptake of macromolecules in the fluid phase of the endocytic vesicle (fluid endocytosis) can be determined with markers such as sucrose or polyvinylpyrollidone (10).

We have studied uptake and intracellular transport in isolated rat hepatocytes of asialoglycoproteins (asialofetuin, asialoorosomucoid, asialotransferrin), mannose-terminated glycoproteins (yeast invertase), and polyvinylpyrollidone. The strategy has been to follow the labeled ligands by means of subcellular fractionation in sucrose or Nycodenz gradients. Marker enzymes have been employed to determine the distribution of lysosomes, plasma membrane, Golgi apparatus etc. To determine the role of various organelles involved in the processing of the ligands specific inhibitors have been used: lysosomotropic drugs, rotenone (decreases energy production), colchicine (disrupts microtubuli), monensin against the Golgi, benzyl alcohol against membrane fusion. Some of our recent data concerning intracellular transport of glycoproteins and the effect of inhibitors on this process will be reviewed here.

INTRACELLULAR TRANSPORT OF ASIALOGLYCOPROTEINS IN RAT HEPATOCYTES

Isolated rat hepatocytes possess about $5x10^5$ asialoglycoprotein receptors on the surface of each cell (11). Binding of ligand is calcium-dependent. This makes it possible to dissociate ligand from the surface receptors by adding EGTA. By measuring surface-bound and internalized ligand as function of time, the rate of internalization may be determined. The half-time for internalization at $37^{o}C$ is about 3 min (11).

We have studied the binding of calcium ions by the isolated asialoglycoprotein receptor from rat hepatocytes (12). The isolated receptor binds 4 calcium ions; the binding exhibits marked positive cooperativity and the association constant at half saturation of the binding was of the order of 10^5 M^{-1}. The isolated receptor was almost saturated with calcium ions at a calcium concentration of 0.1 mM. The binding capacity of isolated hepatocytes for asialoglycoproteins increased, however, even when the calcium concentration was increased above this level. This may be explained by the exposure of increasing numbers of functional receptors on the surface of the cell with increasing membrane potential, since the membrane potential has been shown to be partially regulated by the calcium concentration in the medium (12).

The internalized ligand starts to be degraded after about 15 min. The lag corresponds to the time needed for the ligand to reach the lysosomes (7). Nearly all of the intracellular asialofetuin in hepatocytes is contained in endocytic vesicles ("endosomes"). The endosomes are recovered at densities between 1.11 and 1.16 g/ml in sucrose gradients. The density of the endosome peak gradually increases on prolonged incubation of the cells; the increase probably reflects processing of the endosomes. Degradation products from ^{125}I-labeled asialoglycoproteins are released rapidly from the cells and the subcellular localization of this ligand therefore does not reflect its intracellular transport. By using asialofetuin derivatized with ^{14}C-sucrose the labeled degradation product (^{14}C-sucrose) is trapped at the intracellular degradation site. In this way the degradative organelles are revealed in a fractionation system by the product of its action. Sucrose-derivatized asialofetuin is not treated differently from ^{125}I-labeled asialofetuin by hepatocytes neither in terms of binding and rate of uptake nor in the rate of degradation (13). Fig. 1 compares the distribution in sucrose gradients of ^{14}C-sucrose asialofetuin (14CSAF) and ^{125}I-labeled asialofetuin, after endocytic uptake in rat hepatocytes. The distributions of 5'nucleotidase and acid phosphatase are shown for comparison. The cells were incubated with the labeled ligands (100 nM) for 30, 60 and 120 min, cytoplasmic extracts were prepared (14) and fractionated in linear sucrose gradients. The distribution patterns were markedly different for the two ligands. The distributions of both 14CSAF and ^{125}I-asialofetuin showed two peaks in the gradients; however, the peak at higher density became much more prominent in gradients with 14CSAF than in those containing ^{125}I-asialofetuin. This peak coincides with the distribution of acid phosphatase and probably represents labeled sucrose accumulated in the lysosomes. This notion was supported by experiments in which acid soluble and acid precipitable radioactivities were determined in the gradient fractions. Fig. 2 shows the results of an experiment in which the cells were incubated with 14CSAF at 37°C. Aliquots of cells were removed after 20, 45 and 90 min, and fractionated by sucrose gradient centrifugation. Acid soluble and acid precipitable radioactivities were determined in the fractions. Again, radioactivities were distributed in two peaks. Acid precipitable activity was almost exclusively in the more buoyant peak and probably reflected ligand in endosomes. Initially, all acid soluble radioactivity was in the upper region of the gradient. When the incubation was prolonged, increasing amounts of acid-soluble radioactivity was found in the denser part of the gradient. The distribution of radioactivity in this region became almost identical to the of a lysosomal enzyme, b-acetylglucosaminidase . The present data suggest that the intracellular degradation of 14CSAF is, in fact, a two-step process. Degradation starts in endosomes and is completed in secondary lysosomes.

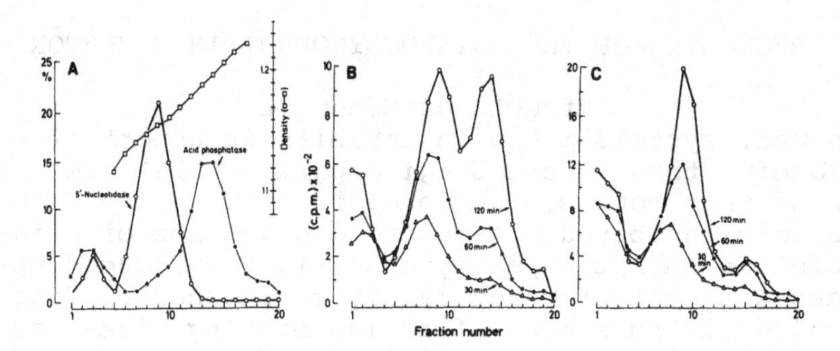

Fig. 1. Distribution patterns of 14CSAF (B) and ^{125}I-asialofetuin (C) after fractionation of hepatocytes by centrifugation in sucrose gradients. The cells were incubated with 100 nM of the ligands for the indicated time intervals. A: Distribution of acid phosphatase and 5'nucleotidase in the gradient containing material from cells incubated for 2 h.

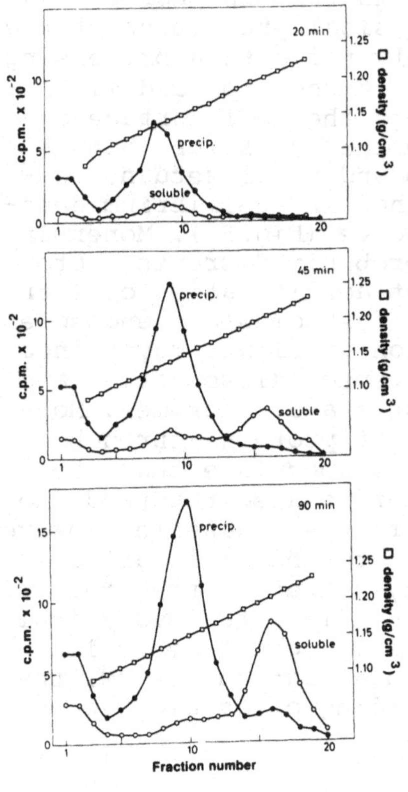

Fig. 2. Distribution of acid precipitable (●) and acid soluble (o) radioactivity after fractionation of hepatocytes in sucrose gradients. The cells were incubated for 20, 45 and 90 min with 100 nM 14CSAF. The recovery of radioactivity (as % of that initially layered on top of the gradient varied between 75 and 85%.

RECYCLING OF THE ASIALOGLYCOPROTEIN RECEPTOR

Effects of Monensin

In many systems monensin inhibits transport of proteins through the Golgi apparatus (15). We found that monensin in concentrations above 25 uM rapidly inhibited uptake and degradation of asialofetuin in rat hepatocytes (16). The reduced uptake was evidently due mainly to reduced binding activity; the number of active binding sites was reduced to 25% of control values 10 to 15 min after the addition of monensin. Monensin also changed the distribution of both acid soluble and acid precipitable radioactivities when it was added to cells that had bound 14CSAF in advance. Subcellular fractionation of control cells taking up labeled asialofetuin have shown that radioactivity is initially in a light particle and then in gradually denser organelles (16). Eventually the median density is about 1.15 g/ml in sucrose gradients. This change in physical properties of the endocytic vesicles probably reflects a processing which leads to separation of receptor and ligand: the receptor is returned to the cell surface and the ligand is directed to the lysosomes. Monensin, added to cells that had bound asialofetuin, completely prevented the change in equilibrium density of the endocytic vesicles (Fig.3). Monensin, a carboxylic ionophore, probably disrupts proton gradients and will prevent acidification of intracellular vesicles such as lysosomes and endosomes. The establishment of proton gradients may be necessary for both ligand-receptor dissociation and for fusion between lysosomes and endosomes. Monensin inhibits processing of proteins through the Golgi and it is tempting to speculate that the asialoglycoprotein receptor is also returned to the cell surface by way of this organelle. However, the return of the receptor to the cell surface probably follows a pathway different from that of secretory proteins, as monensin effectively inhibits secretion of VLDL (very light density lipoprotein) at much lower concentrations (1 uM) than those needed to reduce binding of asialoglycoproteins (10 uM).

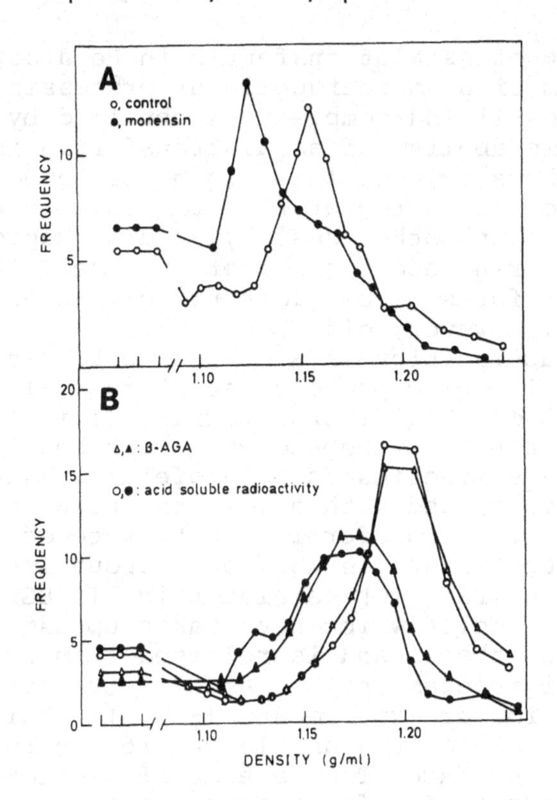

Fig. 3. Sucrose density gradient fractionation of hepatocytes incubated with or without monensin (25 uM) after uptake of 14CSAF. Following preincubation the cells were washed and reincubated in presence or absence of monensin. Cells were fractionated after 60 min and acid precipitable (A) and acid soluble (B) radioactivity were determined in the fractions along with b-acetylglucosaminidase (triangles). Closed symbols: monensin treated cells. Radioactivities and enzyme activities are presented as % of total recovered activity in the gradient.

Recycling of Asialotransferrin in Hepatocytes

More information on intracellular processing of the receptor-ligand complex was obtained by studying the metabolism of asialotransferrin in isolated rat hepatocytes. This ligand is bound to the asialoglycoprotein receptor in the same manner as asialofetuin and other desialylated glycoproteins; the transferrin receptor is not involved. The intracellular fates of asialofetuin and asialotransferrin are, however, different.

By fractionation on a column containing immobilized asialoglycoprotein receptors (17), a minor fraction ("asialotransferrin 3") is obtained which is taken up by hepatocytes with substantially the same kinetics as asialofetuin. When the cells are incubated with a low concentration (1 nM) of asialotransferrin 3, the rate of degradation of the ligand is low, only about 1/4 of the corresponding rate for asialofetuin. If EGTA is added to the cells which have taken up asialotransferrin, the ligand is released from the cells, while negligible amounts of asialofetuin are released in similar experiments (Fig. 4). This indicates that some of the asialotransferrin is found on the cell surface. The release of asialotransferrin proceeds with first-order kinetics, and the half-life is 20 min.

However, it may be shown that only about 1/5 of the cell-associated asialotransferrin is actually on the cell surface (and thus EGTA releasable) at any given moment (Fig. 4). The amount of surface-bound asialotransferrin may be determined by taking parallell samples and treating one set with EGTA; the difference in cell-associated ligand before and after EGTA treatment gives the amount of ligand on the cell surface. To a separate suspension, EGTA may be added when the amount of cell-associated asialotransferrin is maximal (Fig. 4); the loss of ligand from the cells is slow, but sure, confirming that only a fraction of the asialotransferrin is actually EGTA-releasable at any given point in time.

We interpret these observations as meaning that asialotransferrin is being continually recycled between the interior of the cell and its surface. Only a small fraction of the intra-

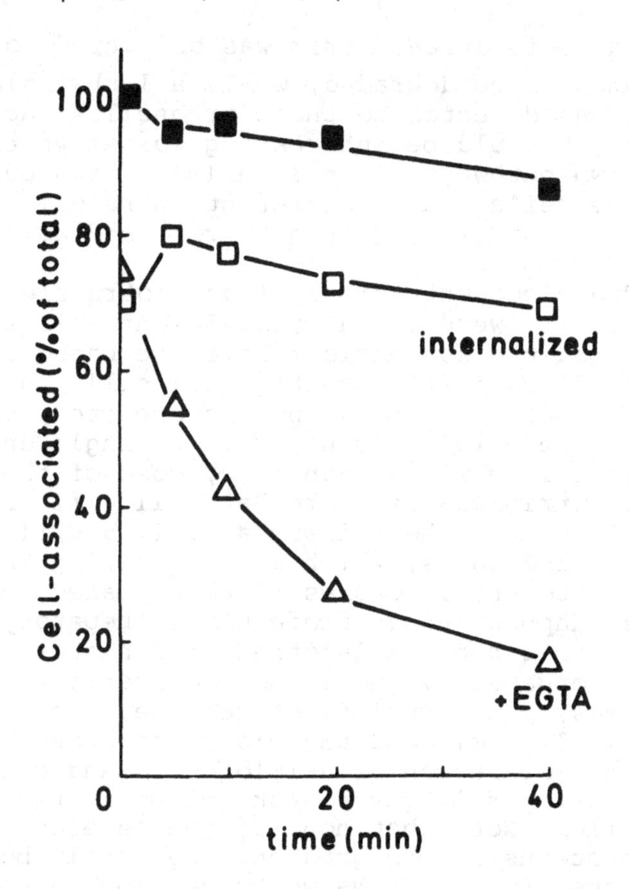

Fig.4. Cell-associated and internalized asialo-transferrin in hepatocytes. The cells were allowed to take up ^{125}I-asialotransferrin for 10 min, then the suspension was chilled and unbound ligand was removed. One portion was re-incubated at 37 °C in the presence of 3 mM EGTA, which inactivates the cell-surface receptors. From these cells asialo-transferrin was gradually released. - The other portion was re-incubated at 37 °C with no additions. Two sets of uptake samples were taken at each time point; one set showed total cell-associated ligand (upper curve), the other set was treated with EGTA before centrifuging through the oil, and the radioactivity in these samples indicated internalized ligand (middle curve).

cellular asialotransferrin was delivered to the lysosomes to be degraded, while all the asialofetuin was directed to these organelles. Accordingly, it would be interesting to see whether these two pathways of intracellular transport would be reflected in different intracellular distributions of asialotransferrin and asialofetuin.

Trace concentrations of asialotransferrin or asialofetuin were added to different cell suspensions, and the suspensions were incubated for 80 min at 37°C. On fractionating the cells in sucrose gradients, it was found, as expected, that most of the asialofetuin was in a single endosome peak at 1.15 g/ml. In contrast, most of the asialotransferrin was recovered at 1.11 g/ml (Fig. 5). We believe that the endosomes at 1.15 g/ml are "mature" endosomes, which are delivering their contents to the lysosomes to be degraded, which is what happens to asialofetuin in hepatocytes. On the other hand, asialotransferrin stays in a lighter organelle, and is not delivered to the lysosomes, but recycles between the light endosomes (1.11 g/ml) and the plasma membrane. Intracellular separation of asialofetuin and asialotransferrin is likely to take place in these organelles. Note that none of the labeled asialoglycoproteins in the gradient are merely bound to fragments of the plasma membrane. The isotonic sucrose solution which is used to wash the cells before fractionation as well as the gradient medium contain sufficient calcium chelators to inactivate the receptors; in other words, all of the labeled ligand in the gradients is contained in membrane-bound vesicles.

In order to examine the kinetics of separation of the two ligands, cells were fractionated after incubations for a few minutes. The ligands were labeled with different isotopes of iodine, and taken up into the same cells, so that any difference between their distributions is real. Again, trace concentrations were used. 6 min after the addition of the ligands, their intracellular distributions were similar but not identical; asialotransferrin as well as asialofetuin were recovered at a density close to 1.11 g/ml.

(Fig. 6 , upper panel). This observation supports our hypothesis that the endosomes at this density are the ones into which the asialoglycoproteins are transferred immediately after internalization; after 6 min of incubation, the degradation of asialofetuin is not yet under way.

After 24 min, the distributions of asialofetuin and asialotransferrin were clearly different. Much more asialofetuin was found in the "mature endosomes"-peak at 1.15 g/ml; this is the material which is scheduled for delivery to the lysosomes and eventual degradation (Fig. 6 ,upper right panel). Thus the time course of intracellular separation of the two ligands is related to the time course of degradation of asialofetuin.

It is unlikely that asialotransferrin and asialofetuin are taken into different endosomes, because they compete for the same receptors. It is possible that asialotransferrin binds to the receptors in a different manner than asialofetuin, because they have different numbers of complex type carbohydrate chains (2 and 3, respectively). The manner of binding to the receptors may serve as a signal which directs the ligand back to the plasma membrane or the "mature" endosomes.

A different mechanism is suggested by the experiments shown in the lower panels of Fig. 6 . It was established early on that the kinetics of degradation of asialotransferrin depends in a very curious manner on the amount of intracellular ligand; to wit, an increase in the amount of asialotransferrin in cells increases not only the absolute amount of asialotransferrin that is degraded, but there is even an increase in the <u>fraction</u> of intracellular ligand that is degraded per unit time (17). To take a rough numerical example: If the amount of intracellular ligand is doubled, the fraction that is degraded per unit time is also doubled (for instance,from 0.2% per min to 0.4% per min); in other words, doubling of intracellular ligand may lead to a four-fold increase in the number of molecules that are degraded per minute. This works in the opposite direction from what one would expect if the degradative system was merely being saturated.

The lower panels of Fig. 6 describe an experi-

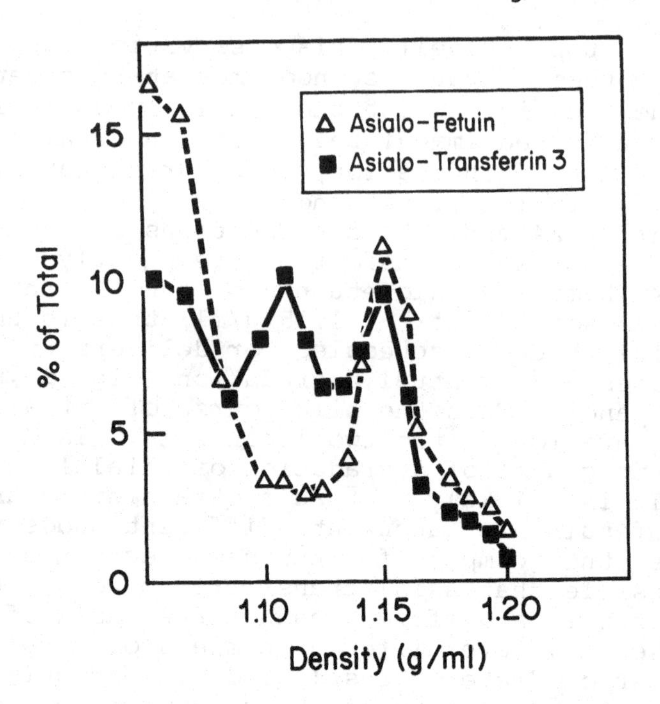

Fig. 5 . Intracellular distributions of asialo-transferrin and asialo-fetuin after prolonged incubation of the hepatocytes. ^{125}I-asialo-transferrin (1.2 nM) was added to one cell suspension and ^{125}I-asialo-fetuin (0.2 nM) to another. They were incubated at 37 °C for 80 min. The cells were chilled, washed, and fractionated by isopycnic centrifuging in a sucrose gradient.

ment which was designed in order to explore this effect. The intracellular distribution of asialo-transferrin was determined at high concentrations of ligand, and compared to corresponding distributions of asialofetuin. In the "high asialotransferrin" cells, the distribution was nearly identical to the distribution of asialofetuin, and it is particularly to be noted that after 24 min, most of the asialotransferrin is bound in the "mature endosomes"-peak at 1.15 g/ml, which corresponds to the observation that in these cells, asialotransferrin is being degraded at a rate

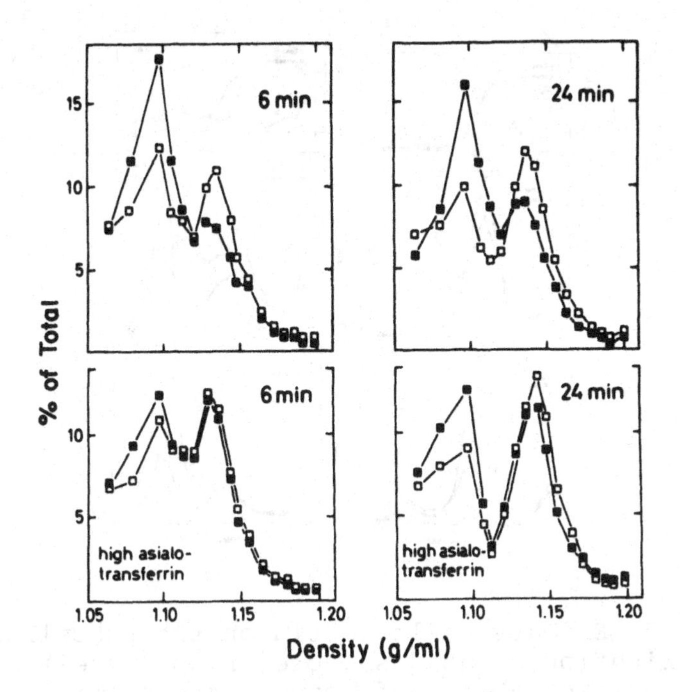

Fig. 6 Intracellular segregation of asialo-transferrin from asialo-fetuin in hepatocytes. Both ^{131}I-asialo-transferrin 3 (■) and ^{125}I-asialo-fetuin (□) were added to two different cell suspensions (0.013 nM and 0.067 nM, respectively). One of the flasks also received 200 nM unlabeled asialo-transferrin 1 ("high asialo-transferrin"). From both flasks, samples were taken after 6 min (left-hand panels) and 24 min (right-hand panels) for subcellular fractionation in sucrose gradients.

which is comparable to that of asialofetuin.

In order to explain the paradoxical effect of high levels of intracellular asialotransferrin on the degradation of this ligand, we would like to postulate first, that the processing of recently formed endosomes takes place by a fusion/budding process; and second, that the affinities of asialotransferrin and asialofetuin for the receptor change in different manners as these vesicles are acidified (18), and/or their content of calcium ions are decreased. The effect of the changes in affinities would be to allow the

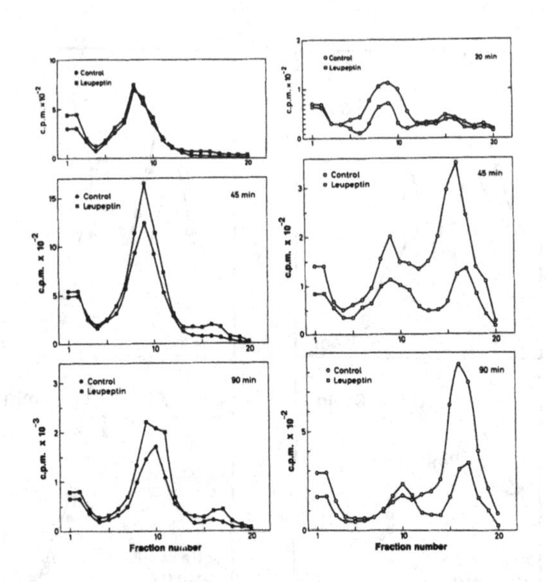

Fig. 7. Effects of leupeptin on the subcellular distribution of acid soluble (right panel) and acid precipitable (left panel) radioactivity after incubating hepatocytes in presence of 100 nM 14CSAF for 20, 45 and 90 min. The cells were fractionated by isopycnic centrifugation in sucrose gradients. Acid soluble and acid precipitable radioactivities were determined in the fractions. The densities of the fractions were as in Fig. 2.

asialotransferrin to remain membrane-bound, while asialofetuin is released into solution.

We envision the fusion/budding process as taking place very much as described by Geuze _et al_ (19). Small vesicles are pinched off from recently formed endosomes; coincidentally, the endosomes fuse in order to economize on membrane material. The end products of the process would be "mature" endosomes ready to fuse with lysosomes, plus smaller vesicles which carry membrane components back to the plasma membrane.

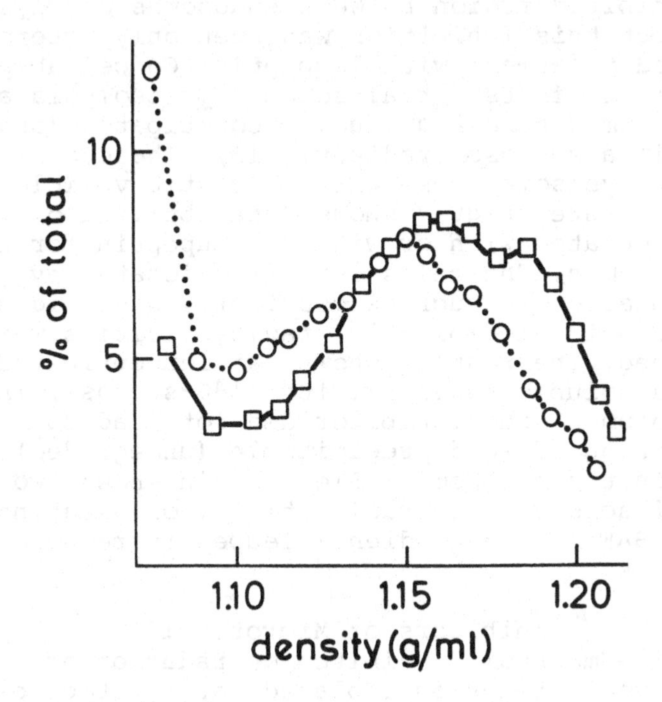

Fig. 8. Intracellular transport of yeast invertase in hepatocytes. A cell suspension received 0.2 nM of ^{131}I-invertase. After 10 min at 37 °C, a sample was removed for fractionation. The rest of the cells were washed to remove extracellular ligand and incubated for another 10 min. O ; intracellular distribution after 10 min incubation with labeled invertase, and □ ; distribution after another 10 min of incubation (as determined by isopycnic centrifugation in sucrose gradients).

EFFECTS OF INHIBITORS ON THE INTRACELLULAR TRANSPORT OF ASIALOGLYCOPROTEINS

Lysosomal Inhibitors

We have shown earlier that chloroquine and ammonium ions inhibit the degradation of asialoglycoproteins in isolated rat hepatocytes. Subcellular fractionation experiments suggested that this inhibition was mainly due to an inhibition of fusion between lysosomes and endosomes (20). Leupeptin also inhibited fusion between endosomes and lysosomes, but this inhibition was seen only after prolonged treatment with leupeptin. Conceivably, leupeptin inhibits intralysosomal proteolysis and the lysosomes become gradually constipated (and denser in a sucrose gradient) (13). These constipated lysosomes fuse with endocytic vesicles at a reduced rate. Fig. 7 shows data obtained on cells incubated with or without leupeptin for 20, 45 and 90 min. The cells were fractionated by centrifugation in sucrose gradients, and acid soluble and acid precipitable radioactivities were determined. The results show that leupeptin reduced acid soluble radioactivity (^{14}C-sucrose) in the lysosomes; the inhibitor did not lead to much accumulation of acid precipitable (undegraded) ligand in the lysosomes. Fig.7 again shows two peaks of acid soluble radioactivity originating from 14CSAF in the gradient; leupeptin reduced both peaks.

Inhibitors of Microtubuli

Colchicine markedly inhibited degradation of asialoglycoproteins in isolated rat hepatocytes (14). Subcellular fractionation experiments suggested that degradation was inhibited indirectly; colchicine prevented the transfer of ligands from endosomes to lysosomes, suggesting that this process was depebdent on intact microtubuli. The inhibitor did not, however, reduce the formation of acid soluble radioactivity (from 14CSAF) in the "light" peak in the gradient (Lill Naess, unpublished).

Metabolic Inhibitors

Many metabolic inhibitors depress uptake and degradation of asialoglycoproteins in rat hepatocytes (Helge Tolleshaug, unpublished). We have particularly studied the effect of rotenone. This inhibitor of oxidative phosphorylation reduced ATP production and at the same time inhibited both uptake and degradation of asialoorosomucoid. Subcellular fractionation of cells that had bound labeled ligand before being incubated with rotenone showed that rotenone, like monensin, prevented the change in density distribution of the endocytic vesicle containing radioactivity. Again, if this change is a reflection of the establishment of a proton gradient then rotenone may prevent it by reducing the ATP supply needed (to create the gradient.

INTRACELLULAR TRANSPORT OF OTHER LIGANDS IN ISOLATED RAT LIVER CELLS

As already pointed out, liver parenchymal cells have several types of binding sites which mediate the uptake of various ligands. The cells could therefore be used to compare binding, uptake and intracellular transport and turnover of ligands taken up by different receptors. Earlier studies have shown, for instance, that lipoproteins such as HDL (high density lipoproteins) are internalized into the hepatocytes at a slower rate than asialoglycoproteins. Once inside the cells the HDL are degraded rapidly (21).A direct comparison in the same experiments of the turnover of lipoproteins and asialoglycoproteins has not yet been done.

Uptake of Mannose-Terminated Ligands

Nonparenchymal liver cells are very active in endocytosing glycoproteins whose carbohydrate side chains terminate in mannose, such as RNAse B, ovalbumin, or artificial conjugates of mannose with bovine serum albumin. The endothelial cells are particularly active in vivo (22), and they retain their activity after isolation and purification.

Most of the mannose-binding capacity of a liver homogenate is actually associated with the hepatocytes (23,24), but uptake into hepatocytes of the usual mannose-terminated ligands has not been demonstrated. In our experiments, we tested a large glycoprotein which contains about 50% mannose in large,branched chains (25), namely yeast invertase. Following injection of ^{131}I-invertase into a rat, 98% of the radioactivity was recovered in the liver, and about half of it was associated with the hepatocytes; on a "per cell" basis, the endothelial cells were the most active ones in endocytosing invertase.

This protein was taken up into suspended hepatocytes by a saturable mechanism which may be inhibited by α-methylmannoside or EGTA. The rate of uptake was much lower than for asialofetuin, which is probably due to the small binding capacity of the cell surface (of the order of 10^3 molecules/cell). The intracellular distribution was different from asialofetuin. After 10 min of incubation with the ligands, the asialofetuin peak was at 1.14 g/ml, while the invertase peak was at 1.15 g/ml, which is the same as the peak of 5'nucleotidase activity. If at this point the extracellular ligand is removed by changing the medium, and the incubation was continued for another lo min, then the asialofetuin peak merely shifted slightly towards higher density, but a new peak appeared in the distribution of invertase. The new peak was at the lysosomal density of 1.19 g/ml (Fig. 8). Thus yeast invertase was transported towards the lysosomes at a much faster pace than asialofetuin, and very likely by a different mechanism.

ACKNOWLEDGEMENT

The competent technical assistance of Kari Holte and Lill Næss is gratefully acknowledged. The work was supported by The Norwegian Council for Science and the Humanities, and by The Norwegian Council on Cardiovascular Disease.

REFERENCES

1. WISSE, E., J. Ultrastructure Res.46,393 (1974)
2. NORUM, K.R., BERG, T., and DREVON, C.A., in L.A. CARLSON and B. PERNOW (Editors), Metabolic risk factors in ischemic caridiovascular disease, Raven Press, New York,1982,p.35.
3. NAGELKERKE, J.F., BARTO, K.P., and VAN BERKEL, T.J.C., in D.L. KNOOK and E. WISSE (Editors), Sinusoidal liver cells, Elsevier Biomedical Press, Amsterdam, 1982, p.319.
4. BLOMHOFF, R., ESKILD, W., and BERG, T. (1983) (submitted).
5. BLOMHOFF, R., HELGERUD, P., RASMUSSEN, M., BERG, T., and NORUM, K.R., Proc. Natl. Acad. Sci. (USA) , 79, 7326,(1982).
6. BERG, T., BLOMHOFF, R., and NORUM, K.R., in D.L. KNOOK and E. WISSE (Editors),Sinusoidal liver cells,Elsevier Biomedical Press,

Amsterdam, 1982, p. 37.
7. TOLLESHAUG, H., BERG, T., NILSSON, M., and NORUM, K.R., Biochim.Biophys.Acta,499,73 (1977)
8. NILSSON, Å, and ÅKESSON, B., FEBS-letters, 51, 219, (1975).
9. HIGA, Y., OSHIRO, S., KINO, K., TSUNO, H., and NAKAJIMA, H., J. Biol. Chem., 256, 12322(1981).
10. OSE, L., OSE, T:, REINERTSEN, R., and BERG, T, Exp. Cell Res.,126, 109 (1980).
11. TOLLESHAUG, H., Int. J. Biochem.,13,45 (1981).
12. BLOMHOFF, R., TOLLESHAUG, H., and BERG, T., J. Biol. Chem. 257, 7456 (1982).
13. TOLLESHAUG, H., and BERG, T., Exp. Cell Res. 134, 207 (1981).
14. KOLSET, S.O., TOLLESHAUG, H., and BERG, T., Exp. Cell Res. 122, 159 (1979).
15. TARTAKOFF, A.M., and BASSALLI, P., J. Exp. Med. 146, 1332 (1977).
16. BERG, T., BLOMHOFF, R., NÆSS, L., TOLLESHAUG, H., and DREVON, C.A., Exp. Cell Res. (In press)
17. TOLLESHAUG, H., CHINDEMI, P.A., and REGOECZI, E., J. Biol. Chem. 256, 6526 (1981).
18. TYCKO, B., and MAXFIELD, F.R., Cell, 28, 643 (1982).
19. GEUZE, H.J., SLOT, J.W., STROUS, G.J.A.M., LODISH, H.F., and SCHWARTZ, A.L., Cell, 32, 277, (1983).
20. BERG, T., and TOLLESHAUG, H., Biochem. Pharmacol., 29, 917 (1980).
21. OSE, L., OSE, T., NORUM, K.R., and BERG, T., Biochim. Biophys. Acta, 620, 120 (1980).
22. HUBBARD, A.L., WILSON, G., ASHWELL, G., and STUKENBROK, H., J. Cell Biol., 83, 47 (1979).
23. MAYNARD, Y., and BAENZIGER, J.U., J. Biol. Chem. 257, 3788 (1982).
24. MORI, K., KAWASAKI, T., and YAMASHINA, I., Arch. Biochem. Biophys. 222, 542 (1983).
25. LEHLE, L., Eur. J. Biochem., 109, 589 (1980).

ON THE TRANSPORT OF SECRETORY PROTEINS FROM THE ENDOPLASMIC RETICULUM TO THE GOLGI COMPLEX

Erik Fries

Dept. of Medical and Physiological Chemistry,

University of Uppsala, Box 575,

S-751 23 Uppsala, Sweden

Early in the evolution of the living cell there must have appeared proteins that spanned the plasma membrane and which mediated the uptake and disposal of specific molecules. Proteins of this kind, that were bound to the cell membrane and that were partially exposed to the exterior of the cell, might also have come to serve other functions, such as degrading macromolecules in the surrounding medium to forms that could be taken up and be used as nutrients by the cell. It is reasonable to assume that it would have been an advantage for those early cells to shed, or secrete, those degrading proteins, thereby making the processing of the nutrients in the medium more efficient (1).

It is also possible that the secretion of the proteins was made more efficient through the formation of an invagination of the cell membrane, which increased the surface available for the transfer of the secretory proteins to the outside of the cell (2,3). As the surface of the invagination increased, direct contact with the cell surface membrane may have been lost and indirect contact had to be established, probably by vesicles that shuttled back and forth. Once the two membranes were separated they could become specialized for different functions and the former invagination became the sole site of the synthesis for secretory proteins - a protoform of the endoplasmic reticulum (ER). This membrane segregation necessitated a sorting mechanism so that the traffic of the transport vesicles would not cause randomization of the membrane components (4). It is now believed that the Golgi complex (GC) , through which all secretory traffic seems to pass, has a central role for this sorting function (5,6). It has been suggested that the lysosomes may also have developed from invaginations of the cell surface membrane (1), and it is interesting to note that proteins that are to be transported to this organelle have to be transported there via another organelle, the endosome, which may have a sorting function similar to that proposed for the GC (7).

The study of the transport of proteins from the ER to the GC is a relatively difficult task since this process cannot readily be manipulated and very little is therefore known on the molecular level about this process. In contrast, the exterior of the cell is readily accessible to experimental manipulation and consequently a great deal is known today about the mechanisms underlying the transport of specific proteins from the outside of the cell to the endosomes (7,8). This process can be summarized as follows : 1) external proteins bind to their respective cell surface receptors, 2) the protein-receptor complexes migrate into patches on the plasma membrane which have a lattice of a specific protein (clathrin) on their cytoplasmic side, 3) these patches invaginate (coated pits) and form vesicles, (coated vesicles), and finally 4) the coated vesicles lose their coat and fuse with the target membrane. If it is true that both the ER and the lysosomes have developed from the cell surface membrane then the interior of these organelles should be topologically equivalent to the exterior of the cell and one would expect to find similarities in the mechanisms by which proteins are transported from the plasma membrane to the endosomes and from the ER to the GC. One of the questions which will be discussed in this paper is whether the transfer of proteins from the ER to the GC is mediated by receptors and clathrin-coated vesicles.

Are Secretory Proteins Bound to Receptors in the ER ?

Early studies of protein secretion, in which isolated ER vesicles were made permeable to macromolecules by the addition of small amounts of detergent, showed that most of the secretory proteins in the vesicles could diffuse out (9). This finding might be taken as evidence that secretory proteins are freely soluble in the ER; however, it is also possible that the unphysiological conditions that were used in the experiments had caused the dissociation of any binding to the inner surface of the ER that had existed in the intact cell. An alternative interpretation of the results could be that binding did exist but that it was not detected because at any given instant only a small fraction of the secretory proteins would be engaged in this binding.

If one assumes that secretory proteins are not bound to receptors in the ER, as part of a transport process, but are transported to the GC by mass transfer (10), one would expect all secretory proteins to be transported at the same rate through a cell (provided that the proteins are not transferred at different rates at other steps of the secretory pathway). In pancreatic acinar cells secretory proteins have been reported to be transported in parallel (11) whereas in hepatocytes, albumin and transferrin have been shown to be transported at different rates (12-14). In order to study this latter phenomenon of non-parallel transport in more detail, we decided to determine the time courses for the intracellular transport of three serum proteins, albumin, transferrin, and the retinol binding protein, in isolated rat hepatocytes. Through the use of pulse-chase experiments, followed by subcellular fractionation and quantitative immuno-precipitation, we were able to establish that these three proteins were transported from the ER to the GC at greatly different rates, their half-lives in the ER being 15, 90, and 130 minutes, respectively (15). In contrast, their turn-over rates in the GC appeared to be comparable, the estimated half-lives being 15 - 30 minutes.

Studies of the intracellular transport of different membrane glycoproteins have indicated that proteins of this kind may also be transported from the ER to the GC at different rates (14,16). On the basis of such observations, Fitting and Kabat (16) recently proposed that proteins destined for the outside of the cell bind to transport receptors in the ER and that proteins with high affinities for the receptors have small pools (relative to their rates of synthesis) and turn over quickly. Conversely, proteins with low affinities have large pools and turn over slowly. Our finding that transferrin (which was synthesized at a much lower rate than albumin) occured in the ER of the hepatocytes at a concentration similar to that of albumin but was transported at a lower rate (15), is consistent with this model.

It has been shown for a number of secretory proteins that a small change of their structure may prevent their secretion. For example, a single aminoacid substitution in an immunoglobulin light chain may cause this protein to accumulate in the ER (17). Although data of this kind strongly suggest that secretory proteins contain structural information that is necessary to effect their transport the possibility cannot be ruled out at present that the altered proteins are not secreted because of low solubility.

Are Secretory Proteins Taken Up in Clathrin Coated Vesicles?

In the late 60's, Jamieson and Palade suggested, on the basis of electron microscopical data, that newly synthesized secretory proteins leave the ER from special regions located close to the GC. In these regions, which they called transitional elements, they could see buds on the ER membrane which they proposed would form vesicles that would carry secretory proteins over to the GC (18). More recently, Yokota and Fahimi, who used an immunocytochemical technique, were able to show in hepatocytes that these buds contained albumin. Interestingly, they also found that lipoprotein particles,which could be recognized without the aid of antibodies because of their large size, did not occur in the same buds as albumin, suggesting that different secretory proteins may be segregated in the ER, possibly through the binding to different receptors (19).

There is as yet no experimental evidence for the existence of clathrin on the transitional elements. However, the fact that coated vesicles have been isolated containing a newly synthesized membrane glycoprotein destined for the surface of the cell which had immature oligosaccharides (20) strongly suggests that coated vesicles mediate the transport of at least membrane proteins from the ER to the GC. However, certain observations indicate that the vesicles that form on the ER are different from those formed elsewhere in the cell. A special staining procedure has revealed rings of bead-like structures specifically at the base of the buds on the transitional elements (21) and the appearance of the vesicles that can be seen between the ER and the GC has been reported to be different from that of coated vesicles at other locations in the cell (22).

Other factors affecting transport rates

It has been observed for a number of glycoproteins that they after their synthesis reach the GC only after a clearly defined lag period (14,23). It has been suggested that this lag period reflects the time required for certain modifications of the oligosaccharides to take place in the ER (13). This hypothesis implies that the carbohydrates play a role in the transport process. However, the fact that at least some glycoproteins will still be transported even if they have been prevented from acquiring their normal oligosaccharides in the ER (14,23) argues against this hypothesis. In our experiments with isolated hepatocytes we observed that newly synthesized transferrin reached the GC fraction only after a 10 - 20 min lag period (15). In these cells the retinol binding protein was transported at a lower rate than was transferrin but for the retinol binding protein no time lag was observed. This observation shows that the time lag was not simply an effect of slow, or inefficient, transport from the ER. It is conceivable that those proteins that appear to be delayed in the ER are transported to the GC via a compartment which cofractionates with the ER and through which other proteins do not pass (14).

Reconstitution of Intracellular Protein Transport.

One standard approach for dissecting a complex cellular process is to study the process in a cell-free system. Some years ago, Jim Rothman and I set out to find conditions under which a newly synthesized membrane glycoprotein would be transferred from the ER to the GC in a cell homogenate. As a model system we used Chinese hamster ovary (CHO)cells infected with vesicular stomatitis virus (VSV). In cells infected with this virus the only membrane protein produced is the viral glycoprotein (G protein) that is incorporated in the membrane of the mature virus particle. G protein is synthesized in the ER and is subsequently transported to the plasma membrane, apparently along a pathway that is identical to that used by the cell´s own surface membrane proteins in uninfected cells (24).

During its synthesis G protein acquires two mannose-rich oligosaccharides that are subsequently processed as G protein passes through the GC (24). This change in oligosaccharide structure provides an indirect means for assaying the arrival of G protein at the GC. Endo-β-N-acetylglucosaminidase H (Endo H) has proved useful in this regard because this enzyme will cleave the mannose-rich precursor oligosaccharide, causing a marked reduction of the apparent molecular weight of G protein, but will not attack the Golgi-processed oligosaccharides (25).

The general approach we took to obtain transport of G protein in a cell-free system was to prepare extracts of VSV-infected CHO cells that had been briefly incubated with 35S-methionine so as to label G protein in the ER. This resulted in an extract in which all of the labelled G protein was sensitive to Endo H. The production of Endo H resistant G protein after incubation of the cell-free extract was then taken as initial evidence that transport to the Golgi complex in the cell-free system had occurred.

To ensure that only events resulting in the transfer of G protein between organelles would be detected by our indirect assay, we used extracts prepared from a mutant of the CHO cells (clone 15B) and an *in vitro* complementation scheme. Clone 15B lacks an enzyme in the GC that is needed to initiate the process that leads to the formation of Endo H resistant G protein (26). G protein is transported normally in infected 15B cells but always remains sensitive to Endo H (27). The appearance of Endo H resistant G protein upon incubation of a mixture of an extract of infected, 35S-methionine labelled 15B cells and an extract of uninfected wild type cells should thus provide strong evidence that transfer of G protein from 15B cell membranes to the GC membranes of the wild type cells had occurred.

However, the use of the protocol just described did not give any indication of transfer. We then argued that the transfer reaction probably consisted of more than one step and that possibly one of the first steps required the integrity of the cellular organization, which was destroyed upon homogenization. To exclude this putative early step in the reconstituted transport reaction we instead used extracts of cells that had been chased for increasing periods of time after the pulse. With this modified protocol, Endo H resistant G protein was obtained and we found that the most efficient transfer was achieved with extracts of 15B cells that had been chased for 10 minutes (28).

The ^{35}S-methionine labelled G protein in the 15B cell extracts that yielded maximal transfer was found to co-fractionate with a marker for the GC upon equilibrium centrifugation in sucrose density gradients (27). This, and other observations (29) suggested that the G protein that could be transferred to the GC of the wild type cells upon incubation was in the GC of the 15B cells. The fact that the efficiency of the transfer reaction decreased rapidly with extracts prepared from cells chased for longer periods of time suggested that transfer could occur only from membranes derived from an "early" part of the GC (28,29). These observations, together with the postulate that proteins are transported through the GC from the side facing the ER (the cis side) to the opposite side (the trans side) (6), suggested to us that we had achieved transfer of G protein from membranes derived from the cis side of the GC to membranes derived from the trans side (29). The possible implication of the existence of such an intra-Golgi transfer process has been discussed elsewhere (5).

The cell-free system that we had obtained will most probably prove useful for the study of the role of the GC in the transport of proteins. However, for the elucidation of the mechanisms for the transport of proteins from the ER to the GC other systems must be designed. It is justified to ask why the viral protein was not transferred between those organelles under the conditions that we used. One reason could be that upon homogenization of the cell the ER was fragmented into small vesicles with only a small fraction of the protein within vesicles derived from the transitional elements. Since transfer to Golgi derived vesicles could occur only from these latter vesicles, the over-all efficiency of the transfer reaction in this crude cell-free system was low. It therefore appears that substantial progress in the elucidation of the mechanisms by which proteins are transported from the ER to the GC will be made only when the transitional elements can be isolated and their components characterized.

ACKNOWLEDGEMENTS

This work was supported by grants from the Swedish Research Council (B 3074-103) and from O.E. and Edla Johansson´s Scientific Foundation.

REFERENCES

1 de Duve and Wattiaux,R. (1966) Annu.Rev.Physiol. 28,435
2 Blobel,G. (1980) Proc.Natl.Acad.Sci. USA 77, 1496
3 Sabatini,D.D., Kreibich,G., Morimoto,T., and Adesnik,M. (1982), 92,1
4 Palade,G.E. (1982) in Ciba Found.Symp. 92, 1
5 Rothman,J.E. (1981), Science (Wash.DC) 213, 1212
6 Tartakoff,A. (1981) Int.Rev.Exp.Pathol. 22, 227
7 Brown,M.S., Anderson,R.G.W. and Goldstein J.L. (1983), Cell 32, 663
8 Pearse,B.M.F. and Bretscher (1981) Annu.Rev.Bioch. 50,85
9 Kreibich,G. Debey,P. and Sabatini,D.D. (1973), J.Cell Biol.58, 436
10 Palade,G.E. (1975) Science(WashDC) 189, 347
11 Tartakoff,A.M., Jamieson,J.,Scheele,G.A. and Palade,G.E. (1975) J.Biol.Chem.250, 2671
12 Peters,T.,Fleischer,B.and Fleischer,S. (1971) J.Biol. Chem. 246, 240
13 Morgan, E.H. and Peters,T. (1971)J.Biol.Chem. 246,3508
14 Strous,G. and Lodish,H.F. (1980) Cell 22, 709
15 Fries,E., Unger,E.,Gustafsson,L. and Peterson,P.E., submitted for publication
16 Fitting,T. and Kabat,D. (1982) J.Biol.Chem.257, 14011
17 Wu,G.E., Hozumi,N. and Murialdo,H. (1983) Cell 33, 77
18 Jamieson,J.D. and Palade, G.E. (1967)J.CellBiol.34, 577
19 Yokota,S. and Fahimi,H.D. (1981)Proc.Natl.Acad.Sci. USA 78, 4970
20 Rothman,J.E. and Fine,R.E. (1980)Proc.Natl.Acad.Sci.USA 77, 780
21 Locke,M and Huie,P. (1976)J.CellBiol. 70, 384
22 Croze,E.M.,Morré,D.M.,Kartenbeck,J.,Franke,W. and Morré D.J (1982) J.Cell.Biol.95,429a
23 Carlson,J. and Stenflo,J.(1982)J.Biol.Chem. 257,12987
24 Lenard,J. (1978) Annu.Rev.Biophys.Bioeng. 7, 139
25 Robbins,P., Hubbard,S. and Wirth,D. (1977) Cell 12,893
26 Gottlieb,C.,Baenzinger and Kornfeld,S.(1975) J.Biol. Chem. 250,3303
27 Li, E. and Kornfeld (1978) J.Biol.Chem.253,6426
28 Fries,E. and Rothman,J.E.(1981)J.CellBiol.90,697
29 Dunphy,W.G., Fries,E., Urbani,L.J. and Rothman,J.E. (1981) Proc.Natl.Acad.Sci.USA 78,7453
30 Rothman,J.E., Fries, E., Dunphy.W.G. and Urbani,L.J. (1982) Cold Spring Harbor Symposia on Quant.Biol. Vol.XLVI, 797

Intracellular Transport of Proteins and Protein Modification

SJUR OLSNES

Norsk Hydro's Institute for Cancer Research and The Norwegian Cancer Society, Montebello, Oslo 3, Norway

Extensive transport of proteins takes place between different intracellular compartments. Protein synthesized for export or for insertion into a number of membraneous compartments are transferred from the endoplasmic reticulum to different parts of the vesicular space where modifications may take place. Various sugars are added by enzymes which first recognize defined structures in the polypeptide chain and then structures in the growing oligosaccharide chains. The different additions take place in different vesicular compartments. Thus sugars like N-acetyl glucosamine present in the proximal part of the oligosacharide chain, are added in the endocytic reticulum, while the more distal sugars, like galactose and sialic acid, are added in the Golgi apparatus. Subsequently the proteins are transported in vesicles to their final target or, in the case of export proteins, to the cell surface where they are released.

On their way out of the cells these proteins meet another streem of proteins which are on their way into the cell. These are proteins which are bound to cell surface receptors and then transported into various vesicular compartments. In some cases as with certain toxins, the proteins are even transferred into the cytosol. Other proteins, like transferrin, are transported in and out of the cell in a cycle as if they were attached to a conveyor belt. Many mitochondrial and peroxisomal proteins are synthesized in the cytoplasm and then posttranslationally transferred into the organelles. Clearly an extensive sorting apparatus must be operating to ensure that the different proteins reach their respective destinations.

The transport of different newly synthesized proteins to the Golgi apparatus occurs at different rates, possibly by a receptor-mediated process. In the Golgi apparatus extensive sorting of proteins destined to the different compartments appears to take place. The chemical basis for this sorting is in most cases not known. Some of the proteins are directed to the lysosomes. In this case the signal recognized by the sorting apparatus is mannose 6-phosphate, which is recognized and bound by a receptor in the membrane. Some of the lysosomal enzymes are released, possibly erronously, into the medium. However,due to the presence of receptors at the cell surface,the enzymes are recognized by their mannose 6-phosphate groups and returned to the lysosomes where the phosphate groups are removed.

A variety of physiological molecules are taken up by receptor-mediated endocytosis and transported to various intracellular locations. Among these are a number of protein hormones, antibodies, desialylated glycoproteins and serum proteins carrying vitamins, metabolites and growth factors. The internalized proteins are first observed in endosomes with low boyand density, but later they are transferred to vesicles with higher density, the lysosomes. During this process the receptor is recycled back to the cell surface ready to bind a new ligand molecule. In many cases the low pH found in endosomes induces release of ligand from the receptor. If this release is inhibited, the receptors may not be recycled back to the cell surface which after a while becomes depleated of receptors.

Several protein toxins are also bound to receptors at the cell surface and internalized by endocytosis. Among these are diphtheria toxin, Pseudomonas aeruginosa exotoxin A, cholera toxin, Shigella toxin and the plant toxins abrin, modeccin, ricin and viscumin. These toxins contain a binding moiety which binds to cell surface receptors and an enzymatically active moiety which penetrates the membrane and inactivates components present in the cytosol. There is now good evidence that the toxins are first endocytosed and then the enzymatically active moiety crosses the membrane.

The reason why toxins must be endocytosed before they can cross the membrane is in most cases not clear. However, in the case of diphtheria toxin and possibly also in the case of modeccin and Pseudomonas aeruginosa exotoxin A, the toxins must reach an acidic environment before they can cross the membrane. In the case of diphtheria toxin the low pH exposes a hydrophobic region present in the B-fragment which then inserts itself into the membrane and somehow assists in the transfer of the A-fragment. Different toxins appear to cross the membrane from different intracellular vesicular compartments indicating that the particular conditions required are only present in defined vesicular compartments. The entry of the toxins is a complex process which requires energy, ion gradients and in some cases, Cl^- and Ca^{2+} in the medium. Possibly an electrical potential across the membrane is also required for entry.

The transport of mitochondrial, chloroplast and peroxysomal proteins from the cytosol into their respective organs resembles that of toxin entry in the sense that in all cases a completed polypeptide is transported. In the case of the mitochondrial and chloroplast proteins in most cases a precursor is made which is then split upon entry into the organelle. Such precursors have not been found in the peroxisomal proteins. The proteins appear to first bind to receptors on the surface of the respective organelle. So far little is known about the vectorial forces involved in the transfer of the proteins across the membranes. The membrane potential is essential for entry of the mitochondrial proteins, but not for binding to the surface of the organelle. Possibly the potential assists in pulling certain domains of the proteins across the membrane. It is in this connection interesting that the entry of a number of colicins into bacteria also depend on a normal membrane potential.

The elucidation of the various signals for correct sorting of proteins, the routing of proteins between the different organelles and the mechanisms and forces involved in protein translocation across membranes is likely to greatly increase our understanding of the organisation and functioning of the eucaryotic cell.

5. PROTEIN GLYCOSYLATION

FUNCTION AND METABOLISM OF MEMBRANE-BOUND DOLICHOL

G. Dallner, C. Edlund, T. Ekström, I. Eggens,
Ö. Tollbom, I.M. Åstrand
Department of Biochemistry, Arrhenius Laboratory, University of Stockholm and Department of Pathology at Huddinge Hospital, Karolinska Institutet, Stockholm, Sweden

Membranes of eucaryotic cells contain glycoproteins and consequently their biosynthesis is of great interest. It is well established that lipid intermediates of dolichol type are obligatory in the assembly of the N-glycosidicly linked oligosaccharide chains (Struck & Lennarz, 1980; Dallner & Hemming, 1981). Membranes, and to a great extent secretory proteins, are glycosylated; a process involving lipid intermediates. The role of glycosylation in the biosynthesis of cell membranes has stimulated interest in the dolichol lipid intermediates during the last ten years.

Despite evidence of dolichol having an important function in glycoprotein synthesis, there are serious doubts as to the importance of this lipid in many biological membranes. Only a minor and sometimes very small fraction of the polyprenol is phosphorylated, that is, in the activated form. This means that mg quantities of this lipid may be present in membranes and this would be unlikely for a precursor substance. Even the phosphorylated lipid is compartmentalized and only a fraction of it seems to participate in sugar transport.

Not only the amount and distribution of dolichol, but also the mechanism of biosynthesis raises questions as to the significance of these lipids in glycoprotein synthesis. The series of condensation reactions leading to the synthesis of polyprenol pyrophosphates are not known in any detail. The possibility that the active intermediate is de- and rephosphorylated has been intensively investigated but the results are not conclusive. The rapid production of the free alcohol during *in vivo* labeling raises the possibility that alternative pathways of biosynthesis may exist; alternative to the sequential additions of the isopentenyl pyrophosphate residues. If this is true, one must begin to consider the possibility that dolichol has several roles in cell function.

Dolichol distribution

Dolichol vary in size from 17 to 22 isoprene residues. Both shorter and longer polyisoprenes have been identified in various tissues but are less common. The α-isoprene in animal tissues is saturated but it appears that a small portion, about 1%, is in unsaturated form. The two isoprene residues at the ω-end are in trans form but the rest of the chain has cis configuration (Hemming, 1974).

Most animal tissue contain dolichols in similar amounts to those recorded for rat tissue (10 - 150 μg/g tissue) as shown in Table 1 (Eggens *et al.*, 1983). Human tissues, however, has a considerably greater polyprenol content. Human spleen, kidney and lung contain dolichol amounts comparable to that of many other animal organs but a number of human tissues have mg quantitities of dolichol. The pituitary gland is especially interesting since dolichol is the largest lipid component here (8 mg/g tissue) even exceeding lecithin.

Intracellular distribution

The only tissue where dolichol distribution has been studied is the liver. In rat liver microsomes, which are the main or exclusive site of glycosylation reactions, have very low dolichol content as related to protein content (Table 2). Lysosomes, outer mitochondrial and Golgi membranes are rich in this lipid. On the other hand, dolichyl-P distribution reflects the glycosylation function since a great proportion of the lipid is phosphorylated in microsomes in opposite to, for example lysosomes. The lipid is also

TABLE 1

Amount of dolichol in rat and human tissues

	Rat	Human
	μg/g tissue	
Liver	43	465
Spleen	113	84
Pituitary gland		8,335
Testis	13	1,493
Adrenals	35	1,371

TABLE 2

Distribution of dolichol and dolichyl-P in liver fractions in control and drug-treated rats

	Dol.	Dol.-P	% Phosph.
	μg per mg protein		of total
Control			
Homogenate	0.196	0.028	13
Microsomes	0.21	0.062	23
Lysosomes	3.9	0.079	2
Supernatant	0.020	0.003	13
Mitochondria	0.91	0.038	4
outer membrane	4.7		
inner membrane	0.18		
Golgi membranes	4.8		
Peroxisomes	0.8		
Nuclei	0.2		
Plasma membranes	1.8		
Phenobarbital-treated			
Microsomes	0.21	0.048	19
Lysosomes	1.3		
N-Nitrosodiethylamine-treated			
Microsomes	0.47	0.047	9
Lysosomes	10.5	0.11	1
Diethylhexylphthalate-treated			
Microsomes	0.22	0.041	16
Lysosomes	7.3	0.15	2

Phenobarbital (80 mg/kg) and N-nitroso-diethylamine (20 mg/kg) were injected intraperitoneally once a day during 5 days. Diethylhexyl-phthalate was included in the diet (2%) and given *ad libitum* during 4 weeks.

present in the supernatant but the nature of the lipid at this location has not been investigated. It is probably present in a protein-bound form and is part of a transport process.

Dolichol content is greatly influenced by various treatment which affect membrane synthesis (Table 2). Phenobarbital, which is a well known inducer of the endoplasmic reticulum and associated drug hydroxylating system, decreases the lysosomal dolichol content, but has no effect on

TABLE 3
Dolichol and dolichyl phosphate in pathological conditions

Exp. Tissue	Amount (μg/g)	Composition (% of total) D17	D18	D19	D20	D21	D22
1. Dolichol							
Rat liver	43	11	42	32	10	5	0
Rat liver hyperplastic nodules	147	8	36	38	15	3	0
Human liver	465	4	12	39	30	11	4
Human liver cirrhosis	52	5	14	41	26	11	3
Human hepatocarcinoma	62	6	33	48	11	1.5	0.5
2. Dolichyl-P							
Rat liver	8.75	7	33	38	15	5	2
Rat liver hyperplastic nodules	5.12	9	34	39	10	5	3

Hyperplastic nodules were produced by treatment of rats with 2-acetylaminofluorene. Samples of human liver, cirrhosis (Leannec type) and primary carcinoma (hepatocellular type) were diagnosed histologically parallel to dolichol measurements on HPLC.

microsomes. The carcinogenic substance N-nitrosodiethylamine considerably increases both microsomal and lysosomal dolichol concentrations. The plasticizer diethylhexylphthalate substantially elevates lysosomal polyprenols without affecting the lipids in microsomes. These examples illustrate the compartmentalization of dolichol in the hepatocyte. It appears that an independent regulation is operating among the different compartments and this independence is also evident between the free alcohol and its phosphorylated counterpart.

Isoprene pattern and malignant transformation

The isoprene pattern of dolichol is more species than organ specific. In rat liver, the 18-isoprenenoid compound is the most prevalent form, but dolichol 19 is also present in high concentrations (Table 3). In human liver the dominating component is the 19-isoprene form and the 20-residue counterpart the next most prevalent. Some important changes appears to occur during malignant transformation. Acetylaminofluorene induced preneoplastic transformation leads to

TABLE 4

Dolichol kinase and dolichol monophosphatase activities in microsomes of preneoplastic nodules

Enzyme	Substrate	Control	Nodules
Dol. kinase	Dol.-11[a]	507	391
	Dol.-20[a]	150	81
Dol.-Pase	Dol. 17-21[b]	3590	5700

[a]Dol.-P formed (cpm/mg prot/10 min)

[b]Dolichol formed (cpm/mg prot/10 min)

hyperplastic nodules containing severalfold increased (particular in microsomes) amounts of dolichol. In contrast, dolichyl-P content is decreased in this process. The preneoplastic condition of the rat, however, does not lead to a change in the isoprenoid pattern. Both human liver cirrhosis and hepatocellular cancer cause a drastic reduction of polyprenol content and the latter also introduces a change in the isoprene distribution. In this malignant process the pattern changes and becomes similar to that in rat.

The importance of the above described changes are not known but it is reasonable to assume that they are associated with the variations appearing during carcinogenesis. Changes in the dolichol content may be in causal relation with the production of the changed glycoproteins appearing in and on the surface of the malignant cell. Dolichol kinase, the enzyme capable of phosphorylating dolichol, shows a significant decrease in activity in microsomes isolated from nodules (Table 4). Dolichol monophosphatase, which catalyzes the production of the free alcohol, increases substantially. The observations of decreased phosphorylation and increased breakdown of dolichyl-P agree well with the finding that in spite of the extensive synthesis of the free alcohol, the amount of active lipid intermediate is decreased. The main importance of such an event is that in several experimental systems dolichyl-P is the rate limiting factor of the glycosyl transferase system. This may also be the case in carcinogenesis since the dolichol-mediated protein glycosylation exhibits a clear decrease.

TABLE 5

Properties of dolichol-Pase and -PPase

	Dol-Pase	Dol-PPase
pH optimum	6,5	8,0
Triton effect	Activation	Inhibition
Inhibitor	NaF	Bacitracin
Product	Pi	Pi(no PPi)
Present	Lysosomes, plasma membranes	
Absent	Inner mitochondrial membranes, peroxisomes	

CTP-kinase and phosphatases

It is generally accepted that the final step in dolichol biosynthesis is α-saturation and dephosphorylation of the appropriate polyisoprenes to produce the active lipid intermediate. Dolichyl-PP is also liberated when the core oligosaccharides are transferred to the protein and regeneration of this lipid requires a pyrophosphatase action. If the free alcohol is produced along the same pathway involving the condensation reactions, dolichyl-P has to be hydrolyzed by a monophosphatase. The free alcohol is then either esterified with a fatty acid which probably represents a storage product or it may be rephosphorylated by a CTP-specific kinase. The key enzymes mediating these processes have not as yet been isolated and are poorly characterized.

There appears to be two enzymic mechanisms for dolichyl-PP and P-dephosphorylation (Table 5) in lysosomes, plasma membranes and, to some extent, in microsomes (Appelkvist et al., 1981). There is no direct evidence for the specificity of these, but enzymes specific for dolichol dephosphorylation must exist. The two "activities" differ in pH optimum and in response to detergent treatment. The monophosphatase is inhibited by NaF while the pyrophosphatase action is blocked by bacitracin.

If the CTP-mediated kinase participates to a substantial extent in dolichol phosphorylation, it can be regarded as playing a key role in the process (Allen *et al.*, 1978). The enzyme is present only in the outer surface of the microsomes and phosphorylates α-saturated polyisoprenes (Ekström *et al.*, 1982). There still remains a question as to whether or not it functions in the synthesis of active lipid intermediates *in vivo*.

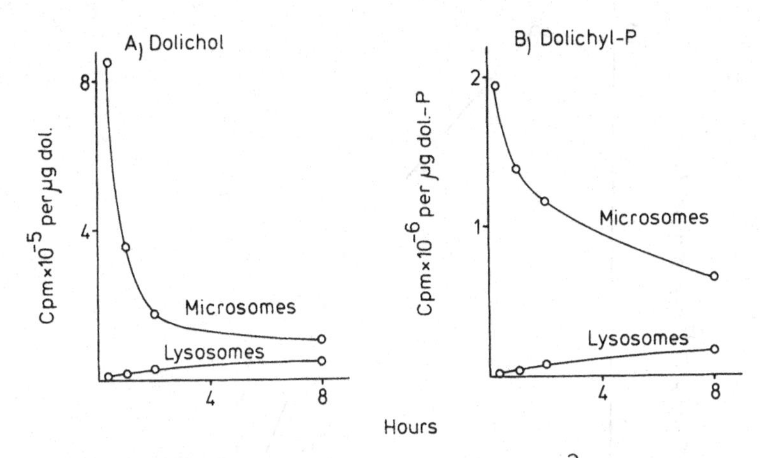

Fig. 3 Time course of the appearance of [^{3}H]undecaprenol in different organs. [^{3}H]Undecaprenol was injected through a gastric tube, rats were decapitated after various time points and the organs were homogenized. Aliquots were taken for determination of radioactivity.

The site of synthesis and transport

The initial step in dolichol synthesis is identical with that of the cholesterol. All the enzymes for biosynthesis of farnesyl-PP, with the exception of the HMG-Coenzyme A reductase, are present in the soluble cytoplasm. It is generally accepted that dolichol synthesis is proceeded by the cis-addition of isopentenyl-PP residues to trans-trans farnesyl-PP. The consecutive addition of the appropriate number of isoprenes leads to the final α-unsaturated polyprenyl-PP. This sequence of events was originally thought to be present in the outer mitochondrial membrane, but high initial incorporation into microsomal dolichol and particulary microsomal dolichyl-P *in vivo* (Fig. 1) contradicts this (Ekström *et al*., 1982). In the initial period, labeling of the mitochondrial dolichol (not shown in the figure) is very low. If *de novo* synthesis of dolichyl-P and dolichol is localized exclusively to the endoplasmic reticulum, the question arises as to how the lipid is transferred to other organelles. This would involve an intensive process as majority of polyprenols, at least in hepatocytes, are situated in membranes other than the ER. We know very little about the half life of the membrane-bound polyprenols and even less about the catabolic reactions involving these

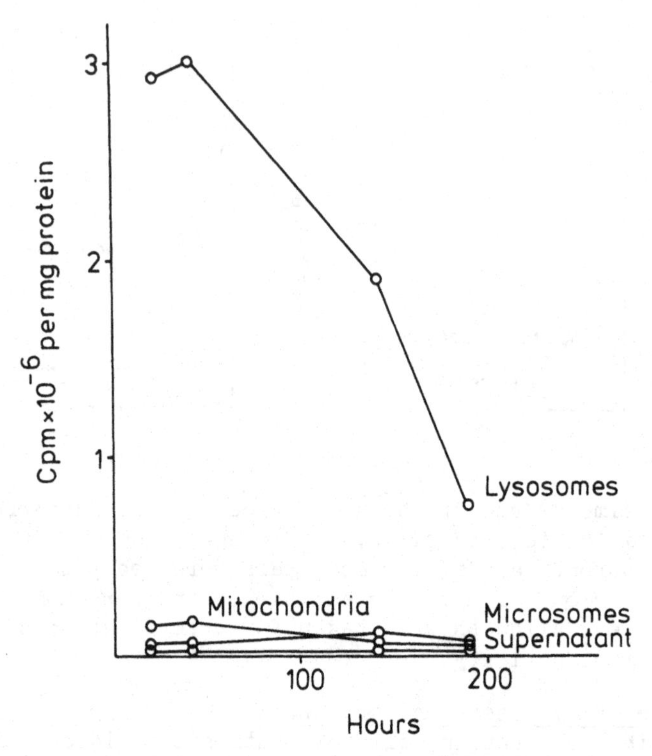

Fig. 2 Incorporation of liposomal [^{3}H]dolichol-11 into liver membranes. Liposomes were injected into the portal vein, livers were fractionated after various time points and radioactivity was determined in the isolated dolichol fraction.

lipids. The transfer process is assumed to be a slow one. It can be seen in Fig. 1 that the labeling of microsomal dolichol, and particularly that of the dolichyl-P, is not in equilibrium with the lysosomal counterpart even after 8 hours. This type of experiment, as well as those involving drug induction, suggest that the intracellular polyprenes are compartmentalized and probably represent functionally separate units.

The possibility also exists that dolichol move from the extracellular compartment via an endocytotic process. When labeled dolichol is incorporated to liposomes consisting of phosphatidylcholine, the lipid can be adminstered to the liver through the portal vein. As expected, the labeled dolichol is rapidly taken up by the liver and after a few minutes appears in the lysosomes as measured by continuously increasing radioactivity. Some radioactivity is also present in the microsomes but there is no increase in activity with time at this location. The fate of the lysosomal dolichol was followed for 200 hours (Fig. 2). There is a decrease of the radioactivity in the lysosomes after 40 hours but at the same time no labeling appears in mitochondria, microsomes or supernatant. It is therefore improbable that extracellular dolichol, taken up by the lysosomes, contributes substantially to the microsomal dolichol but it is probably enzymatically broken down.

Dietary uptake

Dolichol, like cholesterol, appears to be synthesized in several organs in sufficient amount to fulfill the requiremen for normal function. The animal diet contains both

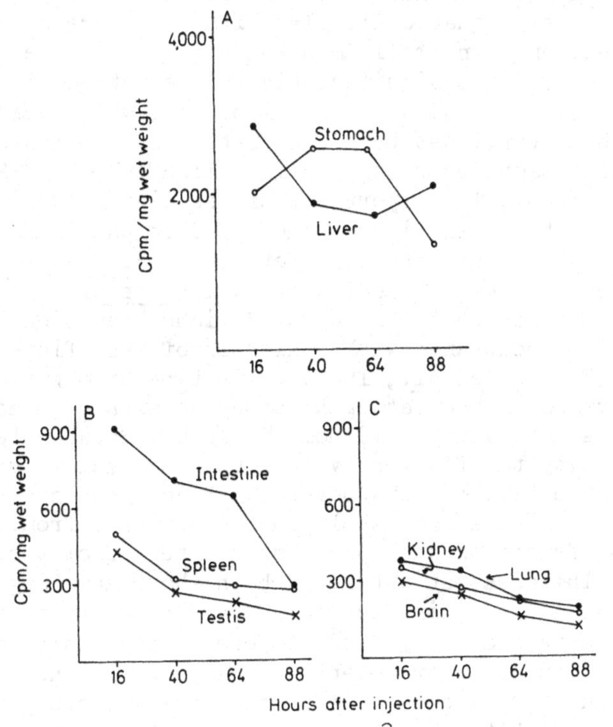

Fig. 1 In vivo incorporation of [^{3}H]mevalonate into dolichol and dolichyl-P of liver microsomes and lysosomes. [^{3}H]Mevalonate (1.25 mCi) was injected into the portal vein and livers were removed at various time points for fractionation and isolation of dolichol and dolichyl-P.

saturated and unsaturated polyprenols. The question arises as to whether or not these lipids are utilized by the body (Chojnacki & Dallner, 1983). When rats were supplied with dietary liposomal polyprenols it was found that short chain isoprenes (11 residues) were taken up and found mainly in the liver (Fig. 3). Long chain polyprenes (19-isoprenes) were not taken up from the gastrointestinal tract. Undecaprenol occurs in high amount in plants which means that this lipid is a part of the human diet. Undecaprenol (α-unsaturated polyprenol-11) is found in liver as a free alcohol and esterified with fatty acid. A sizeable portion of the polyprenol becomes α-saturated after uptake and a lesser portion is phosphorylated. The dietary polyprenol is modified to active lipid intermediate capable of participating in oligosaccharide biosynthesis. The main problem what remains here is that dolichyl-P with 11 isoprene residues is present only in small amounts in liver. If the dietary uptake contributes significantly to the intracellar dolichyl-P pool, the short chain polyprenol, which were taken up by the liver, must be a substrate for a series of condensation reactions resulting in dolichyl-P of 18-20 isoprene residues. The importance of the dietary dolichol also depends on the half-life of the lipid in the intracellular compartments (a property not yet studied).

Enrichment of the intracellular dolichol pool

One approach to the study of dolichol function is by enrichment of membranes with dolichols of specific chain length (Ekström *et al.*, 1982). This type of experiment can be performed by incubating hepatocytes with egg lecithin liposomes containing dolichol (Table 6). Hepatocytes have the capacity to efficiently take up liposomal material. After incubation of hepatocytes with dolichol containing liposomes, labeled dolichol can be recovered from various membrane fractions and to a lesser extent from the supernatant. The greatest uptake of both short and long chain dolichols is found in microsomes. A large part of these dolichols is phosphorylated. Interestingly, this dolichyl-P can function as sugar acceptor in glycosyl transfer reactions and even to some extent stimulate protein glycosylation. Although there is no evidence that the incorporated dolichol is associated with the microsomal membrane in the same manner as the endogenous lipid intermediate, this process provides a unique possibility to study the importance of specific dolichols in membrane.

TABLE 6

Transfer of liposomal [^{3}H]dolichol(C95) and (C55) into isolated hepatocytes

Fraction	Type of Dol.	Dol.	Dol.-P[6]
300 g Pellet	C95	51.9	5.1
	C55	26.9	12.2
Mitochondria	C95	46.0	5.9
	C55	20.5	9.1
10.000 g Pellet	C95	139.4	10.5
	C55	101.9	24.8
Microsomes	C95	197.4	27.7
	C55	115.5	50.1
Supernatant	C95	51.8	2.6
	C55	33.6	2.0

The hepatocytes were incubated with dolichol(C95) or dolichol(C55) containing liposomes for 90 min at 37°C. Dolichol and dolichyl-P were separated by chromatography on DEAE-Sephadex.

TABLE 7

Incorporation of [^{3}H]mevalonate into dolichol, dolichyl-P and dolichyl-PP of microsomes in isolated liver cells

	Control	
	cpm	%saturated
Polyprenol 18	7,479	17
Polyprenol 19	6,792	17
Polyprenyl-P-18	5,726	7
Polyprenyl-P-19	5,056	6
Polyprenyl-PP-18	2,845	2
Polyprenyl-PP-19	3,023	3

Cells were incubated with [^{3}H]mevalonate (1 mCi/25 ml) for 1 min.

Mechanism of biosynthesis

Details of the events which occur during condensation are not known and *in vivo* experiments indicate the possibility of more than one pathway in the biosynthetic process. *In vivo* administration of $[^3H]$mevalonate results in the initial phase mainly unsaturated polyprenols, polyprenyl-P and-PP (Table 7). α-Saturation is increased with time in all three lipids indicating that the cell has the capacity to α-saturate both the phosphorylated derivatives and the free alcohol. The distribution of radioactivity further indicates that almost equal parts of the tree lipids are produced. Thus, it is not possible to define a single pathway for biosynthesis of dolichyl-P and it is likely that different pathways are operating for the production of free alcohols and phosphorylated isoprenes.

The functional importance of dolichyl-P

Dolichyl-P is present in both rough and smooth microsomes as well as in other intracellular membranes. The presence of dolichyl-P, however, does not prove its participation in glycosylation processes (Eggens & Dallner, 1982). Incubation of rough microsomes with UDP-GlcNAc or GDP-mannose results in glycosylation of dolichyl-P, dolichyl-PP-oligosaccharide and protein (Table 8). Inhibitors (tunicamycin or amphomycin) completely eliminate lipid carrier

TABLE 8

Glycosylation in rough microsomes

Substr.	Treatment, inhibitor	Dol-P	Dol-PP-O	Protein
		% of control		
UDP-GlcNAc	None	100	100	100
- " -	Tunicamycin	5	17	55
- " -	DABS	35	40	46
GDP-mannose	None	100	100	100
- " -	Amphomycin	2	21	56
- " -	DABS	61	39	58

[1] 10 μg/ml incubation medium

[2] 400 μg/ml incubation medium

TABLE 9

Glycosylation in smooth microsomes

Substr.	Treatment, inhibitor	Dol-P	Dol-PP-P	Protein
		% of control		
UDP-GlcNAc	None	100	100	100
- " -	Tunicamycin[1]	16	26	99
- " -	DABS	46	51	89
GDP-Mannose	None	100	100	100
- " -	Amphomycin[2]	5	29	98
- " -	DABS	76	89	89

[1]10 µg/ml incubation medium

[2]400 µg/ml incubation medium

interaction with sugars and to some extent reduces protein glycosylation. Diaminobenze-sulfonate (DABS) interacts with protein components on the outer cytoplasmic surface of microsomes and interferes with lipid and protein glycosylation.

The situation is quite different with smooth microsomes (Table 9). Tunicamycin and amphomycin as well as DABS inhibits or decreases glycosylation of the two lipid intermediates, but has not effect on protein glycosylation. Dolichyl-P in smooth microsomes has a functional role which differs from that described in rough microsomes where it functions in establishement the core oligosaccharide. It is not known why dolichyl-P at this location is interacting with the local transferases nor whether this interaction is an *in vivo* event or is important in biosynthesis.

Conclusions

Dolichol is present in all or almost all eucaryotic membranes but its function is described for only a minority of these. In the phosphorylated form, it is an obligatory intermediate in the biosynthesis of the N-glycosidically bound oligosaccharides. Its distribution and biosynthesis indicates a compartmentalization and a structural and functional heterogeneity. It is very possible that dolichol is an important membrane component and regulate membrane properties and functions similar to other neutral- and phospholipids.

This work was supported by grants from the Swedish Medical Research Council, the Swedish Council for Planning and Coordination of Research and the Magnus Bergwall Foundation.

REFERENCES

Allen, C.H. Kalin, J.R., Sack, J. & Verizzo, D. (1978). Biochemistry, 17, 5020-5026.

Appelkvist, E.L., Chojnacki, T. & Dallner, G. (1981). Biosci. Rep. 1, 619-626.

Chojnacki, T. & Dallner, G. (1983). J. Biol. Chem. 258, 916-922.

Dallner, G. & Hemming, F.W. (1981). In Mitochondria and Microsomes (Lee, C.P., Schatz, G. & Dallner, G., eds), pp. 655-681, Addison-Wesley, Reading.

Eggens, I., Chojnacki, T., Kenne, L. & Dallner, G. (1983). Biochim. Biophys. Acta, 751, 355-368.

Eggens, I. & Dallner, G. (1982). Biophys. Acta, 686, 77-93.

Ekström, T., Chojnacki, T. & Dallner, G. (1982). J. Lipid Res. 23, 972-983.

Ekström, T., Eggens, I. & Dallner, G. (1982). FEBS Letters, 150, 133-136.

Hemming, F.W. (1974). In Biochemistry of Lipids (Goodwin, T.W., ed) vol. 4, pp 39-97, Butterworths, London.

Struck, D.K. & Lennarz, W.J. (1980). In The Biochemistry of Glycoproteins and Proteoglycans (Lennarz, W.J., ed), pp. 35-73, Plenum Press, New York.

GLYCOSYLTRANSFERASES IN PROTEIN GLYCOSYLATION

ROBERT L. HILL, THOMAS A BEYER,
KEITH R. WESTCOTT, ALLEN E. ECKHARDT
AND ROBERT J. MULLIN

DEPARTMENT OF BIOCHEMISTRY
DUKE UNIVERSITY MEDICAL CENTER
DURHAM, NORTH CAROLINA 27710 U.S.A.

INTRODUCTION

Glycosylation of glycoproteins in eucaryotic cells is initiated on nascent polypeptide chains in the endoplasmic reticulum and completed post-translationally in the lumen of the Golgi apparatus. Recent interest in the details of protein glycosylation stems from our presently incomplete knowledge of the structure, functions and mechanisms of biosynthesis of oligosaccharides in glycoproteins and other glycoconjugates.

Oligosaccharide Structure

Two major types of oligosaccharides exist in glycoproteins, the asparaginyl, or N-linked oligosaccharides, and the seryl/threonyl, or O-linked oligosaccharides (Table I). Each of these types of oligosaccharide has a core structure, which is the branched pentasaccharyl sequence containing three mannose and two N-acetylglucosamine residues in N-linked oligosaccharides, and the disaccharyl sequence, Galβ1,3GalNAc, in O-linked oligosaccharides (Table I). A great diversity of oligosaccharide structures is generated, however, by addition of other monosaccharides of different kinds and in different linkages to the core structures. Thus, N-linked oligosaccharides may have two, three or four branches that extend from the two mannose residues in the core oligosaccharide. Branching of O-linked oligosaccharides also occurs, and may, in some cases, be as complex structurally as that in N-linked oligosaccharides (Slomiany and Slomiany, 1978).

The structural diversity of oligosaccharides is potentially enormous. Only ten monosaccharides exist in mammalian oligosaccharides, but over 800 different, isomeric disaccharide sequences can be generated from the 10 monosaccharides, thus the possible tri-, tetra, penta- and oligosaccharide sequences that may occur is extraordinarily large.

Oligosaccharide Functions

The functions of oligosaccharides in glycoconjugates appear to be many, but one type of function that predominates the thinking in the field, is their role as recognition markers (Neufeld and Ashwell, 1980; Ashwell and Harford, 1982). Thus, oligosaccharides are thought to contain structural information that controls various cellular processes. The structural information inherent in oligosaccharide groups is transmitted from one molecule to another, or from one cell or subcellular organelle to another, by interaction of an oligosaccharide with specific oligosaccharide-binding proteins, or lectins. Several specific examples of oligosaccharide-lectin interactions are listed in Table II. Among the most thoroughly studied processes of this type is carbohydrate-mediated endocytosis of glycoproteins by hepatic receptors formed by lectins. Three different lectins are currently known to be involved in hepatic endocytosis of glycoproteins. The best understood lectin binds galactose or N-acetylgalactosamine at the non-reducing termini of oligosaccharides, whereas another binds N-acetylglucosamine or mannose and a third, fucose or galactose (Ashwell and Harford, 1982). The major function of these lectin/receptor systems that is known at present is in the turnover of plasma glycoproteins, but other functions may be found in the future.

The other lectin-oligosaccharide receptor system that has been studied in detail is involved in the sorting (Pearse and Bretscher, 1981) of lysosomal enzymes from among the many proteins synthesized in the

TABLE I

The oligosaccharide structures of porcine mucin (Aminoff et al., 1979) human glycophorin (Thomas and Winzler, 1969) and human transferrin (Spik et al., 1975) as illustrations of typical N- and O-linked oligosaccharides. The core structures shared by many oligosaccharides are enclosed.

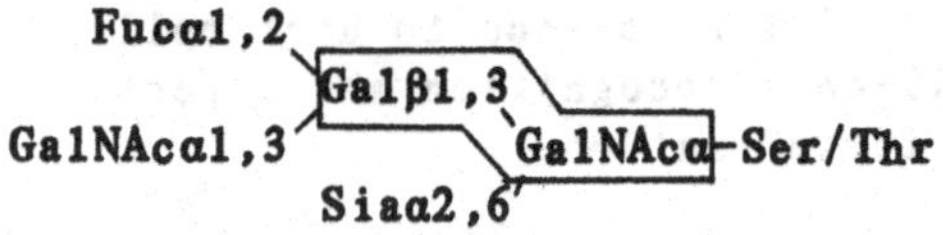

```
Fucα1,2
       \Galβ1,3\
       /        \
GalNAcα1,3       GalNAcα-Ser/Thr
                /
        Siaα2,6
```

porcine mucin

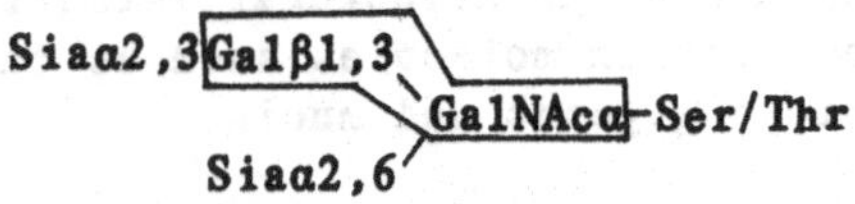

```
Siaα2,3Galβ1,3\
               \
                GalNAcα-Ser/Thr
               /
       Siaα2,6
```

human glycophorin

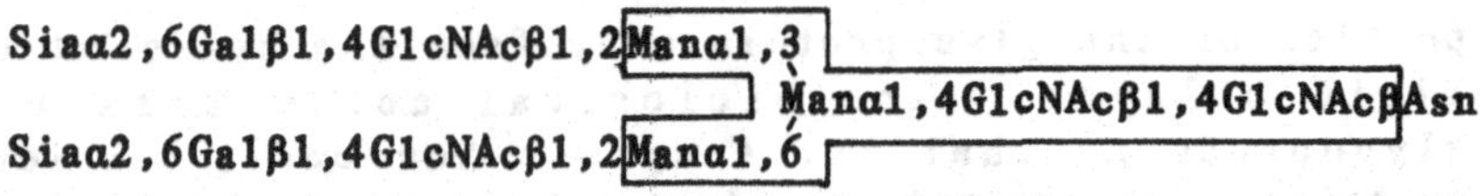

```
Siaα2,6Galβ1,4GlcNAcβ1,2Manα1,3
                               \
                                Manα1,4GlcNAcβ1,4GlcNAcβAsn
                               /
Siaα2,6Galβ1,4GlcNAcβ1,2Manα1,6
```

human transferrin

endoplasmic reticulum and the delivery of lysosomal enzymes to lysosomes. These processes depend upon the interaction of mannose 6-phosphate residues in lysosomal enzymes with a mannose 6-phosphate specific lectin. This system is considered more fully in another chapter of this volume.

Some oligosaccharides have structural roles. Thus, such complex glycoproteins as the proteoglycans (Rodén, 1980) and the mucins (Hill et al., 1977), which contain upward of two-thirds carbohydrate by weight, have properties that are markedly dependent on their carbohydrate content and composition. In membrane-bound glycoproteins with a lower carbohydrate content, the N- or O-linked oligosaccharides may also serve structurally as head groups of these amphiphiles and provide a very hydrophilic domain that can aid in orientation of a

TABLE II

Some Functions of Oligosaccharides in Glycoproteins

1. **Recognition Markers**

   Receptor mediated endocytosis.
   Sorting of lysosomal enzymes.
   Viral adherence to host cells.
   Bacterial adherence to host cells.
   Cell-cell recognition and aspects of fertilization.

2. **Structural Roles**

   Maintain the structural integrity of certain molecules such as the proteoglycans and mucins.

   Serve as hydrophilic head groups in glycoproteins and glycolipids.

portion of the glycoprotein away from the hydrophobic lipid bilayer. The biological activities of glycoproteins that are enzymes, hormones, carrier proteins, etc. are not dependent on the oligosaccharide content of the molecule.

Oligosaccharide Biosynthesis

Different glycosyltransferases act sequentially in oligosaccharide biosynthesis to incorporate one monosaccharide at a time into the growing oligosaccharide chain (Beyer et al., 1981). Two major types of glycosyltransferase are known. The first catalyzes transfer of a monosaccharide from a nucleotide-monosaccharide donor substrate to hydroxyl groups in an acceptor substrate to form glycosides in reactions of the following general form.

Nucleotide-monosaccharide + HO-acceptor ->
Monosaccharide-O-acceptor + nucleotide (1)

The second catalyzes transfer of a monosaccharide from a dolicholphosphoryl-monosaccharide to an acceptor in a similar type of reaction,

Dolicholphosphorylmonosaccharide + HO-acceptor ->
Monosaccharide-O-acceptor + dolicholphosphate (2)

Only nine different nucleotide-monosaccharides and two different dolicholphosphorylmonosaccharides are utilized by mammalian glycosyltransferases, but the number of acceptor substrates is very large. With but few exceptions (Prieels et al., 1981), each of the different transferases utilize only one acceptor substrate, thus, one transferase is required to synthesize each of the different glycosidic bonds in the oligosaccharides of glycoconjugates.

The activities of several glycosyltransferases have been demonstrated in crude tissue preparations, although some activities that must be responsible for synthesis of the variety of different glycosidic linkages known in glycoconjugates have not been examined. Only a few glycosyltransferases have been purified to homogeneity, but on the basis of our knowledge of these few enzymes, some conclusions about the general properties of the glycosyltransferases may be made (Beyer et al., 1981). All transferases are present in very small amounts in tissues, and affinity chromatography is likely essential for their purification. They are membrane-bound glycoproteins, behaving either as integral or peripheral membrane proteins. The enzymic activities of those studied are modulated by detergents or lipids, and some appear to contain a structural domain for binding the lipid bilayer. Some transferases are metalloenzymes, whereas others have no known cofactors. The glycosyltransferases, with the exception of glycogen synthase and lactose synthase, are not regulated by any of the many known molecular mechanisms, such as allosteric control or covalent modification.

The pathway of biosynthesis of the pentasaccharide of porcine submaxillary mucin (Fig. 1) illustrates the synthesis of O-linked oligosaccharides. If the five transferases involved act in the order shown, the pentasaccharide is formed. It is clear, however, that if one transferase acts out of sequence, a product may be formed that cannot be converted ultimately to the pentasaccharide (Beyer et al., 1981). The formation of

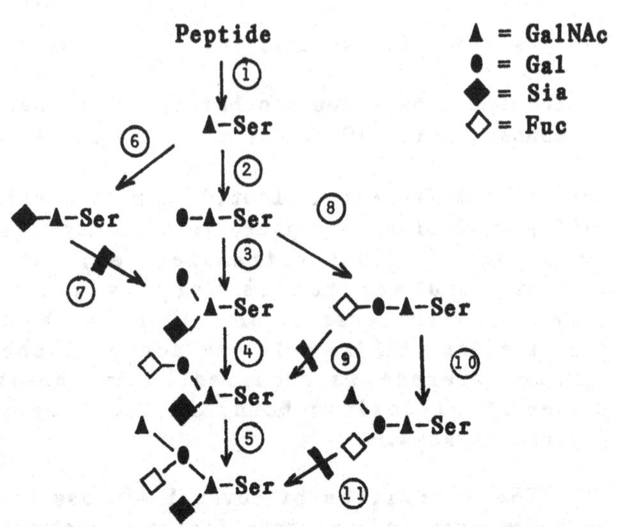

Fig. 1 The biosynthesis of porcine submaxillary mucin. Synthesis of the pentasaccharide proceeds in the order, reactions 1, 2, 3, 4, 5. Other reactions to the right or left give 'dead-end' products that cannot be converted to the pentasaccharide. The bars on reactions 7, 9, and 11 indicate that the reaction is very slow or cannot occur.

such 'dead-end' products may account in part for the structural microheterogeneity of oligosaccharides.

The synthesis of N-linked oligosaccharides (Fig. 2) is more complex than that of O-linked oligosaccharides, and can be considered to occur in four steps (Hubbard and Ivatt, 1981). 1) Synthesis of a dolicholpyrophosphoryl oligosaccharide by the sequential actions of different glycosyltransferases; 2) transfer of the oligosaccharide from the dolichol derivative to the polypeptide chain; 3) processing, or removal of non-reducing terminal monosaccharides from the oligosaccharide; and 4) final elongation and completion of the oligosaccharide by the sequential action of glycosyltransferases. The details of synthesis of tri- and tetra-antennary oligosaccharides are not known, but likely are similar to those of a biantennary chain.

1. **Oligosaccharide Synthesis on Dolicholpyrophosphate.**

Dolichol phosphate (Dol-P) reacts with UDP-GlcNAc to give dolicholpyrophosphoryl GlcNAc, which after 13 consecutive glycosyltransferase reactions gives the following dolicholpyrophosphoryl oligosaccharide.

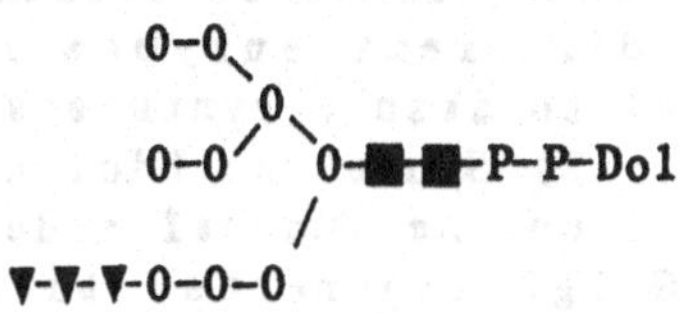

```
        0-0                      0 = Mannose
           \
            0
           / \
        0-0   0-■-■-P-P-Dol      ■ = N-acetylglucosamine
             /
    ▼-▼-▼-0-0-0                  ▼ = Glucose
```

2. **Transfer of Oligosaccharide to Nascent Polypeptide Chain.**

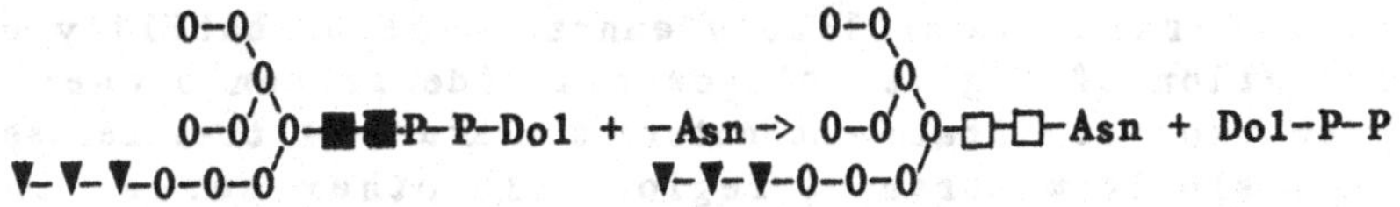

```
      0-0                            0-0
         \                              \
          0                              0
         / \                            / \
      0-0   0-■-■-P-P-Dol + -Asn ->  0-0   0-□-□-Asn + Dol-P-P
           /                              /
▼-▼-▼-0-0-0                     ▼-▼-▼-0-0-0
```

3. **Processing.**

The oligosaccharide on the protein is processed by the sequential actions of glycosidases to give the following structure.

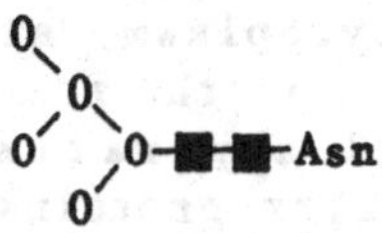

```
    0
     \
      0
     / \
    0   0-■-■-Asn
       /
      0
```

4. **Elongation and Completion of Chain.**

The processed polypeptide is completed after removal of two mannose residues by the action of GlcNAc, Gal (●) and Sia (◆) transferases to give the following biantennary chain.

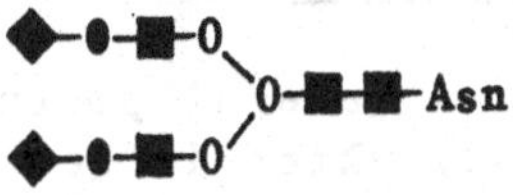

```
◆-●-■-0
       \
        0-■-■-Asn
       /
◆-●-■-0
```

Fig. 2 The Steps in the Pathway of Biosynthesis of N-linked Oligosaccharides.

REGULATION OF OLIGOSACCHARIDE BIOSYNTHESIS

Several factors may aid in regulation of oligosaccharide biosynthesis. 1) There appears to be a high degree of organization of glycosyltransferases and other enzymes involved in oligosaccharide synthesis within the lumen of the endoplasmic reticulum-Golgi apparatus. Thus, many different enzymes may be physically separated so as to assure synthesis of the correct oligosaccharide. 2) Since the biosynthetic enzymes are membrane-bound on the lumenal side of the endoplasmic reticulum-Golgi apparatus, the lipid constituents of the lipid bilayer may exert effects on the catalytic activity of the enzymes, as well as provide the matrix for their organization in membranes. 3) The strict, acceptor substrate specificities of the transferases most likely enhance the probability of formation of a given oligosaccharide sequence when a precursor substrate encounters a mixture of transferases in a single membranous region. In other cases, the substrate specificity may prevent formation of an oligosaccharide sequence. 4) Some evidence suggests that the glycosyltransferases may form multienzyme complexes within the membranes of the endoplasmic reticulum-Golgi apparatus, although direct isolation of such complexes has not been achieved. 5) The nucleotide-monosaccharide donor substrates for the glycosyltransferases are synthesized primarily in the cytoplasm, although one, CMP-sialic acid, is synthesized in the nucleus. Recent evidence suggests that Golgi apparatus contains nucleotide-monosaccharide carrier proteins that aid in transport of the donor substrates from the cytoplasmic to the lumenal side of the Golgi apparatus. Factors that effect the transport systems may also influence oligosaccharide synthesis. In the subsequent sections of this report, each of the forgoing potential regulatory mechanisms will be considered.

Organization of Biosynthetic Enzymes

During the past few years a general picture of protein glycosylation has emerged, which is generally consistent with the intracellular membrane traffic among subcellular organelles as revealed by morphological studies (Farquhar, 1983). Glycosylation of N- and O-linked oligosaccharides commences on the nascent polypeptide chain in the rough endoplasmic reticulum. During N-linked synthesis the oligosaccharide is transferred from the dodicholpyrophosphoryl derivative (Das and Heath, 1980) to an asparaginyl residue in the sequence Asn-X-Ser/Thr (Fig. 2) whereas in O-linked synthesis, N-acetylgalactosamine is transferred from UDP-GalNAc to a seryl or threonyl residue (Strous, 1979). The glycoprotein then migrates through a poorly understood intermediate region between the rough endoplasmic reticulum and Golgi membranes. Processing (Fig. 2) of N-linked oligosaccharides and final elongation and completion of both the N- and O-linked oligosaccharides then takes place on transit through the Golgi apparatus. The completed glycoproteins are eventually sorted on the trans side of the Golgi membranes and packaged for secretion from the cell or for transport to other subcellular organelles. Thus, the requisite glycosyltransferases for glycosylation as well as other enzymes required for posttranslational modification of oligosaccharides appear to reside in different locations within the membranous network of the rough and smooth endoplasmic reticulum and the Golgi apparatus.

The best evidence for separate location of enzymes in the membrane preparations comes from studies on Golgi membranes. Thus, N-acetylglucosamine β1,4 galactosyltransferase (Beyer et al., 1981), which acts late in elongation and completion of oligosaccharide chains, exists in the lightest fractions of Golgi membranes separated on sucrose density gradients (Roth and Berger, 1982). In contrast, mannosidase I, which acts in processing of N-linked oligosaccharides, resides in a more dense fraction of the Golgi membranes (Dunphy et al., 1981). A more extensive study of the Golgi membranes in a mouse lymphoma cell line, as shown in Fig. 3, shows that several different enzyme activities exist in different fractions of membranes obtained by sucrose gradient centrifugation (Goldberg and Kornfeld, 1983). Thus, N-acetylglucosaminylphosphotransferase (Reitman and Kornfeld, 1981), one of the first enzymes to act in lysosomal enzyme processing, was found in the densest membranes and the galactosyl transferase, which acts late in oligosaccharide synthesis was found in the lightest membranes. Two enzymes, phosphodiester glycosidase

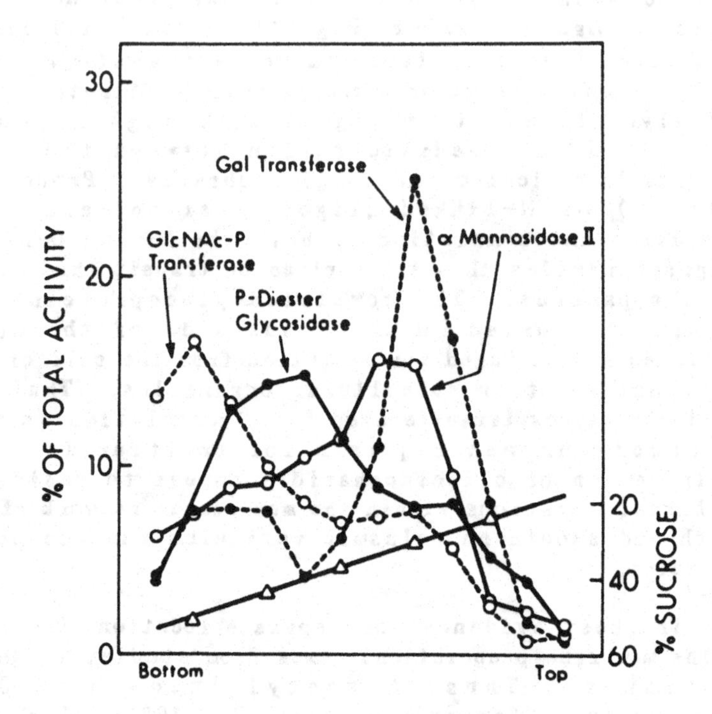

Fig. 3 Distribution of enzymes of oligosaccharide biosynthesis in Golgi membranes separated by sucrose density gradient centrifugation. N-acetylglucosaminylphosphotransferase (o--o), phosphodiester glycosidase (●-●), α-mannosidase II (●---●) and galactosyltransferase (Δ--Δ). From Goldberg and Kornfeld, 1983, with permission.

(Varki and Kornfeld, 1981) and mannosidase II (Tabas and Kornfeld, 1978), which act after the phosphotransferase but before the galactosyltransferase were found in two other fractions of intermediate density. N-Acetylglucosamine α1,6 fucosyltransferase (Wilson et al., 1976), N-acetylglucosamine transferase I (Oppenheimer and Hill, 1981) and N-acetylglucosamine transferase IV (Cummings et al., 1982), which act late in oligosaccharide synthesis, are located in fractions slightly more dense than the galactosyltransferase and in the same membrane fraction as mannosidase II.

The foregoing observations suggest that there must be some mechanism, yet unknown, for allowing the different, newly synthesized Golgi enzymes to reach a specific region of Golgi membranes. Studies on the lipid composition of rat liver membranes show that there is a cholesterol concentration gradient extending from the endoplasmic reticulum (low) to the Golgi to the plasma membrane (high). The different cholesterol contents may provide, in part, the different densities of Golgi membranes separated by sucrose density centrifugation, but it is not known whether cholesterol or any other lipid aids in localization of different transferases in different regions of the Golgi apparatus.

Effects of Phospholipids on Glycosyltransferases

Lipids are known to modulate the activity of membrane bound enzymes (Sanderman, 1978), and recent studies with bovine milk N-acetylglucosamine β1,4 galactosyltransferase (Mitranic and Moscsarello, 1980), rat liver glucuronyltransferase (Hochman and Zakim, 1983) and porcine submaxillary galactose α1,3 sialyltransferase (Sadler et al., 1979) show this to be true for glycosyltransferases. The effects of lipids on the catalytic activity of the α2,3 sialyltransferase provide an interesting model for examining the possible regulation by lipids of late stages of oligosaccharide biosynthesis. This sialyltransferase has been purified to homogeneity and consists of two non-interconvertable forms, designated A and B. The two forms can be separated by gel filtration in Triton X-100, and have Stokes radii of 51Å (A) and 31Å (B). Both forms, however, have molecular weights of about 50,000 as estimated by gel electrophoresis in sodium dodecylsulfate. Form A binds a single micelle of Triton X-100, whereas form B does not bind detergent, accounting for the differences in the Stokes radii of the two forms. Since both forms have the same substrate specificity and identical specific activities in Triton X-100, the B form has been proposed to be a proteolytic fragment of the A form and is thought to lack a small lipid binding domain that anchors the A form to detergents or lipids. Indeed, the two forms are very closely related structurally as judged by peptide mapping (Wolf et al., 1983). This proposal receives support from the additional observation that the activity of the A form is reduced by about 60% in the absence of Triton X-100 but is regained on readdition of Triton X-100 at concentrations above its critical micelle concentration. The activity of the B form is unaffected by Triton X-100.

Further studies (Westcott and Hill, 1982) have shown that the activity of the sialyltransferase is modulated by other detergents as well as by phospholipids. Different types of nonionic detergents affected the A form of the transferase differently. Under standard assay conditions Triton X-100 stimulated the activity of form A about three fold to give the same activity as the B-form in the absence of Triton X-100. Other types of polyoxyethylene ether detergents were less effective in stimulating the A form and octyl-β-D-glucoside and deoxycholate inactivated the enzyme. Moreover, of several lysophosphatides tested, only lysophosphatidyl choline (LPC) stimulated activity and the extent of stimulation was about that obtained with Triton X-100. Lysophosphatidylethanolamine had no effect on activity and lysophosphatidylserine and lysophosphatidylglycerol inhibited the A form of the enzyme. Lauroyl LPC did not stimulate the activity of form A, although myristoyl LPC and palmitoyl LPC gave about 2.5 fold stimulation and stearoyl LPC and oleoyl LPC gave about 3 fold stimulation. The B form was unaffected by LPC irrespective of the fatty acyl chain length, although like the A form, it was inhibited by lysophosphatidyl glycerol and octylglucoside. The A form of the sialyltransferase, but not the B form, could also be incorporated into phosphatidylcholine vesicles as well as into liposomes formed by the mixed phospholipids isolated from porcine submaxillary gland Golgi membranes. Liposome bound transferase activity was about that observed on stimulation with LPC.

Kinetic studies reveal that Triton X-100, lysophosphatidylcholine and phosphatidylcholine vesicles affect both the K_m and V_{max} of the transferase (Table III). The K_m for CMP·sialic acid is decreased 3 fold by both Triton X-100 and LPC whereas V_{max} is increased over

TABLE III

The Effects of Detergents on the Kinetic Properties of the A and B Forms of Galactose α2,3 Sialyltransferase

Kinetic Parameter	A (no detergent)	A (1% Triton X-100)	A (1% LPC)	B
V_{max} (μmol-min^{-1}-ml^{-1})	0.046	0.097	0.091	0.071
K_m Galβ1,3GalNAc (mM)	0.25	0.28	0.28	0.30
K_m CMP-Sia (μM)	12.3	4.38	4.34	4.82

2 fold by both detergents. The K_m for the acceptor substrate (Galβ1,3GalNAc) is unaffected.

The structural basis for the activating effects of lipids remains to be established, but the foregoing results suggest that lipids, on reaction with a lipid-binding domain, change the transferase structurally so as to increase its affinity for CMP-sialic acid and enhance its turnover of substrates (V_{max} effect). In the absence of lipid, the lipid-binding domain exerts a negative influence on the activity, since the B form, lacking the domain, is locked into an activity state equal to that of the A form bound to activating lipids.

Substrate Acceptor Specificities of Glycosyltransferases

If several glycosyltransferases are present in the same membranous compartment and can compete with one another for the same substrate, then it is difficult to understand how oligosaccharides of a single desired structure on a glycoprotein are synthesized with great fidelity. Studies with several highly purified glycosyltransferases illustrate this problem (Beyer et al., 1981). For example, the β1,4 galactose α2,6 sialyltransferase and the N-acetylglucosamine α1,3 fucosyltransferase, both of which act in the terminal stages of oligosaccharide synthesis utilize the same acceptor substrate as follows (Beyer et al., 1979).

CMP-Sia + Galβ1,4GlcNAc->Siaα2,6Galβ1,4GlcNAc + CMP (3)

GDP-Fuc + Galβ1,4GlcNAc->Galβ1,4(Fucα1,3)GlcNAc + GDP (4)

The products of both reactions could potentially be substrates for the other transferase, but in fact, Siaα2,6Galβ1,4GlcNAc is not an acceptor for the fucosyltransferase and Galβ1,4(Fucα1,3)GlcNAc is not an acceptor for the sialyltransferase. Thus, the two enzymes cannot act sequentially, and the action of one prevents the action of the other. In other cases, the action of one transferase enhances the rate of another. In this example, two sialyltransferases which use the same acceptor substrate catalyze the following reactions.

CMP-Sia + Galβ1,3GalNAc -> Siaα2,3Galβ1,3GalNAc (5)

CMP-Sia + Galβ1,3GalNAc -> Galβ1,3(Siaα2,6)GalNAc (6)

The products of reactions 5 and 6 are acceptor substrates for the other sialyltransferases, as follows.

CMP-Sia + Galβ1,3(Siaα2,6)GalNAc ->

Siaα2,3Galβ1,3(Siaα2,6)GalNAc + CMP (7)

CMP-Sia + Siaα2,3Galβ1,3GalNAc ->

Siaα2,3Galβ1,3(Siaα2,6)GalNAc + CMP (8)

But the inital rate of reaction 8 is 4 to 5 fold that of reactions 5 and 6 at the same enzyme concentrations, and 100 to 200 fold that of reaction 7. Thus, the prior action of the 2,6 sialyltransferase inhibits the subsequent action of the 2,3 transferase, but the prior action of the 2,3 transferase enhances considerably that of the 2,6 transferase.

It is possible that the competing effects of two glycosyltransferases in the same membranous compartment are circumvented by differences in the concentration of the two enzymes. This is unlikely to always be the case, however, since glycoproteins are known to have oligosaccharide structures that reflect the apparent activities of two competing transferases. For example, human lactoferrin contains the typical N-linked oligosaccharide shown in Table I, with one non-reducing branch terminating in the Siaα2,6Galβ1,4GlcNAc sequence and the other in the Galβ1,4(Fucα1,3)GlcNAc sequence. In addition, porcine submaxillary mucin contains each of the 'dead-end' products listed in Fig. 1. Thus, the amounts of Siaα2,6GalNAc, Fucα1,2Galβ1,3GalNAc and Fucα1,2(GalNAcα1,3)Galβ1,3GalNAc have been estimated to be 12, 5 and 15%, respectively, of the total oligosaccharides in this mucin (Aminoff et al., 1979).

The substrate specificities of the glycosyltransferases that add monosaccharides at the non-reducing termini of N-linked oligosaccharidses also can give rise to different sequences in the two arms of the biantennary chains (Beyer et al., 1979). This is illustrated by the extent of incorporation of fucose and sialic acid into the glycopeptides of asialotransferrin (Table I) by the two fucosyl and two sialyltransferases listed in Table IV. Thus, all four enzymes have a high degree of preference for incorporating monosaccharides into the arm in the sequence Galβ1,4GlcNAcβ1,2<u>Manα1,3</u>Man- rather than the arm in the sequence Galβ1,4GlcNAcβ1,2<u>Manα1,6</u>Man-. This suggests that the α1,3 linked arm is accomodated into the substrate binding region of the catalytic site better than the α1,6 linked arm. Indeed, this type of effect may account for the fact that the β1,4 galactose α2,6 sialyltransferase cannot transfer sialic acid from CMP-Sia into all of the potential Galβ1,4GlcNAc sites of asialoorosomucoid (Paulson et al., 1977). Only about 50% of the sites can be sialylated by the α2,6 transferase <u>in vitro</u>, but after sialyation of 50% of the sites, the β1,4(3) galactose α2,3 sialyltransferase, which is not as discriminating in its preference for the α 1,3 linked arm as the α2,6 sialyltranferase (Table IV), readily sialylates the remaining 50% of the Galβ1,4GlcNAc sites (Weinstein et al., 1982).

It remains to be established how many transferases show a preference for glycosylation of the α1,3 arm of N-linked oligosaccharides. It may be that only sialyl and fucosyltransferases show this high degree of specificity, since the N-acetylglucosamine β1,4 galactosyltransferase (Table IV) appears to galactosylate the α1,6 linked arm equally as well as the α1,3 linked

TABLE IV

The extent of glycosylation of different branches of the biantennary oligosaccharides of transferrin glycopeptides.

Transferase	Extent of Total Glycosylation of Branch (%)		
	α1,3	α1,6	α1,3 + α1,6
α2,6 Sialyl (1)	86	4	10
α2,3 Sialyl (2)	49	30	21
α1,2 Fucosyl (3)	49	34	17
α1,3 Fucosyl (4)	53	24	23
β1,4 Galactosyl (5)	34	40	26

The acceptor substrate for transferases (1) to (4) was

Galβ1,4GlcNAcβ1,2Manα1,3╲
Manβ1,4GlcNAcβ1,4GlcNAcβAsn
Galβ1,4GlcNAcβ1,2Manα1,6╱

and for transferase (5) the same structure devoid of terminal galactose. The linkages synthesized were as follows: (1) Siaα2,6Gal...; (2) Siaα2,3Gal...; (3) Fucα1,2Gal...; (4) Galβ(Fucα1,3)GlcNAc...; (5) Galβ1,4GlcNAc. From T.A. Beyer and R. L. Hill, unpublished observations.

arm in the glycopeptides of asialoagalactotransferrin.

Differential glycosylation of oligosaccharides at their non-reducing termini and the formation of 'dead end' products appear to be rather extensive in glycoproteins, and reflect a lack of regulation of transferase activity rather than the fine control that is observed with so many biosynthetic enzymes. These effects also provide a mechanism for generation of the microheterogeneity in oligosaccharide structure that has been observed for years and perplexed many investigators. Structural microheterogeneity still remains a puzzle, however, when the functions of some glycoproteins are considered. It is not a particularly difficult problem to rationalize for glycoproteins such as the mucins, since it appears that the oligosaccharides in these molecules provide the viscoelastic gel properties that are required for the mucins to serve as biological lubricants. Indeed, it is the sialic acid content of the oligosaccharides that gives the mucins their gel-like quality and high viscosity in solution, and as long as a substantial fraction of the oligosaccharides in a mucin are sialylated (about 40-50% in porcine submaxillary mucin), then the mucins will have their required biological properties and functions. If the oligosaccharides are to serve as recognition markers, however, and require specific oligosaccharide structures for binding to lectins, then microheterogeneity is more difficult to understand. Perhaps the interaction of oligosaccharides with lectins is a threshold type phenomenon, as seems to be the case for the hepatic lectin that binds galactosides and N-acetylgalactosides. Thus, strong binding of an oligosaccharide with lectin occurs only after several monosaccharides of the requisite type are available in a glycoprotein. This would permit oligosaccharides to express themselves as recognition markers in the face of considerable structural microheterogeneity.

Multienzyme Complexes

Multienzyme complexes provide considerable control over the nature of the final product they form, since the product of each enzyme in the complex would not be expected to dissociate before being acted upon by the next enzyme in the complex. Glycosyltransferases have been suggested to form multienzyme complexes, and the fact that they exist together in the same membranous region of the Golgi apparatus, as discussed above, indicates that they have the potential to form such complexes. Were such complexes to exist, then the problem of different transferases acting on the same acceptor substrate and thereby preventing subsequent actions of the transferases required to form the desired oligosaccharide structure, would be circumvented. On the other hand, of the several glycosyltransferases purified to homogeneity, none has been found to exist in a complex.

The best evidence that glycosyltransferases associate to form a complex was obtained with the core protein xylosyltransferase and the galactosyltransferase that act consecutively to form the Galβ1,4Xylα-O Ser sequence in the biosynthesis of chondroitin sulfate (Schwartz and Rodén, 1975). Thus, the galactosyltransferase was found to specifically adsorb to a xylosyltransferase that was adsorbed to a core protein-agarose affinity matrix. The galactosyltransferase did not, however, adsorb to the core protein-agarose matrix in the absence of xylosyltransferase. Moreover, the galactosyltransferase could be desorbed from the xylosyltransferase saturated core protein-agarose matrix by a nonionic detergent and salt, but not by salt alone. The galactosyltransferase also adsorbed to xylosyltransferase coupled to cyanogen bromide activated agarose. These observations indicate that the two enzymes have the capacity to associate and can be dissociated by neutral detergents that are used in purification of the enzymes. It is possible that the detergents which are used to solubilize the transferases also disrupt any transferases that are associated together within the membrane, thereby preventing isolation of any complexes that may exist *in vivo*.

Athough the foregoing studies on the transferases of chondroitin sulfate synthesis do not establish unequivically that multiglycosyltransferase complexes exist, other studies purporting to demonstrate such complexes are much less conclusive. Indeed the report that a complex is formed between N-acetylglucosaminyltransferases I and II (Mendicino et al., 1981) was subsequently shown to be incorrect (Oppenheimer et al., 1981). Other studies (Ivatt, 1981) in which both a galactosyl and a N-acetylglucosaminyltransferase were incorporated into the same egg white lecithin liposomes, showed that these two enzymes could synthesize N-acetyllactosamine quite efficiently, as would be expected for a complex of the two enzymes. These studies did not, however, establish that the two enzymes formed a complex. Additional studies of this kind with several different transferases that would be expected to form more complex oligosaccharide structures may give more conclusive evidence for complex formation.

Transport of Sugar Nucleotides

CMP-N-acetylneuraminic acid is synthesized in the nucleus whereas the other nucleotide sugars required for oligosaccharide biosynthesis are formed in the cytoplasm (Coates et al., 1980). Since the glycosyltransferases are on the lumenal side of the endoplasmic reticulum-Golgi apparatus, some mechanism must exist for transport of the nucleotide sugars from the cytoplasm to their sites of utilization. Recent studies (Sommers and Hirschberg, 1982) strongly support the view that transport is mediated by sugar nucleotide carrier proteins, thus oligosaccharide synthesis could be regulated in part by factors influencing the transport mechanisms.

The transport of CMP-N-acetylneuraminic acid and GDP-fucose has been examined in 'right side out' vesicles prepared from rough and smooth endoplasmic reticulum and Golgi apparatus. Such vesicles have the same orientation as *in vivo*, with the lumen of the vesicle corresponding to the lumen of the endoplasmic reticulum and the Golgi membranes. The transport of CMP-N-acetylneuraminic acid and GDP-fucose was found to occur only in Golgi-derived vesicles, in accord with the fact that sialyl and fucosyltransferases act late in oligosaccharide synthesis in Golgi membranes. The transport into vesicles is temperature-dependent, saturable and inhibited by analogs of the sugar nucleotides. That carrier proteins mediate transport is strongly suggested by the fact that protease treatment of the Golgi vesicles also inhibits nucleotide sugar transport. CMP-N-acetylneuraminic acid and GDP-fucose are transported intact and are utilized by glycosyltransferases within the lumen for glycosylation of endogenous glycoprotein acceptors, although considerable hydrolysis into nucleotide and monosaccharide occured under the experimental conditions used. The transport of one sugar nucleotide did not compete with transport of another sugar nucleotide, suggesting that transport is mediated by different proteins, each specific for a given nucleotide sugar. The V_{max} for CMP-N-acetylneuraminic acid transport is about 10 times that of GDP-fucose, although the K_m for both was about the same (K_mCMP-NeuAc = 2.5 μm and to K_mGDP-Fuc = 7.5 μm).

Studies (Kuhn et al., 1980) with rat mammary gland Golgi membranes suggested that UDP-galactose transport was also mediated by a carrier protein, but evidence for facilitated transport of the other nucleotide sugars is lacking. It will be of interest, however, to learn whether transport occurs in vesicles other than those of Golgi, since the rough and smooth endoplasmic reticulum are sites where the early stages of glycosylation occur that require GDP-mannose, UDP-glucose, UDP-N-acetylglucosamine and UDP-N-acetylgalactosamine.

References

Aminoff, D., Gathmann, W.D. and Baig, M.M. (1979) J. Biol. Chem., 214, 8909-8913.

Ashwell, G. and Harford, J. (1982) Annu. Rev. Biochem., 51, 531-554.

Beyer, T.A., Rearick, J.I., Paulson, J.C., Prieels, J.-P., Sadler, J.E., and Hill, R.L. (1979) J. Biol. Chem. 254, 12531-12541.

Beyer, T.A., Sadler, J.E., Rearick, J.I., Paulson, J.C. and Hill, R.L. (1981) Adv. Enzymol., A. Meister, ed., 52, 23-175.

Coates, S.W., Gurney, Jr., T., Sommers, L.W., Yeh, M. and Hirschberg, C.B. (1980) J. Biol. Chem. 255, 9225-9229.

Cummings, R.D., Trowbridge, I.S. and Kornfeld, S. (1982) J. Biol. Chem. 257, 13421-13427.

Das, R.C. and Heath, E.C. (1980) Proc. Natl. Acad. Sci., USA, 77, 3811-3815.

Dunphy, W.G., Fries, E., Urbani, I.J. and Rothman, J.E. (1981) Proc. Natl. Acad. Sci., USA, 78, 7453-7457.

Farquhar, M.G. (1983) Federation Proc. 42, 2407-2413.

Goldberg, D.E. and Kornfeld, S. (1983) J. Biol. Chem. 258, 3159-3164.

Hill, H.D., Reynolds, J.A., and Hill, R.L. (1977) J. Biol. Chem. 252, 3791-3798.

Hochman, Y. and Zakim, D. (1983) J. Biol. Chem. 258, 4143-4146.

Hubbard, S.C. and Ivatt, R.J. (1981) Annu. Rev. Biochem., 50, 555-583.

Ivatt, R.J. (1981) Proc. Natl. Acad. Sci., USA, 78, 4021-4025.

Kuhn, N.J., Wooding, F.B.P. and White, A. (1980) Eur. J. Biochem. 103, 377-385.

Mendicino, J., Chandrasekaran, E.V., Anumula, K.R. and Davila, M. (1981) Biochemistry, 20, 957-976.
Mitranic, M.M. and Moscarello, M.A. (1980), Can. J. Biochem., 58, 809-814.
Neufeld, E. and Ashwell, G. (1980) in 'The Biochemistry of Glycoproteins and Proteoglycans', W.J. Lennarz, ed., Plenum Press, New York, pp. 241-266.
Oppenheimer, C.L., Eckhardt, A.E. and Hill, R.L. (1981) J. Biol. Chem. 256, 11477-11482.
Oppenheimer, C.L. and Hill, R.L. (1981) J. Biol. Chem. 256, 799-804.
Paulson, J.C., Rearick, J.I. and Hill, R.L. (1977) J. Biol. Chem. 252, 2363-2371.
Pearse, B.M.F. and Bretscher, M.S. (1981) Annu. Rev. Biochem. 50, 85-101.
Prieels, J.-P., Monnom, D., Dolmans, M., Beyer, T.A., and Hill, R.L. (1981) J. Biol. Chem. 256, 10456-10463.
Reitman, M.L. and Kornfeld, S. (1981) J. Biol. Chem. 256, 4275-4281.
Rodén, L. (1980) in 'The Biochemistry of Glycoprotein and Proteoglycans', W.J. Lennarz, ed., Plenum Press, New York, pp. 267-371.
Roth, J. and Berger, E.G. (1982) J. Cell Biol. 92, 223-229.
Sadler, J.E., Rearick, J.I., Paulson, J.C. and Hill, R.L. (1979) J. Biol. Chem. 254, 4434-4443.
Sanderman, Jr., H.(1978) Biochim. Biophys. Acta 515, 209-237.
Schwartz, N.B. and Rodén, L. (1975) J. Biol. Chem. 250, 5200-5207.
Slomiany, A., Slomiany, B.L. (1978) J. Biol. Chem. 253, 7301-7306.
Sommers, L.W. and Hirschberg, C.B. (1982) J. Biol. Chem. 257, 10811-10817.
Spik, G., Bayard, B., Fournet, B., Strecker, G., Bouquelet, S., and Montreuil, J. (1975) FEBS Lett., 50, 296-299.
Strous, G.J.A.M. (1979) Proc. Natl. Acad. Sci., USA, 76, 2694-2698.
Tabas, I. and Kornfeld, S. (1978) J. Biol. Chem. 253, 7779-7786.
Thomas, D.B. and Winzler, R.J. (1969) J. Biol. Chem. 244, 5943-5946.
Varki, A. and Kornfeld, S. (1981) J. Biol. Chem. 216, 9937-9943.

Weinstein, J., de Souza-e-Silva, U. and Paulson, J.C. (1982) J. Biol. Chem. 257, 13845-13853.

Westcott, K.R. and Hill, R.L. (1982) Fed. Proc. 41, 662.

Wilson, J.R., Williams, D. and Schachter, H. (1976) Biochem. Biophys. Res. Commun., 72, 909-916.

Wolf, C., Westcott, K.R., and Hill R.L., unpublished observations.

SYNTHESIS AND POSSIBLE ROLE OF O- AND N-LINKED OLIGOSACCHARIDES OF YEAST GLYCOPROTEINS

W. Tanner, A. Haselbeck, H. Schwaiger,
E. Arnold and P. Orlean
Institut für Botanik der Universität
Regensburg
8400 Regensburg, FRG

All dolichol-dependent glycosyl transfer reactions of O- and N-glycosylation in yeast proceed at the ER. The mannosyl transferase forming Dol-P-Man from GDP-Man and Dol-P also translocates the mannosyl residue across a liposomal membrane and transfers it to internal GDP. When $[^{3}H]$GDP-$[^{14}C]$Man is used as external substrate in this reaction, only ^{14}C-radioactivity is internalized into the liposome.

Yeast vacuolar enzymes like carboxypeptidase do not require the carbohydrate portion as sorting signal to be segregated into the vacuole. N-glycosylation, however, is obligatorily required for growth and division: cells do not enter S-phase, when N-glycosylation is prevented. α-Factor, which arrests mating type _a_ cells in G1, also is a potent inhibitor of glycoprotein formation.

1. INTRODUCTION

Although an ever increasing number of biologically interesting molecules turn out to be glycoproteins, we still do not have a clue as to what the crucial general importance of glycosylation is (1,2). Probably the most exciting discovery related to this problem, the mannosyl-6-phosphate sorting signal of lysosomal enzymes (3), seems now to be restricted to fibroblasts (4-7).

We have recently obtained evidence that during the Saccharomyces cerevisiae cell cycle, protein N-glycosylation is required for G1/S-phase transition (2,8,9). These results will be summarized in this paper. First, however, the state of knowledge of glycoprotein synthesis in yeast will be reviewed, especially the problem of transmembrane translocation of activated glycosyl moieties.

2. RESULTS AND DISCUSSION

Glycoprotein Synthesis

Dolichyl monophosphate mannose, first described from yeast cells (10,11), serves as mannosyl donor for O- and N-linked oligosaccharides of yeast glycoproteins. The O-linked carbohydrate chains of S.cerevisiae glycoproteins are four to five mannoses long (12,13) and they are synthesized by the following reaction sequence (14,15):

$$\text{GDP-Man} + \text{Dol-P} \xrightleftharpoons{Mg^{++} \text{ or } Mn^{++}} \text{Dol-P-Man} + \text{GDP}$$

$$\text{Dol-P-Man} + \text{Protein(Ser/Thr)} \xrightarrow{Mg^{++} \text{ or } Mn^{++}} \text{Protein(Ser/Thr)-Man} + \text{Dol-P}$$

$$\text{GDP-Man} + \text{Protein(Ser/Thr)Man} \xrightarrow{Mn^{++}} \text{Protein(Ser/Thr)-Man-Man}_{1-4} + \text{GDP}$$

An analoguous reaction sequence has been shown to proceed in a number of fungal cells (16,17), but a participation of dolichyl intermediates in the biosynthesis of O-linked saccharides in any other group of organisms has so far not been reported.

N-glycosylation in yeast cells proceeds in an identical manner (17-19) to core glycosylation in mammalian cells (20,21). Only the trimming is reduced to the removal of the 3 glucosyl units (17,19) and of one mannosyl residue (22). After trimming, the asparagine linked oligosaccharide, therefore, is of the formula $(GlcNAc)_2Man_8$ and this then gets extended to the large polymannose chains as found for example in invertase (13,17,19).

The intracellular localization of the glycosylation reactions is also clear by now. All dolichol dependent steps, and only these, occur at the endoplasmic reticulum (23-25). This is also true for O-glycosylation in

Table 1

Incorporation of [2-^{3}H]mannose into N- and O-linked carbohydrate chains of sec 18 sells at permissive and non-permissive temperature. Cells, grown at 25°C, were preincubated for 30 min at 25 or 37°C, whereupon [2-^{3}H]mannose was added. Experimental details are given in ref. 25.
M_1 = mannose, M_2 = mannobiose, M_3 = mannotriose,
M_4 = mannotetraose, M_5 = mannopentaose.

Carbohydrate Fraction	Incubation temperature	
	25°C	37°C
	counts/min x 10^{-2}	
N-linked oligosaccharides	7 543	434
O-linked sugars (total)	411	1 060
M_1	64 (15%)	934 (88%)
M_2	197 (48%)	120 (11%)
M_3	73 (18%)	6 (1%)
M_4	52 (13%)	0
M_5	25 (6%)	0

S.cerevisiae, where only the first mannose is attached to the protein, when the transfer of that protein to the Golgi is prevented. This is observed when the sec 18 (temperature-sensitive, secretory) mutant obtained by Schekman and co-workers (26) is radiolabelled in vivo with 2-[^{3}H]mannose at the non-permissive temperature of 37°C (Table 1). Although some O-linked disaccharide is formed at 37°C, the radioactivity in the monosaccharide increases more than ten-fold (and 6-fold in relative terms) whereas that in the disaccharide is decreased considerably. Thus, the attachment of the first O-linked mannosyl unit most likely is a cotranslational event at the ER (27). The asparagine-linked core-oligosaccharide is also transferred at the ER (23,24), whereas the extension of the O-linked mannose and the N-linked $(GlcNAc)_2Man_8$ proceeds at the Golgi (23); no evidence for a mannosyl transfer at or outside the cytoplasmic membrane is available (23,25). Trimming of the glucosyl units and of the single mannose residue most likely proceeds at the ER (17,22, Schekman, pers. communication).

The question, whether the dolichyl intermediates somehow are transmembrane transport vehicles, has been a long standing one. N-glycosylation of protein proceeds on the luminal side of the ER and also the immediate precursor, the Dol-PP-$(GlcNAc)_2Man_9Glc_3$ is oriented towards the ER lumen (28-30). In addition, it seems likely that some of the glycosyl transferases catalyze the glycosylation of dolichylphosphate and its transmembrane translocation simultaneously (1,29,31). However, whereas Lennarz et al. (29) believe that the Dol-PP-$(GlcNAc)_2$ disaccharide already faces the ER lumen, we prefer the hypothesis that only the Dol-P dependent extensions of the Dol-PP-oligosaccharide take place in the ER lumen (31). In this way, a clear separation of the sugar nucleotide dependent synthesis of the Dol-PP-heptasaccharide on the cytoplasmic ER surface from the Dol-P requiring extension steps on the luminal side would be achieved. Evidence for a Dol-P dependent translocation of mannosyl residues through liposomal membranes has been reported (31). However, these experiments could not conclusively exclude the possibility of a GDP/GDP-Man exchange reaction. To rule out this possibility a double-labelling experiment with $[^3H]$GDP-$[^{14}C]$Man has now been carried out. As shown in Table 2, the mannosyl transferase translocated $[^{14}C]$mannosyl residues (without $[^3H]$GDP) through Dol-P containing membranes and attaches them to internal non-radioactive GDP according to

$$[^3H]\text{GDP-}[^{14}C]\text{Man}_{out} + \text{GDP}_{in} \xrightarrow{Mg^{++}} [^3H]\text{GDP}_{out} + \text{GDP-}[^{14}C]\text{Man}_{in}$$

("Out" and "in" refers to outside and inside the liposomes).

Table 2

Dolichyl phosphate dependent mannosyl transfer from external $[^3H]$GDP-$[^{14}C]$Man to GDP within liposomes. Liposomes containing Dol-P and purified yeast Dol-P-Man synthase were prepared according to ref. 31. The liposomes preloaded with GDP were incubated with doubled labelled GDP-Man for 2 h, then separated from external medium by the method of Penefsky, H.S. (1977) J.Biol.Chem. 252, 2891-2899.

$[^3H]$GDP-$[^{14}C]Man_{outside}$			Dol-P-Man formed			$GDP-Man_{in}$ formed		
counts/min		$^{14}C/^3H$	counts/min		$^{14}C/^3H$	counts/min		$^{14}C/^3H$
^{14}C	^3H		^{14}C	^3H		^{14}C	^3H	
102600	28100	3.6	6918	13	516	716	2	358

In the artificial test system (31, and Table 2) the formation of internal GDP-Man is used to measure mannosyl translocation. Although this reaction might also occur in vivo, for example to form sugar nucleotides in the lumen of the Golgi cisternae, the Dol-P-Man facing the ER-lumen is thought to undergo either of two reactions (Fig. 1): it donates the sugar either to a Ser/Thr-residue of a protein to be O-glycosylated or to a Dol-PP-$(GlcNAc)_2Man_5$ to form Dol-PP-$(GlcNAc)Man_9$.

The Role of Protein Glycosylation in the Life Cycle of a Yeast Cell

S.cerevisiae contains glycoproteins located in the cell wall such as invertase, and intracellular ones like the vacuolar carboxypeptidase Y. An inhibition of N-glycosylation by tunicamycin prevents formation of external inver-

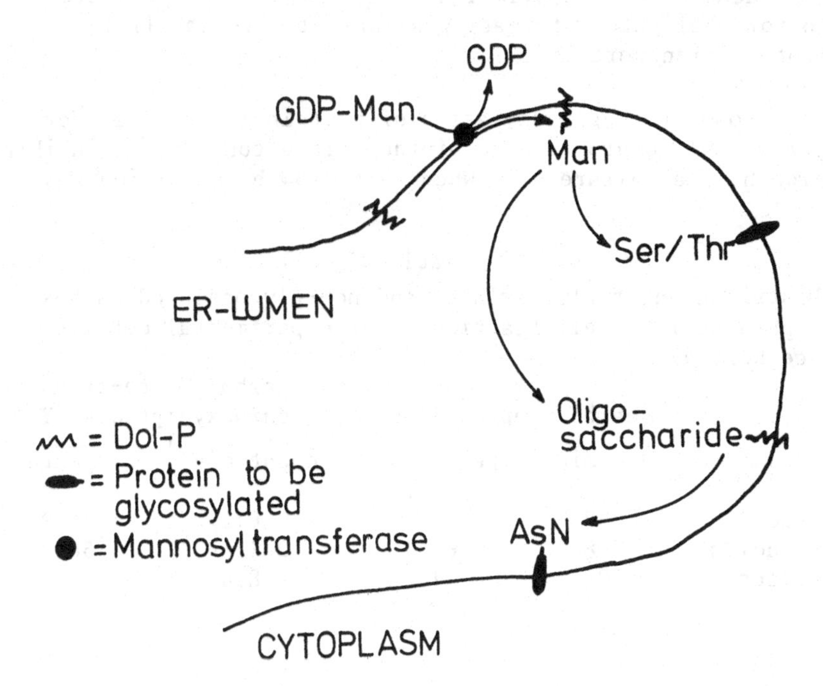

Fig. 1. Formation of Dol-P-Man as intraluminal mannosyl donor for O- and N-glycosylation.

tase (32). Yeast cells growing on sucrose stop growth in the presence of 2-4 μg of tunicamycin per ml of medium (1). Since, however, the same growth inhibition was observed with glucose as substrate (1), the inhibitory effect of tunicamycin could not have been due to the prevention of invertase synthesis.

The yeast vacuole on the other hand is an organelle functionally equivalent to the animal cell lysosome (33). Vacuolar enzymes like carboxypeptidase Y are glycoproteins and even contain phosphomonoester and phosphodiester groups within the oligomannose moiety (4,34). In view of the mannosyl-6-phosphate sorting signal of lysosomal enzymes (3,35) it was natural to speculate that this was also a signal in the case of yeast vacuolar enzymes. In the presence of tunicamycin, carbohydrate-free carboxypeptidase Y is synthesized (36). Where is this protein localized in the cell? In Table 3 data from reference (4) are summarized and show that carbohydrate-free enzyme is still put into the vacuole. N-glycosylation, therefore, is not required to form obligatory targeting signals for intracellular protein transport.

However, N-glycosylation of protein is required for a yeast cell to grow and to divide, since tunicamycin inhibits growth of a culture (1). When this growth inhibition was

Table 3
Distribution of glycosylated and non-glycosylated carboxypeptidase Y in cell fractions (for experimental details see Ref. 4).

	carboxypeptidase Y		carbohydrate-free carboxypeptidase Y	
	count min^{-1}	percentage	count min^{-1}	percentage
cytosol	768	18.3	612	16.9
vacuoles	2 598	62.1	2 126	58.7
pellet	894	21.4	854	23.4

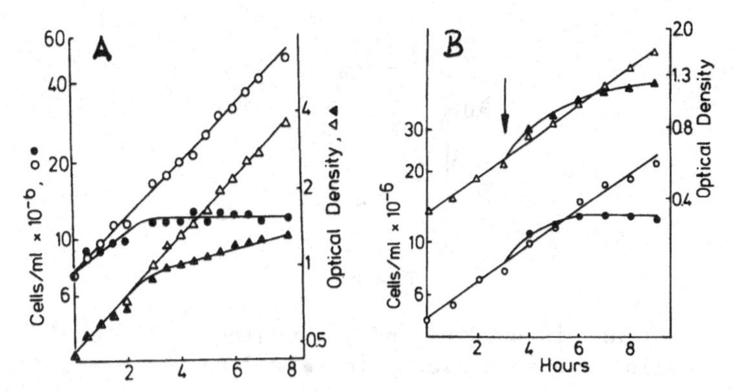

Fig. 2. A: Effect of tunicamycin (●,▲) on growth of S.cerevisiae (2); abscissa = hours. B; Growth of alg1-1 at 25°C (△,o) and 37°C (▲,●); ,△,▲ = O.D. (8)

studied more carefully (2) it became obvious that tunicamycin caused a first cycle arrest (Fig. 2A). Only those cells which had already initiated budding and DNA-replication doubled after tunicamycin was added. Such cells completed their cell cycle, and, after one generation time, the whole population was arrested in G1. S.cerevisiae cells in G1 do not have a bud (37), and in the presence of tunicamycin, the budding index drops from 60 to almost 10 % (Fig. 3). The very same pattern of growth inhibition was observed, when instead of using tunicamycin, a temperature-sensitive N-glycosylation mutant was grown at the non-permissive temperature of 37°C (8, and Fig. 2B). This alg 1-1 mutant is defective in mannosylation of the Dol-PP-$GlcNAc_2$ to the corresponding trisaccharide (38). Shifted to 37°C, the mutant culture stopped growth after one generation time and all cells showed G1 morphology, i.e. only about 10 % of the cells had buds (8). From such results, which do not seem to be restricted to yeast (39), it was postulated that at least one protein carrying an N-linked oligosaccharide is required for G1/S phase transition in the cell cycle.

S.cerevisiae mating type a cells can be synchronized in G1 with the peptide pheromone α-factor (40). Since α-factor was known not to inhibit protein synthesis (41) it was expected that the postulated glycoproteins pile up in the presence of α-factor, and that after removal of the pheromone, cells would enter S phase even under conditions in which N-glycosylation is inhibited. This, however, was not the case (Fig. 4). Cells stayed in G1 after relief from

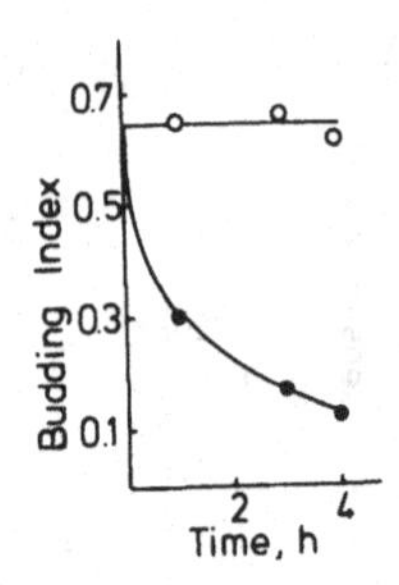

Fig. 3. Decrease in budding index (budding cells/total cells) due to tunicamycin (• = 2 μg/ml; o = control)

α-factor stop, when, subsequently, tunicamycin was present or the alg1-1 mutant was shifted to 37°C (2,8, and Fig. 4). Therefore, it had to be postulated that α-factor also inhibits the formation of the glycoprotein(s) in question or at least its (their) glycosylation. To test this postulate, the effect of α-factor on the incorporation of [^{14}C]glucosamine into polymers in vivo was investigated. As shown in Table 4, α-factor indeed strongly inhibits the formation of N-acetylglucosamine-containing glycoproteins. The fractionation scheme applied is described in greater detail in reference 9. The water soluble and the SDS extractable material separates into 7 and 13 radioactive bands respectively on SDS-polyacrylamide gels. After radiolabelling in the presence of α-factor, the intensity of the bands which can subsequently be displayed by fluorography is much reduced -

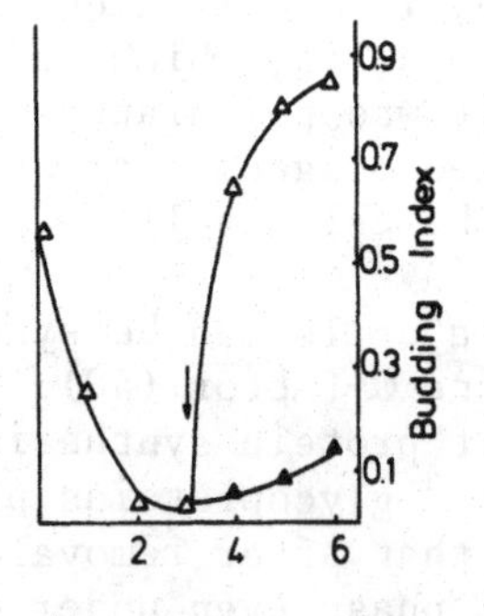

Fig. 4. Effect of tunicamycin on α-factor treated cells (2). α-Factor is removed after 3 h (▵ = control; ▴ = 3 μg/ml tunicamycin added at 3 hrs); abscissa = hours

Table 4

Effect of α-factor on the distribution of radiolabelled glucosamine and phenylalanine among cell fractions of S.cerevisiae X 2180-1a. The cultures radiolabelled for 4 h with [1-^{14}C] glucosamine or [^{14}C]phenylalanine were harvested and fractionated as detailed in ref. 9.

Radiolabelled precursor	Incubation	Total counts/min x 10^{-3} Fraction		
		water-soluble	SDS-soluble	pellet*
[1-^{14}C]glucosamine	control	166.6	116.9	619.2
	+ α-factor	22.2	15.2	449.8
[^{14}C]phenylalanine	control	623.3	396.3	50.2
	+ α-factor	696.8	441.1	74.7

* The radioactivity in chitin was found to be 325 counts/min x 10^{-3} (control) and 335 counts/min x 10^{-3} (α-factor treated cells).

to the same extent as to which the total incorporation of [^{14}C]glucosamine into the two soluble fractions is inhibited (9). The non-solubilized material is probably partly cell wall glycoprotein, formation of which is also decreased in the presence of α-factor, and partly chitin, which is not affected at all (Table 4). Thus, the glycoprotein(s) postulated to be necessary for G1/S phase transition are indeed not accumulated in the presence of α-factor, since this peptide itself is a severe inhibitor - most likely indirectly, since it will interact with a receptor on the cell surface (42) and not penetrate - of glycoprotein formation in yeast cells. This effect is mating type specific: no inhibition is observed in α-cells or in a/α diploids (9).

In summary then, these results with S.cerevisiae α-factor strongly support the conclusions drawn from N-glycoslyation inhibitor studies and from those with temperature sensitive glycosylation mutants. Whenever N-glycosylation is inhibited, cells do not stop growth immediately, but continue almost uninhibited until they reach the G1 phase in their cell cycle. Cells in G1 at the time N-glycosylation is shut off, or reaching G1 after N-glycosylation is halted are no longer able to enter S phase. α-Factor - as it turns out now - a strong physiological inhibitor of glycoprotein synthesis completely conforms to this pattern.

We thank Drs. L. Lehle and N. Sauer, as well as I. Hauner, G. Seebacher, and F. Klebl for help and advice. The work reported from this laboratory has been supported by the Deutsche Forschungsgemeinschaft (SFB 43).

REFERENCES

1. Tanner, W., Haselbeck, A., Schwaiger, H. and Lehle, L. (1982) Phil.Trans.R.Soc.Lond. B 300, 185-194
2. Arnold, E. and Tanner, W. (1982) FEBS Lett. 148, 49-53
3. Kaplan, A., Achord, D.T. and Sly, W.S. (1977) Proc.Natl. Acad.Sci. U.S.A. 74, 2026-2030
4. Schwaiger, H., Hasilik, A., von Figura, K., Wiemken, A. and Tanner, W. (1982) Biochem.Biophys.Res.Commun. 104, 950-956
5. Owada, M. and Neufeld, E.F. (1982) Biochem.Biophys.Res. Commun. 105, 814-820
6. Jessup, W. and Dean, R.T. (1982) Biochem.Biophys.Res. Commun. 105, 922-927
7. Waheed, A., Pohlmann, R., Hasilik, A., von Figura, K. and van Elsen, A. and Leroy, J.G. (1982) Biochem.Biophys. Res.Commun. 105, 1052-1058
8. Klebl, F., Huffaker, T. and Tanner, W. (1983) Exp.Cell. Res., submitted
9. Orlean, P., Seebacher, G. and Tanner, W. (1983) FEBS Lett., in press
10. Tanner, W. (1969) Biochem.Biophys.Res.Commun. 35,144-150
11. Jung, P. and Tanner, W. (1973) Eur.J.Biochem. 37, 1-6
12. Sentandreu, R. and Northcote, D.H. (1969) Carbohydr.Res. 10, 584-585
13. Ballou, C.E. (1976) Adv.Microb.Physiology 14, 93-158
14. Babczinski, P. and Tanner, W. (1973) Biochem.Biophys. Res.Commun. 54, 1119-1129
15. Sharma, C.B., Babczinski, P., Lehle, L. and Tanner, W. (1974) Eur.J.Biochem. 46, 35-41
16. Bretthauer, R.K. and Wu, S. (1975) Arch.Biochem.Biophys. 167, 151-160
17. Lehle, L. (1981) in: Encyclopedia of Plant Physiology New Series, Vol. 13 B; eds. W. Tanner and F.A. Loewus, pp 459-483, Springer-Verlag Berlin Heidelberg New York
18. Lehle, L., Schulz, I. and Tanner, W. (1980) Arch.Microbiol. 127, 231-237
19. Parodi, A.J. (1981) in: Yeast cell envelopes: biochemistry, biophysics and ultrastructure, vol. 2, ed. W.N. Arnold, pp 47-64, Boca Raton: CRC Press, Inc.
20. Robbins, P.W., Hubbard, S.C., Turco, S.J. and Wirth, D.F. (1977) Cell 12, 893-900
21. Li, E., Tabas, I. and Kornfeld, S. (1978) J.Biol.Chem. 253, 7762-7770
22. Byrd, J.C., Tarentino, A.L., Maley, F., Atkinson, P.H. and Trimble, R.B. (1982) J.Biol.Chem. 257, 14657-14666

23. Esmon, B., Nowick, P. and Schekman, R. (1981) Cell 25, 451-460
24. Marriott, M. and Tanner, W. (1979) J.Bacteriol. 139, 565-572
25. Haselbeck, A. and Tanner, W. (1983) FEBS Lett., in press
26. Novick, P., Field, C. and Schekman, R. (1980) Cell 21, 205-215
27. Larriba, G., Elorza, M.V., Vallanueva, J.R. and Sentandreu, R. (1976) FEBS Lett. 71, 316-320
28. Rothman, J.E. and Lodish, H.F. (1977) Nature 269, 775-780
29. Lennarz, W.J. (1982) Phil.Trans.R.Soc.Lond. B 300, 129-144
30. Snider, M.D., Huffaker, T.C., Couso, J.R. and Robbins, P.W. (1982) Phil.Trans.R.Soc.Lond. B 300, 207-223
31. Haselbeck, A. and Tanner, W. (1982) Proc.Natl.Acad.Sci. U.S.A. 79, 1520-1524
32. Kuo, S.-C., and Lampen, J.O. (1974) Biochem.Biophys.Res. Commun. 58, 287-295
33. Matile, Ph. (1978) Ann.Rev.Plant Physiol. 29, 193-213
34. Hashimoto, C., Cohen, R.E., Zhang, W.-J. and Ballou, C.E. (1981) Proc.Natl.Acad.Sci. U.S.A. 78, 2244-2248
35. Neufeld, E. and Ashwell, G. (1980) In: The biochemistry of glycoproteins and proteoglycans, ed. W.J. Lennarz, pp 241-266, Plenum Press, New York and London
36. Hasilik, A. and Tanner, W. (1978) Eur.J.Biochem. 91, 567-575
37. Hartwell, L.H. (1974) Bacteriol. Rev. 38, 164-198
38. Huffaker, T.C. and Robbins, P.W. (1982) J.Biol.Chem. 257, 3203-3210
39. Nishikawa, Y., Yamamoto, Y., Kaji, K. and Misui, H. (1980) Biochem.Biophys.Res.Commun. 97, 1296-2303.
40. Bücking-Throm, E., Duntze, W., Hartwell, L.H. and Manney, T.R. (1973) Exp.Cell.Res. 76, 99-110
41. Throm, E. and Duntze, W. (1970) J.Bacteriol. 104, 1388-1390
42. Fujmura, H., Skimizu, T., Yoshida, K. and Yanagishima, N. (1983) FEBS Lett. 153, 16-20

BIOSYNTHESIS AND TRANSPORT OF LYSOSOMAL ENZYMES

A. Hasilik, R. Pohlmann, F. Steckel, V. Gieselmann, K. von Figura, R. Olsen* and A. Waheed**

Physiologisch-Chemisches Institut der Universität Münster, Waldeyerstrasse 15, D-4400 Münster, West-Germany,
*Institute of Medical Biology, University of Tromsö, N-9001 Tromsö, Norway and **Department of Chemistry, Purdue University, West Lafayette, Indiana 47907, USA

I. INTRODUCTION

Lysosomes are digestive organelles found in most eucaryote cells. Typically they are endowed with a large set of mostly hydrolytic enzymes able to break down in concert nearly all biological polymers or complex molecules, which happen to be delivered into these organelles. The digestion depends on acidic conditions within the lysosomes - the internal pH is lower than that in the extralysosomal space by about 2 units (1,2) - and on the delivery into the lysosomes of both the substrates and the enzymes. The substrates are delivered into the lysosomes from the cell interior as well as from exterior. Traditionally,lysosomes were considered to take care of material interiorized by the cells. For recent reviews on endocytosis pathways in eucaryotes the reader may consult ref. 3-5. Today, the participation of lysosomes in the degradation of endogenous proteins is firmly established, too (ref. 6 and references therein). Lysosomal enzymes also participate in the breakdown of extracellular matrix. Some cells like human fibroblasts constitutively secrete a significant fraction of newly synthesized lysosomal enzymes (7). Certain leucocytes are secreting large amounts of lysosomal enzymes if stimulated properly (8). In this contribution the transport of newly synthesized lysosomal enzymes into the lysosomes will be discussed without attempting to review all pertinent data comprehensively. Special emphasis will be given to synthesis, modification and recognition of lysosomal enzymes in cultured human fibroblasts.

II. SYNTHESIS AND GLYCOSYLATION

With few exceptions lysosomal enzymes are glycoproteins(9). They are supposed to be synthesized in the rough endoplasmic reticulum. Detailed studies on porcine spleen cathepsin D in the laboratory of G. Blobel (10,11) have shown that this enzyme is synthesized as a precursor with a "signal" sequence. By its length (20 amino acids) and hydrophobicity (7 leu residues) this sequence resembles "signal" sequences of presecretory proteins (11). In a few other laboratories mRNA-programmed synthesis of lysosomal enzymes has been demonstrated: cathepsin D from mouse spleen (12), ß-glucuronidase from rat preputial gland (12) and ß-hexosaminidase from human fibroblasts and human placenta (13). In our laboratory mRNA from the human promyelocyte cell line HL-60 was used for in vitro translation of myeloperoxidase. (In leucocytes myeloperoxidase is located in azurophilic granules, which are considered to be primary lysosomes, see ref. 14.) In all mentioned cases the translation products synthesized in the absence of membranes and isolated with the aid of specific antibodies were by about 2000 daltons larger than the corresponding unglycosylated polypeptides isolated from cells (see Table next page). Partially unglycosylated polypeptides can be prepared by treating the lysosomal enzyme precursors from pulse labeled cells with endo-ß-N-acetylglucosaminidase H. The treatment with this enzyme removed rather specifically high-mannose and hybrid oligosaccharides. After the treatment only the asn-linked N-acetylglucosamine residues are left at the cleavage points (15). Alternatively, unglycosylated polypeptides can be isolated from cells treated with tunicamycin. This antibiotic blocks the oligosaccharide transfer in the rough endoplasmic reticulum (16,17). In vivo unglycosylated analogs of glycoproteins are often rapidly degraded (18,19) and this may limit the use of tunicamycin. The yield of radioactivity in various lysosomal enzymes was lowered considerably in the presence of tunicamycin (11,20). In promyelocytes treated with tunicamycin unglycosylated myeloperoxidase could not be identified; the apparent labeling of the enzyme was strongly inhibited in the presence of the antibiotic.

The glycosylation (rev. in 21-23) of asparagine residues takes place in the rough endoplasmic reticulum. Several studies on secretory glycoproteins have shown that the glycosylation occurs on nascent polypeptide chains (21,24-26). Currently it is believed that there is no difference between

LYSOSOMAL ENZYME POLYPEPTIDES SYNTHESIZED IN VIVO AND IN VITRO

Enzyme synthesized	in vivo		in vitro		References
	glycosylated	unglycosylated	- microsomes	+ microsomes	
Cathepsin D					
porcine kidney	46	n.d.	43	46	10
mouse spleen	n.d.	n.d.	48	50	12
rat liver	50	47 (TM)	48	50	12
ß-Glucuronidase					
rat preputial gland	n.d.	n.d.	70	72	12
rat liver	72	68 (TM)	70	72	12
ß-Hexosaminidase					
α-chain	67	63 (EH)	65	n.d.	13
ß-chain	63	57 (EH)	59	n.d.	13

the secretory and the lysosomal glycoproteins in the mechanisms of the initial glycosylation. Indeed, labeled cathepsin D (27) and ß-hexosaminidase (13) isolated from fibroblasts incubated for only a few minutes with a radioactive amino acid contain N-glycosidically linked oligosaccharides, which are sensitive to endo-ß-N-acetylglucosaminidase H.

III. MODIFICATION OF THE OLIGOSACCHARIDES, GENERAL ASPECTS.

Probably, the $Glc_3Man_9GlcNAc_2$ oligosaccharide is universally used in the glycosylation of asn residues (21-23). After the transfer the oligosaccharide may be subject to diverse modifications. Concomitantly with the modifications, the glycosylated product moves through the transitional elements of the endoplasmic reticulum and the Golgi apparatus (28). At present it is not clear what is the influence of the modification reactions on this transport. Some modifications are specific to lysosomal enzymes. The lysosomal enzyme molecules finally leaving the Golgi apparatus may contain diverse oligosaccharides including those of the high-mannose type of various length, some of them carrying phosphomono- and phosphodiester groups, of the complex ($-GlcNAc_2Man_3GlcNAc_2Gal_xNANA_yFuc$) as well as of hybrid types. Further diversification due to hydrolysis in lysosomes has been the main obstacle for isolation and characterisation of oligosaccharides from lysosomal enzymes. Nevertheless, an extensive study on oligosaccharides released from rat liver microsomal and lysosomal ß-glucuronidase by endo-ß-N-acetylglucosaminidase H is available now (29). The neutral oligosaccharides found in the microsomal enzyme contained 5-9 mannose residues, the oligosaccharides in the lysosomal enzyme contained only 5 mannose residues. It is not clear what is the contribution of intralysosomal hydrolysis to this difference.

With the aid of metabolic labeling it has been possible in some systems to use trace amounts of material for structural studies on phosphorylated oligosaccharides in single lysosomal enzymes, in human ß-hexosaminidase and cathepsin D (30) and in mouse ß-glucuronidase (31,32) or in total glycoproteins of mouse lymphoma (33). In the studies on human fibroblast lysosomal enzymes 10 mM NH_4Cl added to the incubation medium was used to block the transport of the newly synthesized enzymes into the lysosomes. The products were isolated from secretions precluding any digestion in lysosomes (30). Human cathepsin D has been shown to contain two glycosylated sites (34). Both of them may carry either the high-mannose or the complex type oligosaccharide. In both sites the high-mannose oligosaccharides may be phosphorylated. Mouse ß-glucuronidase contains three glycosylation sites per subunit and each of them is partially phosphorylated (32).

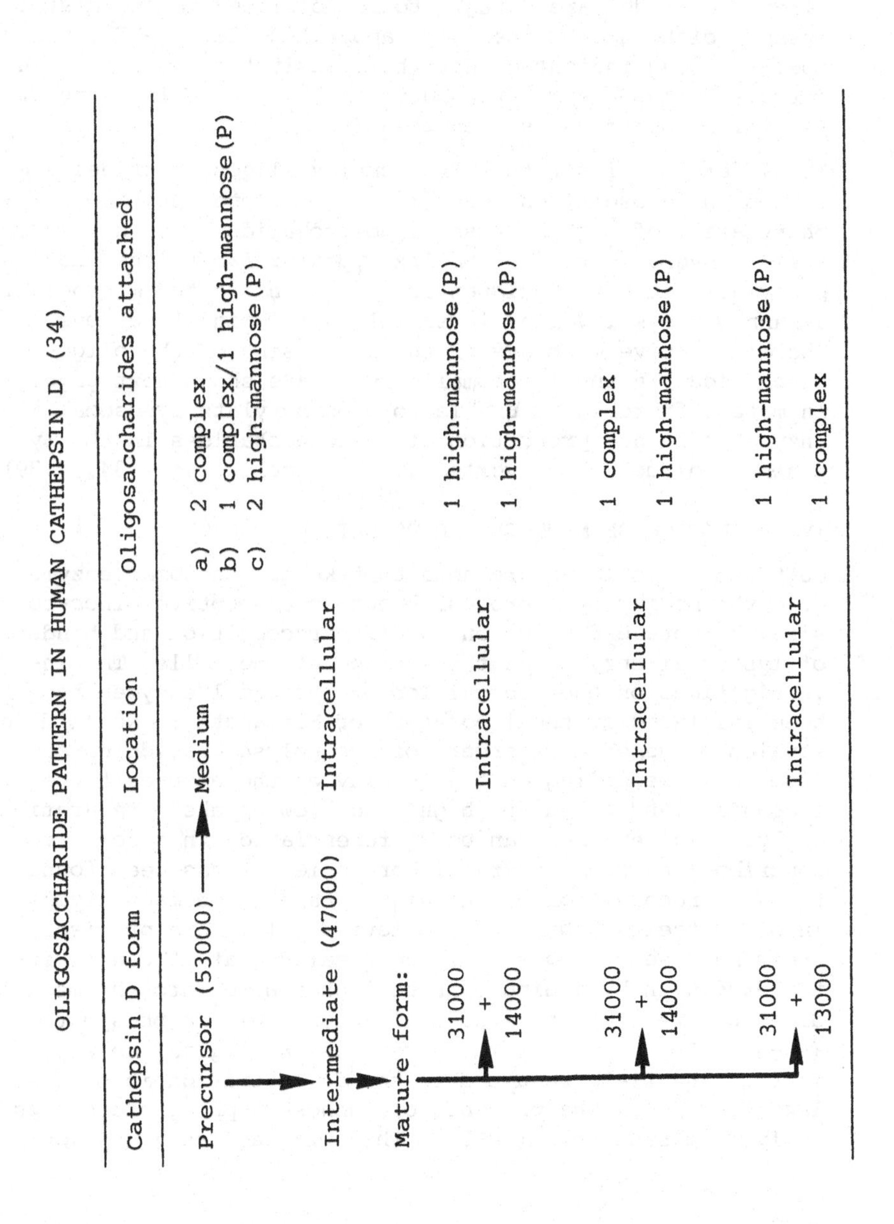

OLIGOSACCHARIDE PATTERN IN HUMAN CATHEPSIN D (34)

Cathepsin D form	Location	Oligosaccharides attached
Precursor (53000) →	Medium	a) 2 complex b) 1 complex/1 high-mannose(P) c) 2 high-mannose(P)
→ Intermediate (47000)	Intracellular	
→ Mature form:		
31000 + 14000	Intracellular	1 high-mannose(P) 1 high-mannose(P)
31000 + 14000	Intracellular	1 complex 1 high-mannose(P)
31000 + 13000	Intracellular	1 high-mannose(P) 1 complex

In a study on cathepsin D in human fibroblasts subjected to a short pulse labeling it was found that the formation of the complex oligosaccharides (i.e. oligosaccharides resistant to endo-ß-N-acetylglucosaminidase H) is initiated within 40-60 min after the synthesis of the protein (27). Since the enzymes involved in the synthesis of the complex oligosaccharides are thought to be localized in the distal (trans) cisternae of the Golgi apparatus (28, 35-37), the above finding indicated that the transit time of cathepsin D from the endoplasmic reticulum to the trans cisternae of the Golgi apparatus is less than 1 h.

In different cells rather few complex oligosaccharides are formed in lysosomal enzymes (30,32). It appears that phosphorylation of high-mannose oligosaccharides precludes their conversion to the complex type. Probably, the phosphorylated mannose residues are resistant to the microsomal α-mannosidases involved in so called trimming reactions. The alternative pathways of the processing of the oligosaccharides in the lysosomal enzymes are shown next page. In mutant fibroblasts unable to phosphorylate lysosomal enzymes a larger proportion of oligosaccharides in the lysosomal enzymes is converted to the complex type (34,38,39).

IV. FORMATION OF MANNOSE 6-PHOSPHATE RESIDUES

Cultured fibroblasts are able to take up lysosomal enzymes from the medium by a process known as adsorptive pinocytosis. The uptake depends on specific recognition and binding of lysosomal enzymes at the surface of the cells. The investigations on the recognition of lysosomal enzymes have been initiated in the laboratory of Elizabeth F. Neufeld in studies on genetic disorders of mucopolysaccharide catabolism (40). Depending on the ability of the enzymes to be recognized and taken up "high" and "low uptake" preparations of lysosomal enzymes can be differentiated. In successive contributions from several laboratories it has been found that the recognition of the high uptake lysosomal enzymes involves their carbohydrate moieties (41), more precisely their mannose residues (42) and finally that it can be interfered with by adding mannose 6-phosphate into the medium or by pretreating the lysosomal enzymes with a phosphatase (43-47). At the end of the seventies the logical development in the field resulted in the demonstration in several laboratories of the presence of mannose 6-phosphate groups in lysosomal enzymes (48-50). This finding has now been

DIVERSIFICATION OF OLIGOSACCHARIDE SIDE CHAINS IN LYSOSOMAL ENZYMES

(Alternatives in trimming-, N-acetylglucosaminyl transferase 1- and N-acetylglucosaminyl phosphotransferase-reactions)

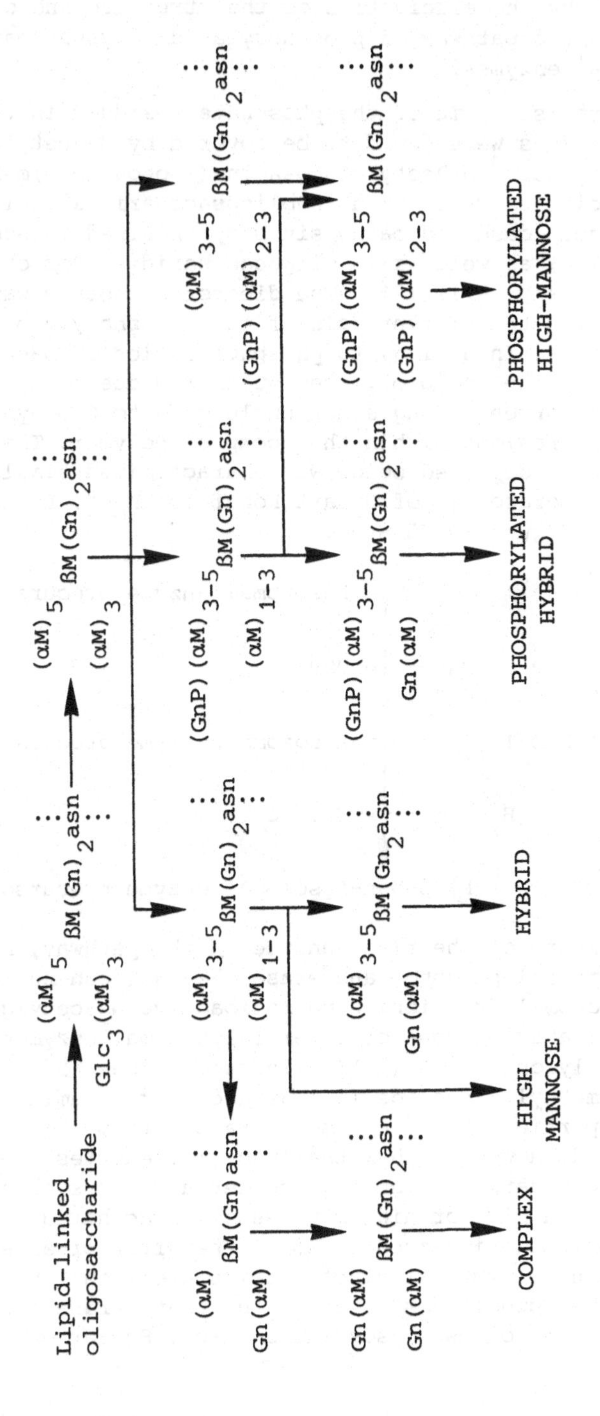

(M = mannose, Gn = N-acetylglucosamine, GnP = N-acetylglucosaminyl 1-phosphate)

followed by the elucidation of the structure and of the biosynthetic pathway of phosphorylated oligosaccharides in lysosomal enzymes.

To a surprise, some of the phosphate residues in the lysosomal enzymes were found to be covered by N-acetylglucosamine residues. N-Acetylglucosamine(1)phospho(6)mannose diester moieties were found in oligosaccharides isolated from ß-hexosaminidase and cathepsin D synthesized in human fibroblasts (30) as well as in oligosaccharides from chinese hamster ovary cells (31). The diester structure was[a] clue in deciphering the biosynthesis of the phosphorylated recognition marker. In analogy to phosphorylation of mannan in yeast (51) the phosphorylated sugar residue is transferred from the corresponding sugar nucleotide to C-6 hydroxyl of a mannose residue within the acceptor polymer. The reaction sequence as depicted below was characterized simultaneously in the laboratories of Stuart Kornfeld in St. Louis and of Kurt von Figura in Münster.

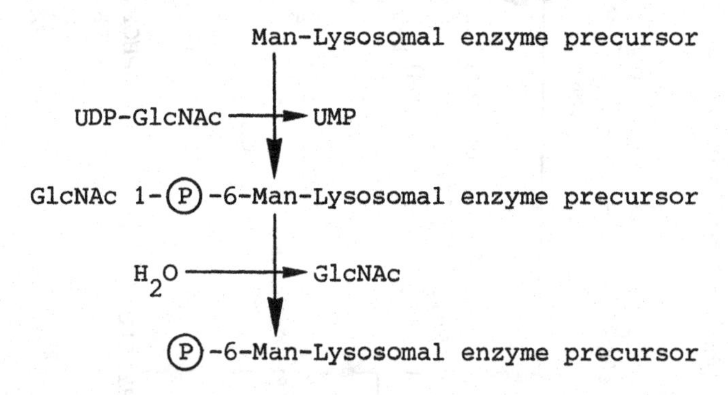

The activity of the first enzyme in the pathway, N-acetylglucosamine-1-phosphotransferase (52-54), can be measured in microsomal fractions with radioactive N-acetylglucosamine as donor and with one of several lysosomal enzymes (52,54-57) or glycopeptides (53), oligosaccharides (55,56), or even with α-methyl mannoside (56) as acceptor. Examinations of the properties of the enzyme extracted from the membranes indicated, however, that the lysosomal enzymes are by two to three orders of magnitude better acceptors than are α-methyl mannoside or high-mannose oligosaccharides (55,56). High-mannose oligosaccharides, either free or as sugar moieties in non-lysosomal glycoproteins are rather poor acceptors. The actual substrates of the transferase are the precursor forms of the lysosomal enzymes. So far unresolved remains the question, what is the structural basis for the selective recognition of the lysosomal enzymes by the transferase? The acceptor activity of the lysosomal enzymes is destroyed by denaturing them with heat (56), a reducing agent or with sodium dodecyl sulfate (unpublished). Probably, native lysosomal enzymes and their precursors contain a signal, which is recognized by N-acetylglucosamine-1-phosphotransferase. In cathepsin D and in ß-hexosaminidase in human fibroblasts about half of the high-mannose oligosaccharides are phosphorylated (30). It appears that not more than one in three oligosaccharides in cathepsin D is phosphorylated, since some of the oligosaccharides are processed to the complex type (34). In ß-glucuronidase synthesized in the murine cell line $P388D_1$ a similar proportion of the oligosaccharides becomes phosphorylated. The phosphate can be linked to most of the α1-2 linked mannose residues in high-mannose oligosaccharides. The two available studies concerning mouse (33) and human (58) ß-glucuronidase slightly differ in assignment of the phosphorylatable residues. A single oligosaccharide may contain one or two phosphate groups (30,33,58). The existence of oligosaccharides containing three phosphate groups has been considered by some authors (30,31,33), while questioned by others (58). In human spleen ß-glucuronidase 85% of phosphorylated oligosaccharides contain one or two phosphates in the diester linkage (58). In mouse lymphoma cells such diesters represented 80% of the total phosphorylated oligosaccharides (59). In several enzymes studied various mixtures of oligosaccharides containing one and two phosphate residues, both free and covered were found. In ß-hexosaminidase and cathepsin D secreted by fibroblasts in the presence of NH_4Cl about 80% of the phosphorylated oligosaccharides contain one or two phosphates in the diester linkage (30).

The second enzyme in the pathway of the recognition marker synthesis, α-N-acetylglucosaminyl phosphodiesterase, cleaves the N-acetylglucosamine residue from the diester thereby generating the actual recognition sites in the lysosomal enzymes. This hydrolase has been solubilized and partially purified from human placenta and from rat liver Golgi fraction (60-62). It is a microsomal enzyme distinct by several other criteria from the lysosomal α-N-acetylglucosaminidase (60-62). This glycosidase reaction may represent the rate limiting step in the formation of the phosphorylated recognition marker in the lysosomal enzymes. Indeed, in several oligosaccharide preparations the diesters were in excess of the monoesters (30,32,58).

The phosphorylating pathway is localized in the Golgi apparatus (60,61,63,64). The two enzymes of the pathway codistribute with α(1-2)mannosidase (64), which is a marker for reactions attributed to the cis part of the Golgi apparatus (28,65), and separate from galactosyl transferase, which participates in the synthesis of complex oligosaccharides and has been demonstrated in the trans cisternae of the Golgi apparatus (28,35,36). Such distribution of the enzymes is in agreement with the kinetic analysis of the modification of the oligosaccharide side chains in ß-glucuronidase in the murine macrophage cell line $P388D_1$. The phosphorylation occured subsequently to the removal of glucose residues from the oligosaccharides and the cleavage of the diesters proceeded concomitantly with that of α(1-2) mannosyl residues (32). In the mouse lymphoma cell line $Pha^{R2.7}$, which is deficient in the activity removing the two inner glucose residues from $Glc_3Man_9GlcNAc_2$ oligosaccharides (glucosidase II) the phosphorylation is impaired. In the glucosylated oligosaccharides phosphate residues were found only in the glucose-free branch (66). The processing of the two branches is to some extent autonomic, as indicated by finding of phosphorylated hybrid type oligosaccharides in glycoproteins of murine macrophage cell line $P388D_1$ (67).

V. TRANSPORT

Originally the recognition and transport of lysosomal enzymes were studied in endocytosis experiments. These studies indicated existence of a specific receptor for phosphorylated lysosomal enzymes at the cell surface of fibroblasts and other cells. Sensitive assays were developed for the receptors and their number in the plasma membrane in cultured human fibroblasts was estimated in the range of 14000-37000 per cell (68,69). However, the receptor was found also in the intracellular membranes and suggested to be involved in the transport of newly synthesized lysosomal enzymes (69,70). The receptor was recently isolated from several cell types (71-73). It behaved like a glycoprotein with a molecular weight of 215000 daltons. Using fractionation and a binding assay in rat liver most of the receptor was found to be associated with the intracellular membranes (74). In mouse lymphoma cells most of the receptor has been found to be occupied (59), whereas the receptor detected recently in coated vesicles prepared from rat liver and calf brain appeared to be largely free of ligands (75).

The binding of lysosomal enzymes to the receptor is independent of calcium or other metal ions (74). The stability of the complexes is optimum at pH 6 to 7 and drops precipitously at pH of less than 5.8 as observed with both cell surface associated (69) and purified receptors (71). The dissociation of the complex at acidic pH may be the physiological means for separation of the transported lysosomal enzymes and probably of many other ligands as well (rev. in 76) from their receptors.

Interactions of various phosphorylated oligosaccharides with the receptors were studied using whole cells (77,78), and with immobilized purified receptors (67,79). In general, presence of uncovered phosphomonoester groups in the oligosaccharides is a prerequisite for binding or uptake, and, strongest interaction is observed with oligosaccharides containing two uncovered mannose 6-phosphate residues. The detailed data presented by Varki and Kornfeld (67) show further that i) removal of certain peripheral mannose residues may increase the binding, ii) among the monophosphorylated oligosaccharides those with the phosphate group located in the branch linked 3' to the ß-mannose have higher affinity to the receptor, iii) phosphorylated oligosaccharides of the hybrid type bind well unless they carry sialic acid residues.

The receptors of lysosomal enzymes appear to be recycled for several rounds of transport (68,69). Certain fraction of the receptor is associated with clathrin-coated areas of membranes (75,80). In rat liver the receptor is enriched 60-fold in coated vesicles over the microsomes (75). Interestingly, in this preparation, the receptor seems to be little occupied with lysosomal enzymes; its capacity to bind a phosphorylated lysosomal enzyme is not increased by washing with mannose 6-phosphate. In the adsorptive pinocytosis of lysosomal enzymes, in which the transport is specifically directed into the lysosomes (81), complexes of the enzymes with the receptor are formed at the cell surface (68,74) and these are found in the "coated pit" areas of the plasma membrane (80). In the case of the transport of newly synthesized lysosomal enzymes in murine lymphoma cells the interaction with the mannose 6-phosphate-sensitive receptor is restricted to enzymes with uncovered phosphate residues (59) and appears not to take place until the lysosomal enzymes passed the galactosyl transferase and sialyl transferase containing region of the Golgi apparatus as judged from the structural analysis of oligosaccharides in the bound enzymes (82).

In contrast, in electron microscopic examination of rat liver cells labeled with an immunoglobulin-gold technique, association of cathepsin D with mannose 6-phosphate receptors in the Golgi apparatus has been observed (83). In this study the receptor was found besides Golgi apparatus in GERL and in plasma membrane but not in lysosomes or endoplasmic reticulum. This latter finding contrasts data obtained with a binding assay applied to subcellular fractions, that were interpreted to indicate a large pool of the receptor in the endoplasmic reticulum (74).

Cytochemistry studies with acid phosphatase indicated that coated vesicles found in the vicinity of the Golgi apparatus are involved in the transport of newly synthesized lysosomal enzymes to multivesicular bodies in rat vas deferens (84) and rabbit macrophages (14). We have found recently, that in a coated vesicle preparation from human placenta lysosomal enzymes can be detected using a combination of immunoprecipitation, electroblotting and immunoglobulin plus anti-immunoglobulin-peroxidase detection. Cathepsin D associated with the vesicles was enriched in the precursor form (85). This observation supports the idea that coated vesicles are involved in the transport of lysosomal enzymes from the Golgi apparatus to lysosomes (14,84,86). It has also been suggested, however, that coated vesicles mediate transport of secretory products from the Golgi apparatus (87).

In human fibroblasts an excessive secretion of lysosomal enzyme precursors can be induced by adding weak bases such as NH_4Cl or chloroquine to the cultivation medium (7). The weak bases interfere with the acidification of lysosomal (1) and presumably also of prelysosomal compartments (76,88-91). In this way the bases may interfere with the recycling of the lysosomal enzyme receptors (69). Several mutant cell lines with impaired receptor function and excessive secretion of lysosomal enzymes were isolated (92-94). In fibroblasts mutations impairing the synthesis of the phosphorylated recognition marker (e.g. mucolipidosis II) result in abnormal secretion of lysosomal enzymes also (48,50,53,95-97). Similarly, inhibition of glycosylation by tunicamycin prevents segregation of lysosomal enzyme precursors and enhances their secretion (11,12,20).

In patients with mucolipidosis II in several organs including liver normal activities of many lysosomal enzymes are found (98 and ref. in 99). The enzymes responsible for the synthesis of the recognition marker, however, are absent in all patient's organs (55,99). This indicates that in certain cells the transport of lysosomal enzymes into lysosomes cannot be entirely dependent on the interaction of the phosphorylated recognition marker with the receptor, i.e. there may be alternative pathways for the transport. Even in fibroblasts, there are differences in the transport of the various enzymes. Inhibition of the transport by the ionophore monensin or by cyanate, both acting mainly at the level of the Golgi apparatus, effects differentially the transport of ß-hexosaminidase and cathepsin D (100,101). In human monocytes and macrophages under certain conditions ß-hexosaminidase is secreted at a higher rate than cathepsin D whereas under other conditions the ratio is reversed (102).

Some mutant cell lines with a defect in receptor for mannose 6-phosphate were found to possess high intracellular levels of lysosomal enzymes indicating that in these cells intracellular mechanisms for delivery of lysosomal enzymes exist that are independent on the mannose 6-phosphate receptor (94). These mutants, however, secrete lysosomal enzymes at an abnormally high rate. Mutants with less than 2% of the receptor retain only about 30% of newly synthesized ß-hexosaminidase and ß-glucuronidase (94). In a search for an alternative packaging of lysosomal enzymes it appears a significant task to estimate the constitutive, non receptor-mediated transport into the lysosomes in cells concerned, because during any vesicle transport a certain fraction of the soluble contents of the Golgi apparatus is expected to end up in the lysosomes. This process, besides crinophagy, may be the cause e.g. of a rather extensive degradation of newly synthesized collagen in human lung fibroblasts (103, 104) and rat hepatocytes (105).

In yeast phosphorylation and formation of mannose 6-phosphate residues in carboxypeptidase Y was observed (106,107). In contrast to lysosomal enzymes of higher eucaryote cells the carbohydrate moiety in carboxypeptidase Y (107) and in alkaline phosphatase (108), another vacuolar enzyme, however, has no effect on the direction of the intracellular transport. The transport of carboxypeptidase Y into the vacuole is little affected when the glycosylation is blocked by tunicamycin (107). Like the lysosomal enzymes in higher eucaryote cells the vacuolar enzymes in yeast share the transport route through the Golgi apparatus with the secretory (periplasmic) proteins (109). The structures directing segregation of the vacuolar and secretory proteins in yeast remain to be discovered. It is possible, that in yeast the formation of the secretory vesicles depends on receptor mediated segregation of secretory and cell wall precursors in the Golgi apparatus, whereas the formation of "primary vacuoles" does not. In sec 7 mutants the terminal Golgi cisterna is found in the proximity of the vacuole at the non-permissive temperature (109). We may speculate, therefore, that the vacuole derives from the body of the terminal cisterna, whereas the secretory vesicles derive from smaller vesicles pinching off the Golgi cisternae. Inhibition by tunicamycin of secretion of invertase and acid phosphatase in yeast spheroplasts (108) also indicates that the carbohydrate may serve as signal in segregation of secretory proteins in the Golgi apparatus.

VI. PROTEOLYTIC PROCESSING

Segregation of the vacuolar enzymes into vacuoles in yeast (107,110) and of lysosomal enzymes into lysosomes in fibroblasts (rev. in 70, 111-113) is correlated with a maturation of the precursor forms. In mouse (114,115) and human macrophages (102) the secreted lysosomal enzymes contain mature forms of lysosomal enzymes.

The maturation is a proteolytic process (11,115,116). It is not yet known, in which compartment the maturation is initiated. It appears to start in a prelysosomal compartment (27). In a Percoll gradient a large amount of the processing intermediate of cathepsin D was found temporarily to be located in fractions with lower density as compared to lysosomes. The processing pattern of endocytosed lysosomal enzyme precursors resembles that of enzymes synthesized and processed within the cell (93,97,117) and it is the structure of the precursor rather than the maturation machinery, which determines the polypeptide pattern of the mature enzymes (93).

The lysosomal enzymes processed by the secretory pathway appear in the medium exclusively in the precursor form (7). This is very clear-cut in fibroblasts. Other cells may secrete also the mature lysosomal enzymes. Human monocytes and related cells secrete various amounts of precursor and mature forms depending on culture conditions (102). The secretion patterns are characteristic for the individual enzymes as if there were various organelles containing lysosomal enzymes in different proportions (100-102). Based on similar observations the existence of different lysosomal organelles has previously been suggested from studies on mouse macrophage lysosomal enzymes (115). In contrast to human macrophages, in mouse macrophages only mature forms of lysosomal enzymes have been found in the secretions (114,115). The precursors of various glucosidases and sulfatases appear to be enzymically active (113,118). In a few cases the precursors appear to be zymogens. Correlations between processing and activation have been reported for human cathepsin D (118), carboxypeptidase Y and several other vacuolar hydrolases in yeast (119,120). Primary structure of porcine cathepsin D resembles closely that of pepsinogen indicating that a similar strategy may be used in the activation of a lysosomal and a secretory proteinase (11).

Certain organelles specialized in secretion or in storage of secretory products resemble lysosomes in several criteria. Thus segregation of peptide hormones in secretory or storage granules is often accompanied with a proteolytic processing of a precursor. Recently, in a mouse pituitary cell line a constitutive secretion of unprocessed ACTH precursor has been observed (121). Further, the storage was abolished by incubating the cells with chloroquine. The treated cells did not process the ACTH precursor and secreted it instead.

Acknowledgement: This work was supported by The Deutsche Forschungsgemeinschaft (SFB 104).

REFERENCES

1. Ohkuma, S. and Poole, B. (1978) Proc. Natl. Acad. Sci. U.S.A. 75, 3327-3331
2. Kakinuma, Y., Ohsumi, Y. and Anraku, Y. (1981) J. Biol. Chem. 256, 10859-10863
3. Goldstein, J. L., Anderson, R. G. W. and Brown, M. S. (1979) Nature 279, 679-685
4. Besterman, I. M. and Low, R. B. (1983) Biochem. J. 210, 1-13
5. Steinman, R. M., Mellman, I. S., Muller, W. A. and Cohn, Z. A. (1983) J. Cell Biol. 96, 1-27
6. Cockle, S. M. and Dean, R. T. (1982) Biochem. J. 208, 795-800
7. Hasilik, A. and Neufeld, E. F. (1980) J. Biol. Chem. 255, 4937-4945
8. Page, R. C., Davis, P. and Allison, A. C. (1978) Int. Rev. Cytol. 52, 119-157
9. Tulsiani, D. R. P., Keller, R. K. and Touster, D. (1975) J. Biol. Chem. 250, 4770-4776

10. Erickson, A. H. and Blobel, G. (1979) J. Biol. Chem.254, 11771-11774
11. Erickson, A. H., Conner, G. E. and Blobel, G. (1981) J. Biol. Chem. 256, 11224-11231
12. Rosenfeld, M. G., Kreibich, D., Popov, D., Kato, K. and Sabatini, D. D. (1982) J. Cell Biol. 93, 135-143
13. Proia, R. and Neufeld, E. F. (1982) Proc. Natl. Acad. Sci. U.S.A. 79, 6360-6364
14. Nichols, B. A., Bainton, D. F. and Farquhar, M. G.(1971) J. Cell Biol. 50, 498-515
15. Malley, F. and Trimble, B. B. (1981) J. Biol. Chem.256, 1088-1090
16. Tkacz, J. and Lampen, J. O. (1975) Biochem. Biophys. Res. Commun. 65, 248-257
17. Lehle, L. and Tanner, W. (1976) FEBS Letters 71,167-170
18. Olden, K., Parent, B. and White, S. L. (1982) Biochem. Biophys. Acta 650, 209-232
19. Gibson, R., Kornfeld, S. and Schlesinger, S. (1980) Trends Biochem. Sci. 5, 290-293
20. von Figura, K., Hasilik, A., Waheed, A. and Klein, U. (1981) Perspect. Inherit. Metabol. Dis. 4, 463-468
21. Hubbard, S. C. and Iwatt, R. J. (1981) Annu. Rev. Biochem. 50, 555-583
22. Struck, D. K. and Lennarz, W. J. (1980) in: The Biochemistry of Glycoproteins and Proteoglycans (W. J.Lennarz, ed.) pp 35-83, Plenum Press, New York
23. Sharon, N. and Lis, H. (1980) in: The Proteins, Vol. 5, 3rd Edn. (H. Neurath and R. C. Hill, eds.) Acad. Press, New York
24. Kiely, M. C., McKnight, G. S. and Schimke, R. T. (1976) J. Biol. Chem. 251, 5490-5495
25. Lingappa, V. R., Katz, F. N., Lodish, H. F. and Blobel, G. (1978) J. Biol. Chem. 253, 8667-8670
26. Bielinska, M. and Boime, I. (1979) Proc. Natl. Acad.Sci. U.S.A. 76, 1208-1212
27. Gieselmann, V., Pohlmann, R., Hasilik, A. and von Figura, K. (1983) J. Cell Biol., in press
28. Rothman, J. E. (1981) Science 213, 1212-1219
29. Mizuochi, T., Nishimura, Y., Kato, K. and Kobata, A. (1981) Arch. Biochem. Biophys. 209, 298-303
30. Hasilik, A., Klein, U., Waheed, A., Strecker, G. and von Figura, K. (1980) Proc. Natl. Acad. Sci. U.S.A. 77, 7074-7078
31. Tabas, I. and Kornfeld, S. (1980) J. Biol. Chem. 255, 6633-6639

32. Goldberg, D. E. and Kornfeld, S. (1981) J. Biol. Chem. 256, 13060-13067
33. Varki, A. and Kornfeld, S. (1980) J. Biol. Chem. 255, 10847-10858
34. Hasilik, A. and von Figura, K. (1981) Eur. J. Biochem. 121, 125-129
35. Roth, J. and Berger, E. G. (1982) J. Cell Biol. 93, 223-229
36. Dunphy, W. G., Fries, E., Urbani, L. J. and Rothman,J.E. (1981) Proc. Natl. Acad. Sci. U.S.A. 78, 7453-7457
37. Quinn, P., Griffith, G. and Warren, G. (1983) J. Cell Biol. 96, 851-856
38. Sly, W. S., Gonzales-Noriega, A., Natowicz, M., Fisher, H. D. and Chambers, J. P. (1979) Feder.Proc.38, 1256
39. Miller, A. L., Kress, B. C., Stein, R., Kinnon, C., Kern, H., Schneider, A. and Harms, E. (1981) J. Biol. Chem. 256, 9352-9362
40. Neufeld, E. F. and Ashwell, G. (1980) in: The Biochemistry of Glycoproteins and Proteoglycans (W.J.Lennarz, ed.) pp 241-266, Plenum Press, New York
41. Hickman, S., Shapiro, L. J. and Neufeld, E. F. (1974) Biochem. Biophys. Res. Commun. 57, 55-61
42. Hieber, V., Distler, J., Myerowitz, R., Schmickel, R.D. and Jourdian, G. W. (1976) Biochem. Biophys. Res.Commun. 73, 710-717
43. Kaplan, A., Achord, D. T. and Sly, W. S. (1977) Proc. Natl. Acad. Sci. U.S.A. 74, 2026-2030
44. Sando, G. N. and Neufeld, E. F. (1977) Cell 12, 619-627
45. Kaplan, A., Fischer,D., Achord, D. T. and Sly, W. S. (1977) J. Clin. Invest. 60, 1088-1093
46. Ullrich, K., Mersmann, G., Weber, R. and von Figura, K. (1977) Biochem. J. 170, 643-650
47. Kaplan, A., Fischer, D. and Sly, W. S. (1978) J. Biol. Chem. 253, 647-650
48. Hasilik, A. and Neufeld, E. F. (1980) J. Biol. Chem.255, 4946-4950
49. Natowicz, M. R., Chi, M. M. Y., Lowry, O. H. and Sly, W. S. (1979) Proc. Natl. Acad. Sci. U.S.A.76,4322-4326
50. Bach, G., Bargal, R. and Cantz, M. (1979) Biochem. Biophys. Res. Commun. 91, 976-981
51. Karson, E. M. and Ballou, C. E. (1978) J. Biol. Chem.253, 6484-6492
52. Hasilik, A., Waheed, A. and von Figura, K. (1981) Biochem. Biophys. Res. Commun. 98, 761-767
53. Reitman, M. L. and Kornfeld, S. (1981) J. Biol. Chem.256, 4275-4281

54. Waheed, A., Hasilik, A., Cantz, M. and von Figura, K. (1981) Hoppe-Seyler's Z. Physiol. Chem. 363, 169-178
55. Waheed, A., Hasilik, A. and von Figura, K. (1982) J.Biol. Chem. 257, 12322-12331
56. Reitman, M. L. and Kornfeld, S. (1981) J. Biol. Chem. 256, 11977-11980
57. Owada, M. and Neufeld, E. F. (1982) Biochem. Biophys. Res. Commun. 105, 814-820
58. Natowicz, M., Baenziger, J. U. and Sly, W. S. (1982) J. Biol. Chem. 257, 4412-4420
59. Gabel, C. A., Goldberg, D. F. and Kornfeld, S. (1982) J. Cell Biol. 95, 536-542
60. Varki, A. and Kornfeld, S. (1980) J. Biol. Chem. 255, 8398-8401
61. Waheed, A., Hasilik, A. and von Figura, K. (1981) J.Biol. Chem. 256, 5717-5721
62. Varki, A. and Kornfeld, S. (1981) J. Biol. Chem. 256, 9937-9943
63. Waheed, A., Pohlmann, R., Hasilik, A.and von Figura, K. (1981) J. Biol. Chem. 256, 4150-4152
64. Pohlmann, R., Waheed, A., Hasilik, A. and von Figura, K. (1982) J. Biol. Chem. 257, 5323-5325
65. Tabas, I. and Kornfeld, S. (1979) J. Biol. Chem. 254, 11655-11663
66. Gabel, C. A. and Kornfeld, S. (1982) J. Biol. Chem. 257, 10605-10612
67. Varki, A. and Kornfeld, S. (1983) J. Biol. Chem. 258, 2808-2818
68. Rome, L. H., Weissmann, B. and Neufeld, E. F. (1979) Proc. Natl. Acad. Sci. U.S.A. 76, 2331-2334
69. Gonzales-Noriega, A., Grubb, J. H., Talkad, V. and Sly, W. S. (1980) J. Cell Biol. 85, 839-852; Fischer, H. D., Gonzales-Noriega, A. and Sly, W. S. (1980) J. Biol.Chem. 255, 5069-5074
70. Sly, W. S. and Fischer, H. D. (1982) J. Cell Biochem. 78, 531-549
71. Sahagian, G. G., Distler, J. and Jourdian, G. W. (1981) Proc. Natl. Acad. Sci. U.S.A. 78, 4289-4293
72. Steiner, A. W. and Rome, L. H. (1982) Arch. Biochem. Biophys. 214, 681-687
73. Sahagian, G. G., Distler, J. and Jourdian, G. W. (1982) Methods Enzymol. 83, 392-396
74. Fischer, H. D., Gonzales-Noriega, A., Sly, W. S. and Morrė, D. J. (1980) J. Biol. Chem. 255, 9608-9615
75. Campbell, C. H., Fine, R. E., Squicciarini, J. and Rome, L. H. (1983) J. Biol. Chem. 258, 2628-2633

76. Brown, M. S., Anderson, R. G. W. and Goldstein, J. L. (1983) Cell 32, 663-667
77. Creek, K. E. and Sly, W. S. (1982) J. Biol. Chem. 257, 9931-9937
78. Natowicz, M., Hallett, D. N., Frier, C., Chi, M., Schlesinger, P. H. and Baenziger, J. U. (1983) J. Cell Biol. 96, 915-919
79. Fischer, H. D., Creek, K. E. and Sly, W. S. (1982) J. Biol. Chem. 257, 9939-9943
80. Willingham, M. C., Pastan, I. H., Sahagian, G. G., Jourdian, G. W. and Neufeld, E. F. (1981) Proc. Natl. Acad. Sci. U.S.A. 78, 6967-6971
81. Rome, L. H., Garvin, A. J., Alietta, M. M. and Neufeld, E. F. (1979) Cell 16, 143-153
82. Goldberg, D. E. and Kornfeld, S. (1983) J. Biol. Chem. 258, 3159-3165
83. Geuze, H. J., Slot, J. W., Strous, G.J.A.M., Hasilik, A. and von Figura, K. (1983) submitted
84. Friend, D. S. and Farquhar, M. G. (1967) J. Cell Biol. 35, 357-376
85. Tümmers, S., Zühlsdorf, M., Hasilik, A. and von Figura, K. (1983) submitted
86. Wehland, J., Willingham, M. C., Gallo, M. G. and Pastan, I. (1982) Cell 28, 831-841
87. Rothman, J. E. and Fine, R. E. (1980) Proc. Natl. Acad. Sci. U.S.A. 77, 780-784
88. Tycko, B. and Maxfield, F. R. (1982) Cell 28, 643-651
89. Forgac, M., Cantley, L., Wiedenmann, B., Altstiel, L. and Branton, D. (1983) Proc. Natl. Acad. Sci. U.S.A. 80, 1300-1303
90. Marsh, M., Bolzan, E. and Helenius, A. (1983) Cell 32, 931-940
91. Bridges, K., Harford, J., Ashwell, G. and Klausner,R.D. (1982) Proc. Natl. Acad. Sci. U.S.A. 79, 350-354
92. Robbins, A. R., Myerowitz, R., Youle, R. J., Murray,G.J. and Neville, D. M. jr. (1981) J. Biol. Chem. 256, 10618-10622
93. Robbins, A. R. and Myerowitz, R. (1981) J. Biol. Chem. 256, 10623-10627
94. Gabel, C. A., Goldberg, D. E. and Kornfeld, S. (1983) Proc. Natl. Acad. Sci. U.S.A. 80, 775-779
95. Reitman, M. L., Varki, A. and Kornfeld, S. (1981) J.Clin. Invest. 67, 1574-1579
96. Hasilik, A., Waheed, A., Cantz, M. and von Figura, K. (1982) Eur. J. Biochem. 122, 119-123

97. Robey, P. G. and Neufeld, E. F. (1982) Arch. Biochem. Biophys. 213, 251-257
98. Gilbert, E. F., Dawson, G., ZuRhein, G. M., Opitz,J.M. and Spranger, J. W. (1973) Z. Kinderheilk. 114,259-292
99. Waheed, A., Pohlmann, R., Hasilik, A., von Figura, K., Elsen, A. and Leroy, J. G. (1981) Biochem. Biophys.Res. Commun. 105, 1052-1058
100. Hasilik, A., Pohlmann, R. and von Figura, K. (1983) Biochem. J. 210, 795-802
101. Pohlmann, R., Krüger, S., Hasilik, A. and von Figura,K. unpublished
102. Imort, M., Zühlsdorf, M., Feige, U., Hasilik, A. and von Figura, K. (1983) Biochem. J., in press
103. Tolstoshev, P., Berg, R. A., Rennard, R. I., Bradley, K. H., Trapnell, B. C. and Crystal, R. G. (1981) J. Biol. Chem. 256, 3135-3140
104. Berg, R. A., Schwartz, M. L. and Crystal, R. G. (1980) Proc. Natl. Acad. Sci. U.S.A. 77, 4746-4750
105. Diegelmann, R. F., Cohen, I. K. and Guzelian, P. S. (1980) Biochem. Biophys. Res. Commun. 97, 819-826
106. Hashimoto, C., Cohen, R. E., Zhany, W.-J. and Ballou, C. E. (1981) Proc. Natl. Acad. Sci. U.S.A. 78,2244-2248
107. Schwaiger, H., Hasilik, A., von Figura, K., Wiemken,A. and Tanner, K. (1982) Biochem. Biophys. Res. Commun.104, 950-956
108. Onishi, H. R., Tkacz, J. S. and Lampen, J. O. (1979) J. Biol. Chem. 254, 11943-11952
1o9. Stevens, T., Esmon, B. and Scheckman, R. (1982) Cell 30, 439-448
110. Hasilik, A. and Tanner, W. (1976) Biochem. Biophys. Res. Commun. 72, 1430-1436
111. Hasilik, A. (1980) Trends Biochem. Sci. 5, 237-240
112. Neufeld, E. F. (1981) in: Lysosomes and Lysosomal Storage Diseases (J. W. Callahen and J. A. Lowden, eds.) pp 131-146, Raven Press, New York
113. Hasilik, A. and von Figura, K. (1983) in: Lysosomes in Biology and Pathology (R. T. Dean and J. T. Dingle, eds.) Vol. 7, Elsevier/North Holland, Amsterdam, in press
114. Skudlarek, M. D. and Swank, R. T. (1979) J. Biol.Chem. 254, 9939-9942
115. Skudlarek, M. D. and Swank, R. T. (1981) J. Biol.Chem. 256, 10137-10144
116. Frisch, A. and Neufeld, E. F. (1981) J. Biol. Chem.256, 8242-8246
117. Steckel, F., Hasilik, A. and von Figura, K. (1983) submitted

118. Hasilik, A., von Figura, K., Conzelmann, E., Nehrkorn, H. and Sandhoff, K. (1982) Eur. J. Biochem. 125,317-321
119. Hemmings, B. A., Zubenko, G. S., Hasilik, A. and Jones, E. W. (1981) Proc. Natl. Acad. Sci. U.S.A. 78,435-439
120. Zubenko, G. S., Park, F. J. and Jones, E. W. (1983) Proc. Natl. Acad. Sci. U.S.A. 80, 510-514
121. Moore, H.-P., Gumbiner, B. and Kelly, R. B. (1983) Nature 302, 434-436

BIOSYNTHESIS OF GLYCOPHORIN A

Carl G. Gahmberg[1], Mikko Jokinen[1] and Leif C. Andersson[2]
Departments of Biochemistry[1] and Pathology[2], University of Helsinki, Helsinki, Finland

Abstract

The major human red cell sialoglycoprotein, glycophorin A, is structurally one of the best known membrane proteins. Until recently, however, nothing was known about its biosynthesis. We found in 1979 that the K562 cell line is erythroid and synthesizes glycophorin A. Glycophorin A mRNA isolated from K562 cells directed *in vitro* the synthesis of a protein with an apparent molecular weight of 19500. That exceeded the molecular weight of the apoprotein with 5000, indicating the presence of a signal peptide. Glycophorin A synthesized *in vitro* in the presence of dog pancreatic membranes had an apparent molecular weight of 37000. This precursor bound to lentil lectin but not to *Helix pomatia* lectin. Pulse-chase experiments *in vivo* showed that initially a 37000 molecular weight protein, which bound to lentil lectin, was synthesized which gradually was replaced by a 39000 molecular weight protein. Glycosylation was completed in about 10 min and the protein appeared at the cell surface after 25 min. Inhibition of glycosylation of the single N-glycosidic oligosaccharide by tunicamycin did not affect the migration of the protein to the cell surface. Using the N-acetylgalactosamine specific lectin from *Helix pomatia* coupled to Sepharose two early N-acetylgalactosamine-containing O-glycosylated glycophorin A precursors with apparent molecular weights of 24000 and 27000 were identified. O-glycosylation took place in the presence of carboxylic ionophore monensin, but the protein stained intracellularly. The relationship between the 37000 and 24000-27000 molecular weight proteins remains unclear but the results indicate that the synthesis of O-glycosylated glycoproteins is more complex than generally believed and may involve early uncharacterized precursors.

Introduction

During the past few years it has become possible to study the biosynthesis of mammalian secretory and membrane glycoproteins in considerable detail (see Blobel, 1980). These proteins are synthesized on membrane-bound ribosomes and in most cases they contain NH_2-terminal peptide extensions, called signal peptides, which are cleaved off co-translationally. Several molecules and components of the endoplasmic reticulum are involved in directing the newly synthesized polypeptides to their right locations at the membrane and into the lumen of the endoplasmic reticulum (Meyer et al., 1982; Walter & Blobel, 1982).

From the rough endoplasmic reticulum the polypeptides move by way of smooth endoplasmic membranes to the Golgi apparatus. Somewhere in this region the final sorting evidently takes place and possibly plasma membrane-destined proteins move in coated vesicles to the membrane. After fusion of these vesicles with the membrane the carbohydrate portions become externally located (Lodish & Rothman,1979).

The N-glycosylation is initiated co-translationally in the rough endoplasmic reticulum by the transfer of a $(glucose)_3$-$(mannose)_9$-$(N\text{-acetylglucosamine})_2$-oligosaccharide from the dolichol pyrophosphate derivative to the growing chain (Hubbard & Ivatt, 1981). The protein-bound oligosaccharide is then partially hydrolyzed by exoglycosidases on its way to its final cellular or extracellular location. Finally, peripheral sugars are added to build up the complete oligosaccharide.

The subcellular locations of the trimming glycosidases and the sugar transferases acting on N-glycosidic oligosaccharides have been studied extensively. The glycosidases are enriched in the rough and smooth endoplasmic reticulum membranes (Tabas & Kornfeld, 1979; Tulsiani et al., 1982) whereas the sugar transferases predominantly are Golgi-located.

The mechanisms and subcellular locations of O-glycosylation to serine and threonine residues have not been studied to a comparable extent. Glycosyl transferases adding the inner N-acetylgalactosamine residues to the polypeptides have been claimed to be located in Golgi membranes (Hanover et al., 1980; Johnson & Spear, 1983). This glycosylation occurs with UDP-N-acetylgalactosamine as the sugar donor, and no lipid intermediates adding N-acetylgalactosamine have been found.

The structure of very few cellular membrane glycoproteins is known in any detail, although the number is rapidly increasing mainly due to the advance in DNA-technology. Glycophorin A, the major sialoglycoprotein of the human red cell, was the first mammalian integral membrane glycoprotein, the primary structure of which was determined (Tomita & Marchesi, 1975). It contains 131 amino acids. The NH_2-terminal portion has all the carbohydrates and is located on the external aspect of the cell. It is followed by a hydrophobic stretch which is located within the membrane. The COOH-terminal portion is confined to the cytoplasm (Fig. 1 A).

The finding of human red cells lacking glycophorin A, En (a-) cells,(Gahmberg et al., 1976; Dahr et al., 1976; Tanner & Anstee, 1976) made the preparation of specific anti-glycophorin A antiserum possible. Crude antiserum was absorbed with the En(a-) membranes and the resulting antiserum was specific for glycophorin A on red cells and their precursors (Gahmberg et al., 1978).

The extensive knowledge on the structure of glycophorin A and the availability of specific antiserum made the protein an obvious candidate for biosynthetic studies. We therefore looked for cell lines synthesizing the protein. We then found that the leukemia cell line, K562, previously considered to be myeloid, in fact expressed glycophorin A (Gahmberg et al., 1979) and could be induced to erythroid differentiation, including the synthesis of hemoglobin (Andersson et al., 1979; Rutherford et al., 1979).

The biosynthetic studies showed that the synthesis of the polypeptide portion and its N-glycosylation resembled those described for viral membrane proteins (Jokinen et al.,

A

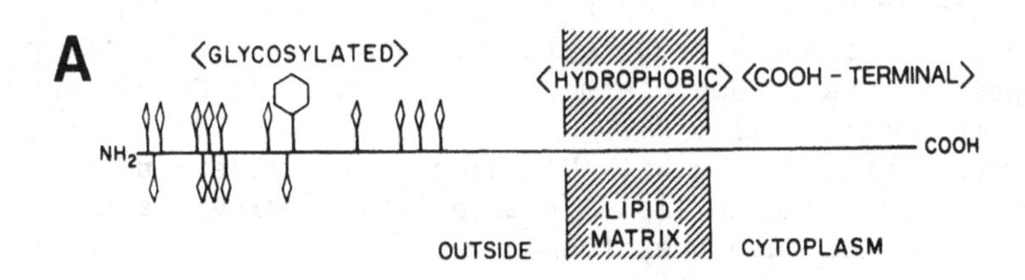

B

NANAα2–3Galβ1–3GalNAc –SER/THR
|
α2–6
NANA

Fig. 1. A. Schematic structure of glycophorin A. The positions of the oligosaccharides are indicated.
B. Structure of major O-glycosidic oligosaccharides.

1979). Other mammalian glycoproteins of this type are synthesized by similar mechanisms.

Glycophorin A differs, however, from most integral membrane glycoproteins in one important aspect. The protein contains 15 O-glycosidic oligosaccharides, most of which have the structure N-acetylneuraminic acid (α2-3)-galactose (β1-3)[N-acetylneuraminic acid (α2-6)]-N-acetylgalactosamine (Fig. 1 B). These oligosaccharides are linked to serine and threonine residues in the NH_2-terminal part of the protein (Fig. 1 A). This large amount of O-glycosidic oligosaccharides should make glycophorin A an excellent model for the synthesis of this type of oligosaccharides.

In this review we describe the synthesis of glycophorin A *in vivo* and *in vitro* with emphasis on the mechanisms of O-glycosylation. Other aspects of the biosynthesis of glycophorin A have been reviewed previously (Gahmberg et al., 1981; Gahmberg et al., in press).

Materials and Methods

All the materials and methods have been described in original articles and the reader is referred to these (Gahmberg et al., 1979; Jokinen et al., 1979; Gahmberg et al., 1980; Jokinen et al., 1981).

Results

Synthesis of glycophorin A in cell-free systems. A mRNA fraction was obtained from K562 cells and translated in a cell-free lysate made from rabbit reticulocytes (Villa-Komaroff et al., 1974). In the absence of membranes a protein with an apparent molecular weight of 19500 was obtained by immune precipitation with anti-glycophorin A antiserum (Fig. 2 B). The molecular weight of the glycophorin A apoprotein can be calculated from the amino acid sequence and is about 14500. This indicates that glycophorin A is synthesized as a larger precursor, most probably containing a signal peptide with an approximate molecular weight of 5000. When protein synthesis took place in the presence of canine pancreatic microsomal membranes a precursor molecule with an apparent molecular weight of 37000 was obtained (Fig. 2 F). This value substantially exceeded that of the apoprotein and strongly indicated that the protein was glycosylated. Glycophorin A contains only a single N-glycosidic oligosaccharide which is located at asparagine-26. This was present in the *in vitro* synthesized protein in an incomplete form because the protein absorbed to lentil lectin-Sepharose, specific for glucose/mannose residues.

In addition, treatment with endo-N-acetylglucosaminidase H lowered its apparent molecular weight with about 2000 (Fig. 2 G). From the apparent molecular weight one can assume that it also must be O-glycosylated, but interestingly the protein did not absorb to *Helix pomatia* lectin-Sepharose, specific for N-acetylgalactosamine residues.

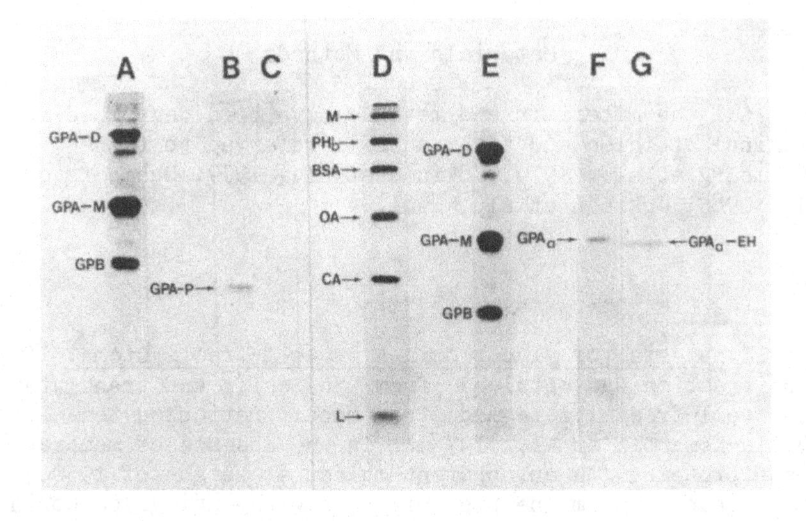

Fig. 2. Synthesis of glycophorin A *in vitro*. A, polyacrylamide gel electrophoresis pattern of periodate/$NaB[^3H]_4$ labeled red cells (Gahmberg & Andersson, 1977). GPA-D = glycophorin A dimer, GPA-M = glycophorin A monomer, GPB = glycophorin B; B, protein immune precipitated with anti-glycophorin A antiserum from proteins synthesized *in vitro* in the absence of membranes; C, with pre-immune serum; D, pattern of standard proteins: M = myosin, PH_b = phosphorylase b, BSA = bovine serum albumin, OA = ovalbumin, CA = carbonic anhydrase, L = lysozyme; E, pattern of periodate/ $NaB[^3H]_4$ labeled red cells; F, protein immune precipitated with anti-glycophorin A antiserum from proteins synthesized *in vitro* in the presence of membranes; G, sample as in F but subsequently treated with endo-N-acetyglucosaminidase H.

Synthesis of glycophorin A in K562 cells. K562 cells were pulse-chase labeled with $[^{35}S]$methionine, lysed in detergent and the extracts immune precipitated with anti-glycophorin A antiserum. A 37000 molecular weight protein was first observed (Fig. 3 A) and during the chase a 39000 molecular weight protein appeared (Fig. 3 G). The

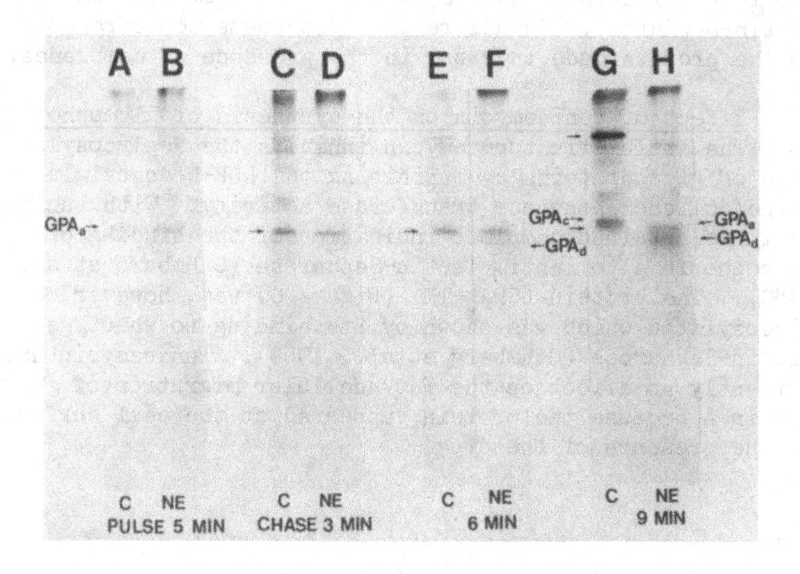

Fig. 3. Pulse-chase labeling of glycophorin A. K562 cells were labeled for 5 min with [^{35}S]methionine and chased as indicated. Detergent extracts were applied to lentil lectin-Sepharose columns and eluates immune precipitated with anti-glycophorin A antiserum. Aliquots of the samples were treated with neuraminidase (NE) after immune precipitation. C = control. GPA_a = glycophorin A precursor; GPA_c = complete glycophorin A; GPA_d = glycophorin A treated with neuraminidase.

37000 molecular weight protein migrated on polyacrylamide gel electrophoresis with the same mobility as the *in vitro* made protein synthesized in the presence of membranes. The 37000 molecular weight protein (GPA_a) was sensitive to endo-N-acetylglucosaminidase H treatment, which showed that it contained an incomplete N-glycosidic oligosaccharide. The addition of sialic acids to the protein occurred after about 9 min, which was observed by using neuraminidase treatment. This treatment resulted in a faster mobility of the highly sialylated proteins on gels (cf. ref. Gahmberg & Andersson, 1982). The GPA_a protein readily absorbed to lentil lectin-Sepharose, but not to *Helix pomatia* lectin-Sepharose. In all respects it seems to correspond to the protein made *in vitro* in the presence of membranes.

Effect of tunicamycin on the synthesis of glycophorin A. The antibiotic tunicamycin inhibits the N-glycosylation of glycoproteins by inhibiting the UDP-N-acetylglucosamine/dolicholphosphate transferase activity. With tunicamycin we obtained complete inhibition of the binding of glycophorin A to lentil lectin-Sepharose (Gahmberg et al., 1980). The protein (GPA_c-TM) (Fig. 4 C) was, however, O-glycosylated which was shown by its binding to wheat germ lectin-Sepharose (Gahmberg et al., 1980). Tunicamycin had evidently no effect on the intracellular migration of glycophorin A because the protein appeared at the cell surface in the presence of the drug.

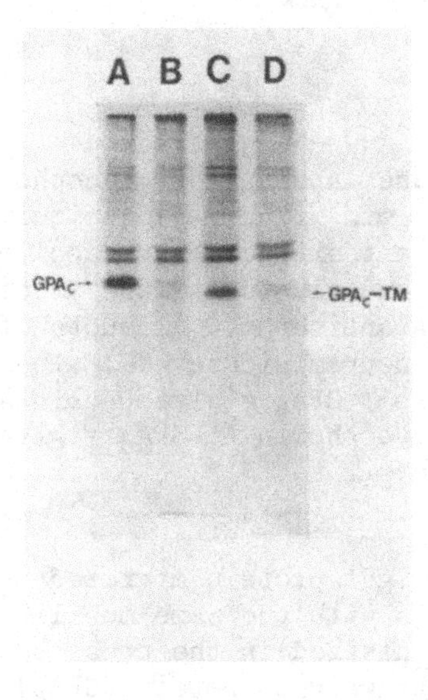

Fig. 4. Effect of tunicamycin on the synthesis of glycophorin A. A, control; B, control obtained from trypsin-treated intact cells; C,glycophorin A obtained in the presence of tunicamycin; D,immune precipitate obtained from trypsin-treated cells grown in the presence of tunicamycin.

Identification of N-acetylgalactosamine-containing precursors of glycophorin A. To identify N-acetylgalactosaminyl residues in glycophorin A we used the lectin from *Helix pomatia* coupled to Sepharose. From the sugar eluates of detergent extracts of [^{35}S]methionine pulse-chase labeled cells we could immune precipitate with anti-glycophorin A antiserum a glycophorin A precursor molecule (GPA_{H1}) (Fig. 5 A,C) with an apparent molecular weight of 24000. This already appeared after a 5 min pulse. During chase its apparent molecular weight increased to 27000 (GPA_{H2}, Fig. 5 C) and finally the complete glycophorin A monomer appeared (GPA_d, Fig. 5 G). These early molecules (GPA_{H1} and GPA_{H2}) did not absorb to lentil lectin-Sepharose.

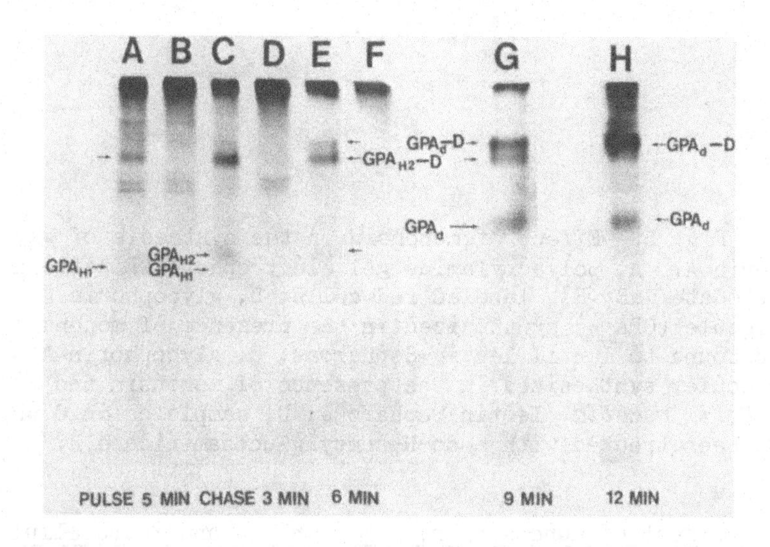

Fig. 5. Pulse-chase labeling of glycophorin A. K562 cells were labeled for 5 min with [^{35}S]methionine and chased as indicated. Detergent extracts were treated for 5 min with neuraminidase to expose N-acetylgalactosamines and applied to *Helix pomatia* lectin-Sepharose columns and immune precipitated with anti-glycophorin A antiserum. GPA_{H1} and GPA_{H2}, precursors of glycophorin A. Other symbols, see legend to Fig. 3.

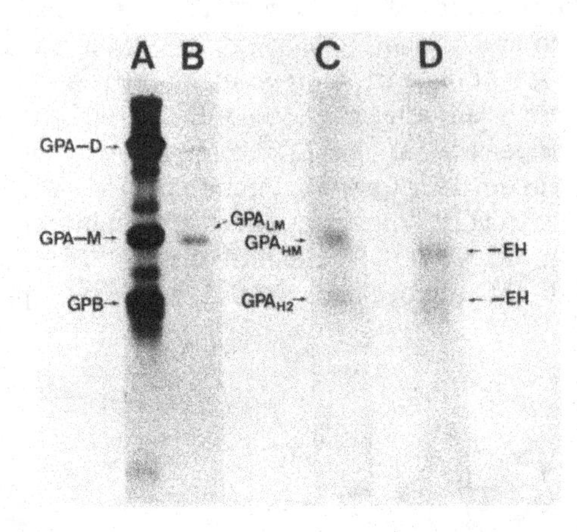

Fig. 6. Effect of monensin on the synthesis of glycophorin A. A, polyacrylamide gel electrophoresis pattern of periodate/$NaB[^{3}H]_4$ labeled red cells; B, glycophorin A molecule (GPA_{LM}) synthesized in the presence of monensin and bound to lentil lectin-Sepharose; C, glycophorin A molecules synthesized in the presence of monensin and bound to *Helix pomatia* lectin-Sepharose; D, sample as in C which has been treated with endo-N-acetylglucosaminidase H.

Effect of monensin on the synthesis and intracellular migration of glycophorin A. The carboxylic ionophore monensin has been used to inhibit intracellular migration of newly synthesized proteins (Tartakoff, 1983). When K562 cells were treated with 10 μg/ml of monensin and pulse-chase labeled with $[^{35}S]$methionine two proteins, GPA_{H2} and GPA_{HM}, accumulated. They bound to *Helix pomatia* lectin-Sepharose and were precipitated with anti-glycophorin A antiserum (Fig. 6 C). Both were sensitive to endo-N-acetylglucosaminidase H treatment (Fig. 6 D). The proteins did not reach the outer cell surface as shown by their insensitivity to trypsin treatment of intact cells (results not shown).

Discussion

It has become clear that in many respects the results obtained concerning the biosynthesis of glycophorin A are applicable to that of other mammalian and viral glycoproteins. Glycophorin A is evidently synthesized with a signal peptide which is cleaved co-translationally. Simultaneously N-glycosylation occurs. It takes about 10 min for the protein to reach the Golgi apparatus and 25 min for it to appear at the external cell surface (Jokinen et al., 1979). Tunicamycin inhibited the synthesis of the N-glycosidic oligosaccharides but had no effect on the O-glycosylation and the intracellular migration of the protein (Gahmberg et al., 1980).

Presently, the most interesting aspect of the biosynthesis of glycophorin A and the greatest benefit to use glycophorin A as a model protein for biosynthetic studies is for studies on the mechanism of O-glycosylation and the subcellular location of this process.

For such investigations glycophorin A is excellent because both the structure of the O-glycosidic oligosaccharides and their locations in the polypeptide chain are known. Furthermore, glycophorin A contains fifteen O-glycosidic oligosaccharides, together constituting more than half of the molecular weight of the protein.

We have previously found that after a short pulse with [^{35}S]methionine a glycophorin A precursor (GPA_a, molecular weight 37000) is obtained which interacts with lentil lectin. This fact and its endo-N-acetylglucosaminidase H sensitivity showed that it contains the single N-glycosidic oligosaccharide. It must also be O-glycosylated because the apparent molecular weight substantially exceeded that of the apoprotein. Most interestingly, the protein did not interact with the N-acetylgalactosamine-specific lectin from *Helix pomatia*. On the other hand we have now found smaller glycophorin A precursors (GPA_{H1} and GPA_{H2}), which interact with this lectin. No doubt, these precursors contain N-acetylgalactosamine residues. The GPA_{H1} was observed after 5 min of labeling and during chase it increased in size to GPA_{H2} and finally to the size of complete glycophorin A. In the presence of monensin GPA_{H2} accumulated and it did not reach the cell surface. Monensin is considered to stop the intracellular migration of glycoproteins somewhere in the Golgi region (Tartakoff, 1983). Using viral glycoproteins no O-glycosylation of these were observed in the presence of the drug which shows that their O-glycosylation was beyond the site of monensin action (Nieman et al., 1983; Johnson & Spear, 1983).

Our pulse chase studies showed that the addition of N-acetylgalactosamines to glycophorin A is an early biosynthetic event and this fact was further strengthened by the results of the monensin experiments, which showed the accumulation of O-glycosylated GPA_{H2}. The relationship between GPA_a and the GPA_{H1} and GPA_{H2} molecules remains unclear. We think that GPA_a is the primary biosynthetic product which contains unknown O-glycosidic structures, which are removed early during biosynthesis and gradually replaced by N-acetylgalactosamines. A good candidate for this unknown structure is mannose. In yeast, serine/threonine linked mannosyl chains are known to occur and these are synthesized through dolichol-P-mannose (see Tanner, this vol.)

It could be important that proteins that finally will be highly O-glycosylated with N-acetylgalactosamines at the reducing ends of the oligosaccharides early during biosynthesis receive hydrophilic side chains at serine/threonine residues. In this way they could achieve a similar conformation as the final protein. Removal of them and simultaneous replacement by N-acetylgalactosamines could be a way to keep the protein in its right shape. Viral proteins containing few O-glycosidic oligosaccharides could achieve their proper configuration without pre-Golgi glycosylation.

Acknowledgements: We thank A. Asikainen and P. Nieminen for technical assistance and B. Björnberg for secretarial help. These studies were supported by the Academy of Finland, the Sigrid Jusélius Foundation, the Association of Finnish Life Insurance Companies and the National Cancer Institute Grant 5 R01 CA26294-03.

References

Andersson, L.C., Jokinen, M. & Gahmberg, C.G. (1979) *Nature* 278, 364-365.

Blobel, G. (1980) *Proc. Natl. Acad. Sci. USA* 77, 1496-1500.
Dahr, W., Uhlenbruck, G., Leikola, J., Wagstaff, W. & Landfried, K. (1976) *Immunogenet.* 3, 329-346.
Gahmberg, C.G. & Andersson, L.C. (1977) *J. Biol. Chem.* 252, 5888-5894.
Gahmberg, C.G. & Andersson, L.C. (1982) *Eur. J. Biochem.* 122, 581-586.
Gahmberg, C.G., Jokinen, M. & Andersson, L.C. (1978) *Blood* 52, 379-387.
Gahmberg, C.G., Jokinen, M. & Andersson, L.C. (1979) *J. Biol. Chem.* 254, 7442-7448.
Gahmberg, C.G., Jokinen, M., Karhi, K.K. & Andersson, L.C. (1980) *J. Biol. Chem.* 255, 2169-2175.
Gahmberg, C.G., Jokinen, M., Karhi, K.K., Kämpe, O.,Peterson, P.A. & Andersson, L.C. *Methods Enzymol.*,in press.
Gahmberg, C.G., Jokinen, M., Karhi, K.K., Ulmanen, I., Kääriäinen, L. & Andersson, L.C. (1981) *Blood Transfusion and Immunohaematology XXIV*, 53-73.
Gahmberg, C.G., Myllylä, G., Leikola, J., Pirkola, A. & Nordling, S. (1976) *J. Biol. Chem.* 251, 6108-6116.
Hanover, J.A., Lennarz, W.J. & Young, J.D. (1980) *J. Biol. Chem.* 255, 6713-6716.
Hubbard, S.V. & Ivatt, R.J. (1981) *Annu. Rev. Biochem.* 50, 555-583.
Johnson, D.C. & Spear, P.G. (1983) *Cell* 32, 987-997.
Jokinen, M., Gahmberg, C.G. & Andersson, L.C. (1979) *Nature* 279, 604-607.
Jokinen, M., Ulmanen, I., Andersson, L.C., Kääriäinen, L. & Gahmberg, C.G. (1981) *Eur. J. Biochem.* 114, 393-397.
Lodish, H.F. & Rothman, J.E. (1979) *Sci. American* 240, 48-63.
Meyer, D.I., Krause, E. & Dobberstein, B. (1982) *Nature* 297, 647-650.
Nieman, H., Boschek, B., Evans, D., Rosing, M., Tamura, T. & Klenk, H.-D. (1983) *EMBO Journal* 1, 1499-1504.
Rutherford, T.R., Clegg, J.B. & Wheaterall, D.J. (1979) *Nature* 280, 164-165.
Tabas, I. & Kornfeld, S. (1979) *J. Biol. Chem.* 254, 11655-11663.
Tanner, M.J.A. & Anstee, D.J. (1976) *Biochem. J.* 153, 271-277.
Tartakoff, A.M. (1983) *Cell* 32, 1026-1028.

Tulsiani, D.R.P., Hubbard, S.C., Robbins, P.W. & Touster, O. (1982) *J. Biol. Chem.* 257, 3660-3668.

Villa-Komaroff, L., McDowell, M., Baltimore, D. & Lodish, H.L. (1974) *Methods Enzymol.* 30, 709-723.

Walter, P. & Blobel, G. (1982) *Nature* 299, 691-698.

PROTEIN GLYCOSYLATION

Carl G. Gahmberg

Department of Biochemistry and
Pathology, University of Helsinki,
Helsinki, Finland.

Most proteins synthesized on membrane-bound ribosomes in mammalian and yeast cells are *glyco*proteins. Usually the oligosaccharides are N- or O-linked to asparagine and serine/threonine residues, respectively. Glycosylation is initiated in the rough endoplasmic reticulum and completed in the Golgi apparatus.

Several enzymes are involved in the synthesis of the oligosaccharides including sugar transferases and glycosidases. Dr. Dallner discussed the important lipid dolichol. Dolichol is used for the initial N-glycosylation of proteins, but only a fraction of the total dolichol is actually involved in glycosylation. The function of the rest is unknown. A $(glucose)_3(mannose)_9(N\text{-acetyl glucosamine})_2$ oligosaccharide is synthesized on dolichol-pyrophosphate and this oligosaccharide is transferred to the acceptor proteins. The protein-bound oligosaccharide is then rapidly partially degraded by exoglycosidases. Then the peripheral monosaccharides are added to complete the oligosaccharides.

Dr. Hill described important results concerning the sugar tranferases adding monosaccharides to the protein backbones on the core structures of oligosaccharides. The transferases must act in a specific order, otherwise a dead-end product is formed. Several transferases are found in a cell because of their high sugar and acceptor specificities. Very few transferases have been purified. They seem to be integral membrane proteins, which need lipid or detergents for optimal activity. Using monoclonal antibodies it is becoming possible to determine their subcellular location and probe their catalytic and recognition sites.

Dr. Tanner has studied N- and O-glycosylation in yeast. Mannose is the polypeptide-linked monosaccharide both in N- and O-linked sugar chains. In both cases this sugar is added through dolichol-P-mannose, whereas the peripheral mannosyl residues are added through GDP-mannose. Using a liposomal model system containing a mannosyl transferase he has got evidence that the transferase acts as a translocator for mannose through the lipid bilayer.

Tunicamycin, known to inhibit the formation of dolichol-P-N-acetyl glucosamine, arrested the cells at the G_1-phase of the cell cycle. This result indicates that glycoprotein carbohydrate is essential for cell division in yeast.

Dr. Hasilik has studied the biosynthesis and posttranslational modifications of lysosomal enzymes. These enzymes are made as large precursors which are cleaved to much smaller proteins. They are N-glycosylated and obtain mannose-6-phosphate residues after removal of peripheral N-acetyl glucosamine residues. The uncovered mannosyl residues are important for the intracellular transport of the enzymes and their uptake from the surrounding medium. In yeast the mannose-6-phosphate residues do not seem essential.

Dr. Gahmberg has studied the biosynthesis and glycosylation of the major human red cell sialoglycoprotein, glycophorin A. This protein contains 15 O-glycosidic oligosaccharides and one N-glycosidic chain. The protein is synthesized with a signal peptide, which is cleaved off co-translationally. Evidently both N- and O-glycosylations take place early during synthesis, but the exact mechanisms of O-glycosylation remain unclear. Using the drug monensin O-glycosylation was obtained, which shows that this type of modification takes place at a pre-Golgi location in the cell.

6. PROTEIN PHOSPHORYLATION

PHOSPHORYLATION OF INITIATION FACTOR eIF-2 AND RIBOSOMAL PROTEIN S6 AND THE REGULATION OF MAMMALIAN PROTEIN SYNTHESIS

M.J. CLEMENS*, S.A. AUSTIN*, J. KRUPPA†, C.G. PROUD‡ and V.M. PAIN§

*CRC Group, Dept. of Biochemistry, St George's Hospital Medical School, London; †Institute of Physiological Chemistry, University of Hamburg; ‡Biological Laboratory, University of Kent; §School of Biological Sciences, University of Sussex.

ABSTRACT

The reversible phosphorylation of eIF-2 and of protein S6 is implicated in the regulation of protein synthesis in several eukaryotic cell types. In the case of eIF-2 it has been shown that phosphorylation impairs the ability of the factor to exchange bound GDP for GTP. This reaction is catalysed by a high molecular weight protein complex, the interaction of which with eIF-2 is modulated by phosphorylation of the latter. GDP is generated by hydrolysis of GTP during 80S initiation complex formation, and also prematurely during 40S complex formation (at least *in vitro*). Since it must be exchanged for GTP before eIF-2 can function catalytically, phosphorylation impairs the ability of eIF-2 to recycle between successive rounds of protein synthesis.

Control of eIF-2 activity is implicated in the regulation of translation by amino acid starvation in cultured cells. Current data suggest a role for GDP exchange in this system also, and eIF-2 specific protein kinase activity is detectable. However, there is no evidence for increased kinase activity in extracts from the starved cells. Recent experiments with temperature-sensitive aminoacyl-tRNA synthetase mutants (in collaboration with Dr J. Pollard) implicate synthetases in the regulatory link between amino acid availability and polypeptide chain initiation.

A number of reports have suggested a role for S6 phosphorylation in 40S subunit-messenger RNP interaction during initiation. However, a study of the kinetics of protein synthesis and changes in S6 phosphorylation during hypotonic salt shock, in the presence or absence of translational inhibitors, leads us to challenge the idea that modification of S6 plays a role in regulation of initiation.

INTRODUCTION

In recent years it has become clear that polypeptide chain initiation is regulated at the translational level under a number of physiological conditions. The mechanisms by which such regulation occurs have been studied in detail only in a few cases, most notably the rabbit reticulocyte lysate system which responds to haem deficiency and other conditions by alterations in the rate of initiation (1-4). Some studies have also been carried out using cultured mammalian cells which rapidly control their rate of protein synthesis at the translational level in response to stimuli such as amino acid starvation (5,6), heat shock (7), salt shock (8) and virus infection (9). It is now known that, at least in the case of the reticulocyte lysate, the phosphorylation of initiation factor eIF-2 provides the mechanism by which initiation is regulated in response to haem supply or low concentrations of double stranded RNA (1-4). Initiation factor eIF-2 is, however, not the only protein involved in translation which can be reversibly modified by phosphorylation; another well studied example is that of ribosomal protein S6 which shows changes in its phosphorylation state in response to a very wide range of conditions experienced by mammalian cells (10,11). It is the purpose of this article to summarise the current state of knowledge concerning the role of phosphorylation of eIF-2 and of protein S6 in the control of translation and to indicate useful experimental systems which will provide further information concerning the functions of these two proteins.

It can be argued that too much attention has been paid in the past to the reticulocyte system, which has proved attractive because of its ease of manipulation and high translational activity *in vitro*. However reticulocytes are by no means typical mammalian cells and the ways in which they regulate their protein synthesis may also not be typical of other cell types. For this reason we have directed our efforts towards analysis of the mechanisms by which mammalian tumour cells respond to amino acid starvation and to hypertonic salt shock in tissue culture, as other examples of rapid translational control occurring in response to environmental changes. The similarities and differences between these systems and the control of reticulocyte protein synthesis will hopefully become apparent in this paper.

RESULTS AND DISCUSSION

Phosphorylation of eIF-2

Studies conducted in several laboratories have now reached a general consensus of agreement concerning the mechanism by which polypeptide chain initiation is regulated by phosphorylation of initiation factor eIF-2. For many years it was unclear as to how the activity of the factor was modulated, primarily because the *in vitro* assays used to examine its activity measured only stoichiometric formation of initiation complexes (12-14). It is now clear that it is the catalytic recycling of eIF-2 between successive rounds of protein synthesis which is sensitive to phosphorylation and our own studies (15-17) and those of others (18-21) have shown that the critical step in this recycling process involves the exchange of guanine nucleotides. The molecule of GTP which is present in the 40S initiation complex, together with eIF-2 and Met-$tRNA_f$, is hydrolysed to GDP and phosphate on or just before the joining of the messenger RNA and 60S subunit to give an 80S initiation complex (12,22,23). The factor eIF-2 is lost from the ribosome at this stage (12,22) and is believed to remain as a complex with the GDP which has been generated. GDP is a potent inhibitor of eIF-2 and has a high affinity for the factor (24) and must therefore be exchanged for a new molecule of GTP before another round of initiation can be carried out. It is now clear that this exchange process is catalysed by another high molecular weight protein, termed RF (21,25), eIF-2B (19) or GEF (for guanine nucleotide exchange factor)(26). It is the ability of this factor

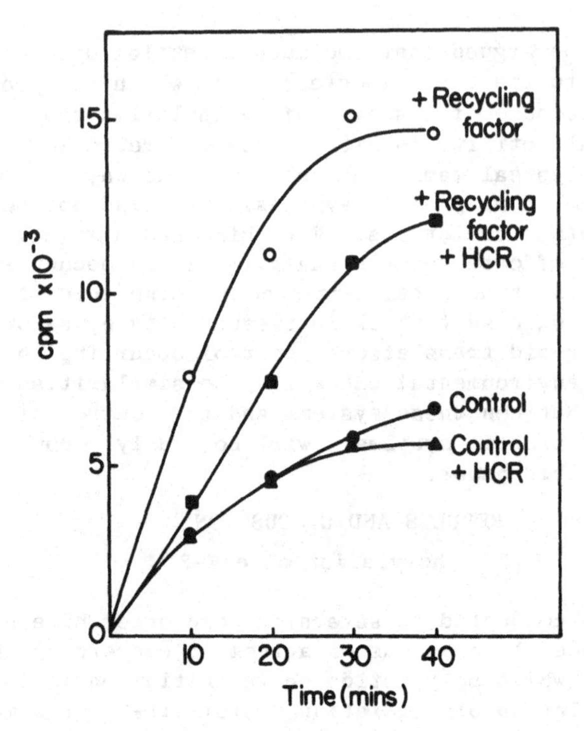

Fig. 1: Stimulation of ternary complex formation on eIF-2 by the recycling factor (GEF)

GTP-dependent binding of [^{35}S]Met-tRNA$_f$ to highly purified rat liver eIF-2 was assayed in the presence and absence of the recycling factor (guanine nucleotide exchange factor - GEF) from Ehrlich ascites cells (26). Where indicated, the eIF-2 was preincubated for 5 min at 30°C with reticulocyte haem controlled repressor (HCR) and 0.2mM ATP. For details see Ref. 16.

to interact with eIF-2 which becomes impaired after phosphorylation of the latter. Such an effect is illustrated in Fig. 1 where the kinetics of [eIF-2.GTP.Met-tRNA$_f$] ternary complex formation can be seen to be stimulated by addition of GEF (purified from Ehrlich ascites cells), in a manner which is partially blocked if the eIF-2 has previously been phosphorylated with the reticulocyte haem controlled repressor (HCR). Phosphorylation of the purified

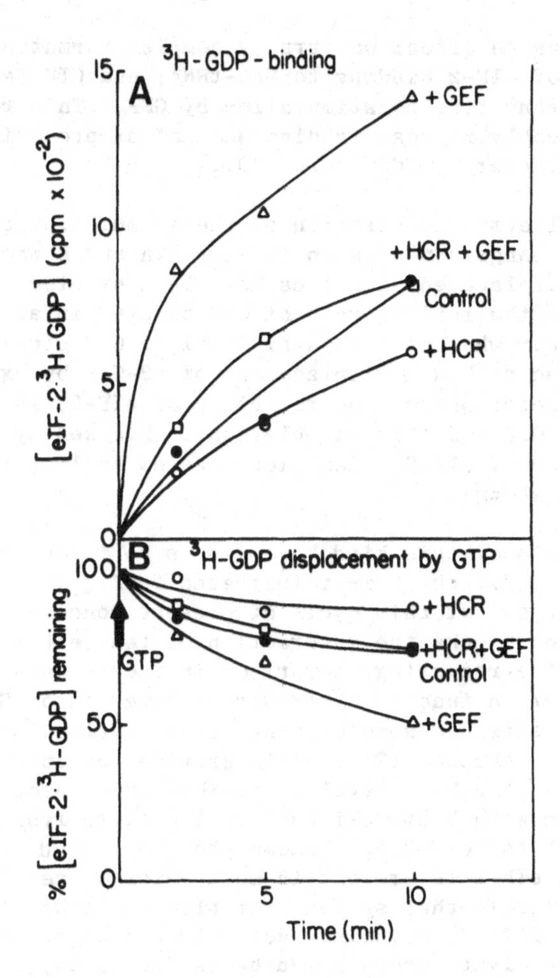

Fig. 2: Stimulation of nucleotide binding and exchange on eIF-2 by the guanine nucleotide exchange factor (GEF)

Panel A shows the kinetics of binding of [^{3}H]GDP to the same preparation of eIF-2 as in Fig. 1. Incubation conditions were identical, except that twice as much eIF-2 and GEF were used, GTP and Met-tRNA$_f$ were omitted and [^{3}H]-GDP was present at 6.7µCi/ml (0.5µM). Panel B shows the displacement of [^{3}H]GDP by 50µM unlabelled GTP, added 20 min after incubation of eIF-2 with [^{3}H]GDP (± HCR). The GEF was added, where indicated, with the GTP. Complexes of eIF-2 with the labelled nucleotide were assayed as described in ref. 15.

eIF-2 alone has no effect on ternary complex formation. The initial rate of eIF-2 binding to Met-tRNA$_f$ and GTP is particularly sensitive to stimulation by GEF. This reaction in effect probably represents displacement of pre-existing GDP from the factor by added GTP (20).

A more direct demonstration of the stimulation of nucleotide exchange can be seen in Figs. 2A and B where GDP binding and displacement kinetics have been studied. Again GEF stimulates the initial rate of GDP binding in an HCR sensitive manner whereas the basal level of GDP binding is barely affected by HCR. Displacement of ^{3}H-GDP by excess GTP, a model reaction for the recycling of eIF-2, is also stimulated by GEF and this stimulation is blocked by phosphorylation of eIF-2. GEF alone has no ability to bind GDP (data not shown).

Experiments of this kind have led us to formulate an 'eIF-2 cycle' (17), which is illustrated in Fig. 3. The least clear aspect of this cycle at present concerns whether phosphorylation blocks the interaction of the [eIF-2.GDP] complex with the recycling factor, or in fact causes accumulation of an inactive [eIF-2.recycling factor.GDP] complex. There is now some evidence accumulating in favour of the latter mechanism (21). This provides an attractive explanation for the high level of inhibition of protein synthesis seen with haem-deficient reticulocyte lysates when only 30-40% of the eIF-2 has become phosphorylated (27,28), assuming that eIF-2 is present in molar excess over the recycling factor in that system. It also explains why addition of purified recycling factor to a haem deficient lysate can reactivate protein synthesis (19,25,29,30).

Control of Protein Synthesis by Amino Acid Starvation

There is good evidence that non-reticulocyte mammalian cells also control their rates of initiation of protein synthesis at the level of eIF-2 activity and 40S initiation complex formation (31). We have studied the responses of cells to deficiencies of single essential amino acids in culture and have shown that extracts from amino acid starved cells are defective in their ability to form initiation complexes on native 40S ribosomal subunits. This defect can be overcome by addition of stoichiometric amounts of eIF-2 *in vitro* (6). The two most important questions concerning

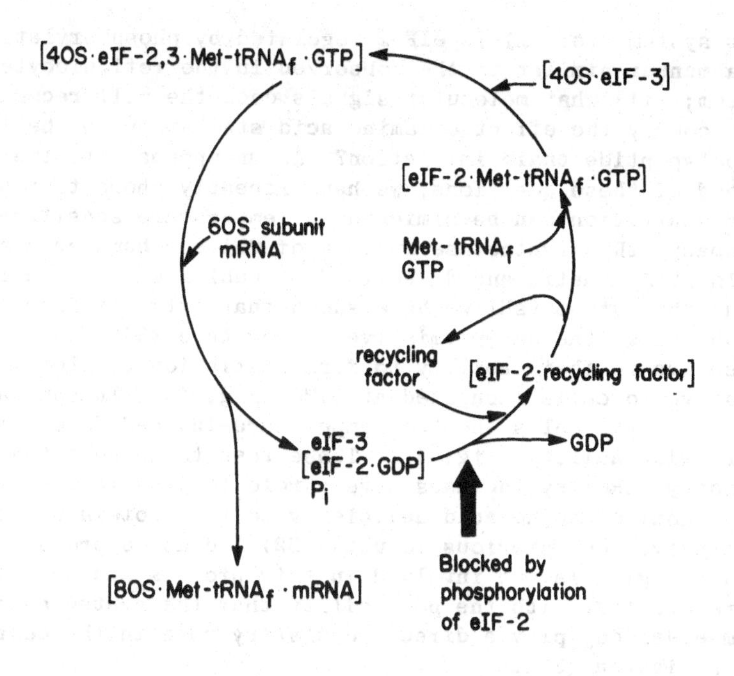

Fig. 3: The eIF-2 cycle and its regulation by phosphorylation

Current evidence indicates that during the process of polypeptide chain initiation release of eIF-2 from the ribosome occurs just before or on joining of the 60S subunit with the [40S.Met-tRNA$_f$.mRNA] initiation complex. Hydrolysis of the GTP associated with this complex also occurs at this stage and it is likely (but not rigorously proven) that the GDP generated by this reaction remains bound to the eIF-2 (the latter has a high affinity for GDP). This GDP is inhibitory and must be exchanged for a new molecule of GTP in a reaction catalyzed by a high molecular weight recycling factor or guanine nucleotide exchange factor (GEF). Phosphorylation of the α subunit of eIF-2 by the reticulocyte HCR or other protein kinases inhibits this nucleotide exchange, causing accumulation of inactive eIF-2. The rate of initiation is thus impaired. Premature hydrolysis of GTP in 40S initiation complexes may also occur under some conditions (ref. 17), requiring a similar exchange reaction to regenerate active eIF-2. Reproduced from ref. 17, by permission of The American Chemical Society.

this system are: (i) Is eIF-2 regulated by phosphorylation in a manner similar to that observed in the reticulocyte system; (ii) what molecular signals does the cell recognise that convey the effect of amino acid starvation to the level of polypeptide chain initiation? As an approach to the second of these questions, we have recently shown that amino acid starvation can be mimicked in temperature sensitive aminoacyl-tRNA synthetase mutants of Chinese hamster ovary cells (S.A. Austin and J. Pollard, unpublished observations). Using the mutant tsH1 we have shown that extracts from cells incubated at the non-permissive temperature (39°C) are defective in their ability to form initiation complexes, relative to cells incubated at 34°C (Fig. 4). In contrast the wild-type cells show no temperature-induced defect in initiation ability (Fig. 4). These results suggest that aminoacyl-tRNA synthetases have a role to play in conveying the signal of amino acid deficiency to the protein synthesis machinery. Our previous results (32) led us to propose that tRNA charging is not involved in this process and we are therefore left with the possibility that the synthetases themselves may play a direct regulatory rôle in the control of initiation (33).

It is clear that eIF-2 activity is modulated by amino acid starvation, not only because addition of this factor to extracts overcomes the difference between fed and starved cells but also because it is possible to demonstrate a difference in activity between eIF-2 from the two sources. Fig. 5 illustrates the kinetics of ternary complex formation using eIF-2 isolated by single step phosphocellulose chromatography from extracts of fed and lysine-starved cells. When ternary complex formation is assayed at low and high magnesium, a difference between the eIF-2 preparations is only observed under the latter conditions. Low magnesium conditions are highly stimulatory for the initial rate of ternary complex formation and no difference is observed in the levels of activity of eIF-2 from fed and starved cells. It has been shown that high magnesium concentrations stabilise eIF-2.guanine nucleotide complexes (20). These results are therefore consistent with the possibility that a higher proportion of eIF-2 is inactive in starved cells because it is complexed with GDP. This could be due to increased phosphorylation of eIF-2, a higher GDP:GTP ratio in the cytoplasm or to defective guanine nucleotide exchange activity. We have attempted to test whether amino acid starvation enhances the phosphorylation

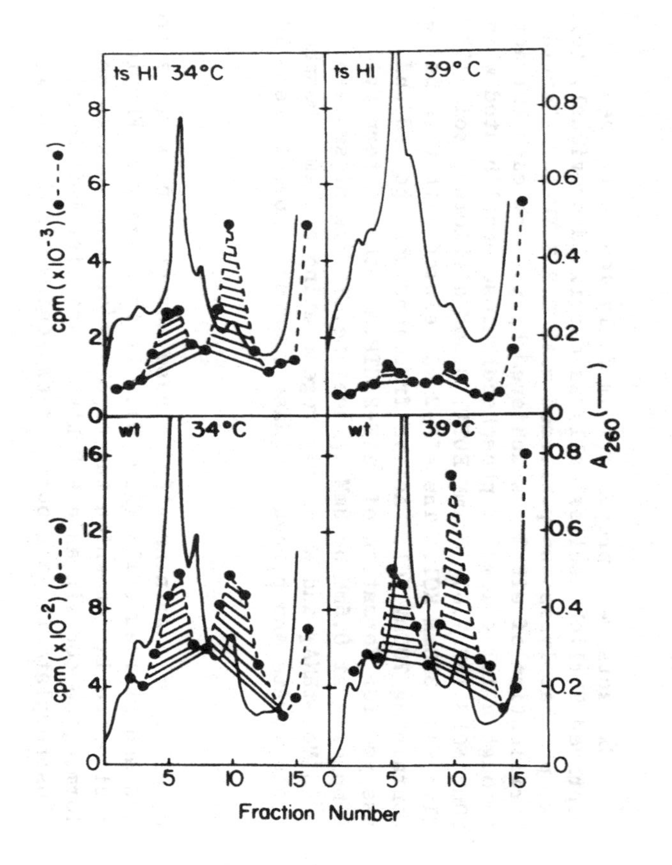

Fig. 4: Formation of 40S and 80S ribosomal initiation complexes in extracts from tsH1 and wild-type Chinese hamster ovary (CHO) cells

Extracts were prepared as described in ref. 6 from CHO cells grown in suspension culture and incubated at either 34°C or 39°C for 30 min. These extracts were then incubated for 2 min at 30°C with [^{35}S]Met-tRNA$_f$ under conditions optimal for protein synthesis and analysed by sucrose gradient centrifugation as described previously (6,32,35). Centrifugation was from right to left. 40S initiation complexes are contained in fractions 8-14 and 80S initiation complexes in fractions 4-7. (——), A_{260}; (●--●), radioactivity in Met-tRNA$_f$. tsH1 is a temperature sensitive leucyl-tRNA synthetase mutant for which 34°C is the permissive temperature and 39°C the non-permissive temperature. wt is the wild-type, which grows equally well at 34°C or 39°C.

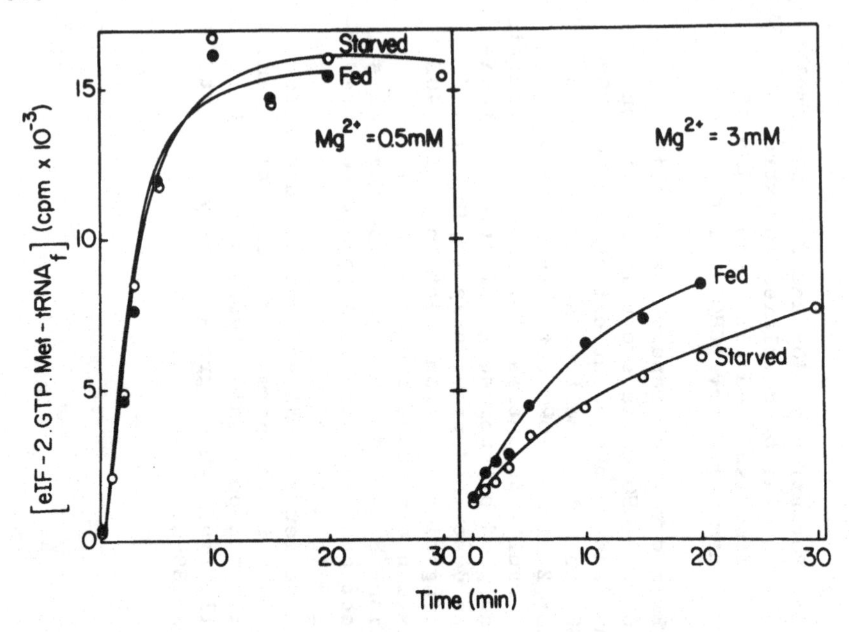

Fig. 5: Kinetics of ternary complex formation by eIF-2 from fed and amino acid-starved cells, at low and high Mg^{2+}

Extracts were prepared from fully fed or lysine-deprived Ehrlich ascites cells as described previously (6), except that lysis was performed with 0.25% NP-40. Partial purification of eIF-2 was achieved by passing each extract through a 1ml column of phosphocellulose equilibrated with 20mM MOPS (pH 7.6), 0.1mM EDTA, 0.5mM dithiothreitol, 10% glycerol, 350mM KCl. The eIF-2 was eluted with this buffer containing 700mM KCl. Active fractions were pooled and assayed for formation of [eIF-2.GTP.Met-tRNA$_f$] ternary complexes at 0.5mM or 3mM Mg acetate, in the presence of [^{35}S]Met-tRNA$_f$ and an energy regenerating system (phosphoenolpyruvate and pyruvate kinase), as described in ref. 16.

of eIF-2 by examining eIF-2 kinase activity in extracts from fed and starved cells (Fig. 6). It is clear that Ehrlich ascites cells, on which the majority of this work is performed, do contain an eIF-2 kinase activity which will phosphorylate the α subunit of exogenous eIF-2 in cell

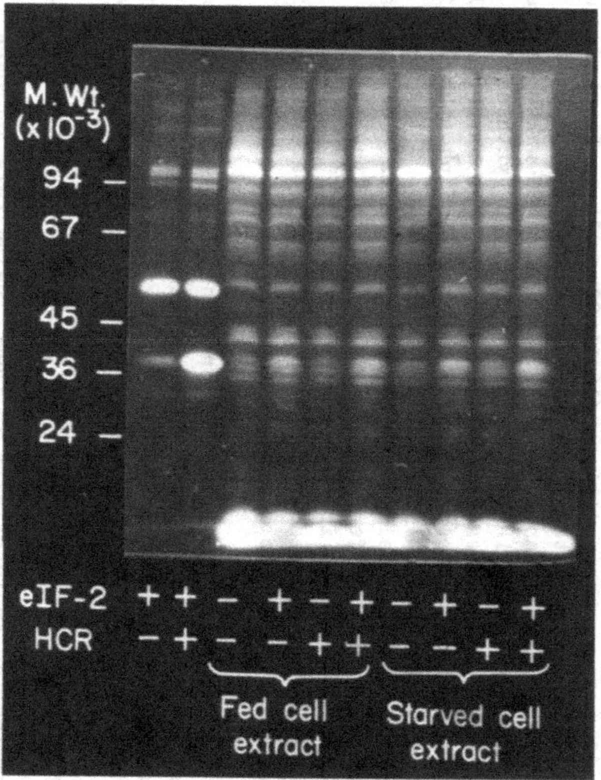

Fig. 6: Protein phosphorylation in extracts from fed and amino-acid starved cells

Extracts from fully fed or lysine-deprived Ehrlich ascites cells were incubated in the presence or γ[^{32}P]ATP (0.1mM, 150μCi/ml) for 10 min at 30°C. Where indicated, highly purified eIF-2 and reticulocyte HCR were added. The samples shown in the left-hand 2 tracks were from incubations containing the same amounts of eIF-2 ± HCR but no cell extract. The incubations were analysed by electrophoresis on a 12.5% SDS polyacrylamide gel followed by autoradiography of the dried gel. Positions of stained molecular weight markers are shown on the left. The α subunit of eIF-2, which is phosphorylated by HCR and in the cell extracts has a molecular weight of 38,000 Daltons. The β subunit of eIF-2 which is phosphorylated by a protein kinase present in the eIF-2 preparation, runs at a molecular weight of 55,000 Daltons in this gel system.

extracts (34,35). However, we have been unable to detect any difference in this kinase activity between fed and starved cells (Fig. 6). This does not necessarily rule out phosphorylation as a mechanism for controlling protein synthesis in this system; what is actually needed is an assay for the extent of phosphorylation of the endogenous initiation factor under different physiological conditions (34). This experiment has yet to be carried out. The relatively low levels of endogenous eIF-2 in these cell extracts make it difficult to assess the phosphorylation of this factor in the experiment shown in Fig. 6 (note, for example, the apparent lack of effect of added HCR on phosphorylation of endogenous eIF-2 in this experiment).

Phosphorylation of Ribosomal Protein S6

Another area which has proved controversial in recent years concerns the biological significance of the phosphorylation of the 40S ribosomal subunit protein S6 in eukaryotic cells. It is clear that multiple species of S6 exist with up to 5 moles of phosphate per mole protein, and that the distribution of the different phosphorylated forms of this protein varies significantly under different cellular growth conditions (10,11,36). In particular, cells which are stimulated from quiescence into growth by addition of serum show enhanced phosphorylation of S6 which has been equated with the increased rate of protein synthesis observed under these conditions (37-39). We have recently reinvestigated the possible link between S6 phosphorylation and the rate of polypeptide chain initiation using mouse myeloma (MPC11) cells. We have taken advantage of the rapid and reversible effect of hypertonic salt on the rate of protein synthesis in cultured cells to study the early kinetics of changes in S6 phosphorylation state in relation to the rate of protein synthesis.

Extraction of ribosomal proteins and their fractionation by two-dimensional gel electrophoresis reveals a large number of components (40) and is particularly useful in that multiple species of S6 can be resolved (Fig. 7). By labelling cells overnight with [^{35}S]methionine it is possible to quantitate the distribution of S6 between the various forms by cutting out individual spots from a 2D gel and determining the radioactivity in them. Using this approach we have quantitated the proportion of S6 in the unphosphorylated

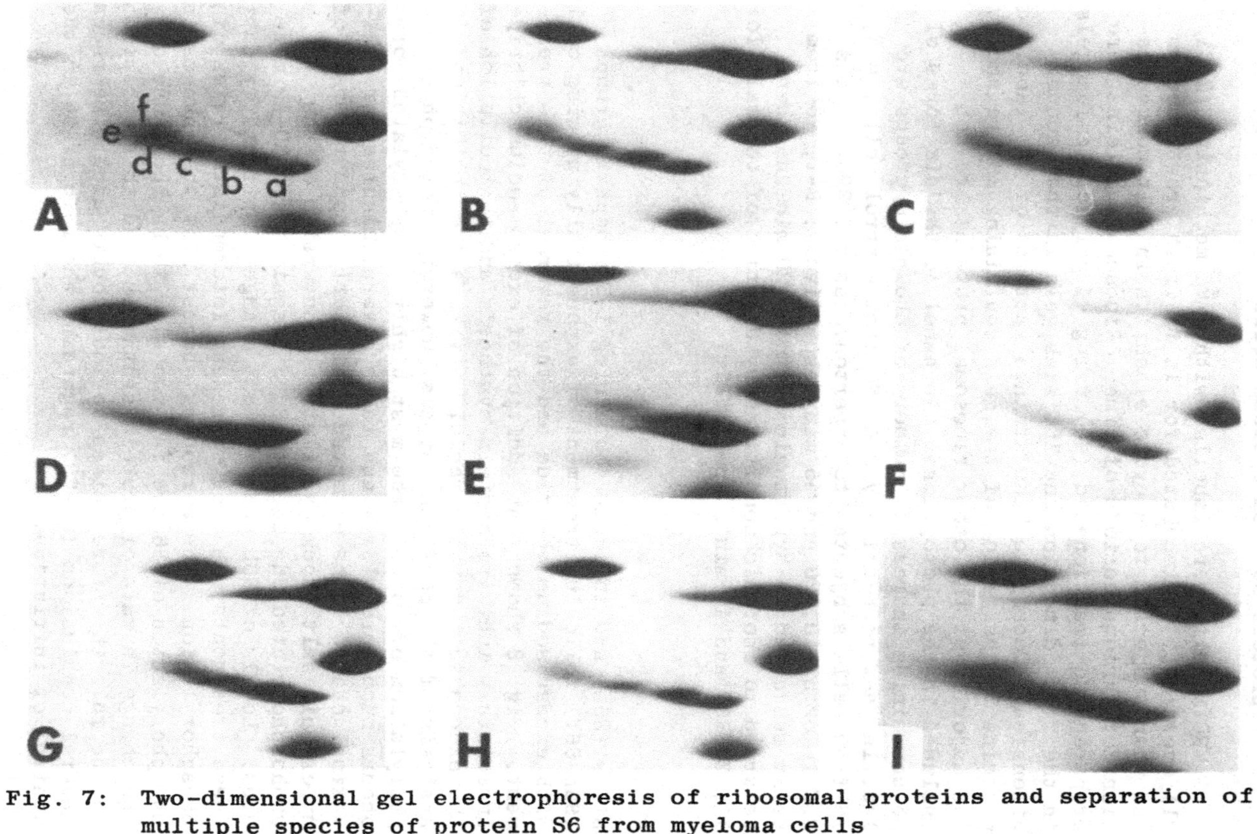

Fig. 7: Two-dimensional gel electrophoresis of ribosomal proteins and separation of multiple species of protein S6 from myeloma cells

Fig. 7: Two dimensional gel electrophoresis of ribosomal proteins and separation of multiple species of protein S6 from myeloma cells

MPC11 cells were labelled with [^{35}S]methionine (2μCi/ml) overnight, preincubated for 1h in fresh growth medium and then subjected to a cycle of shifts in salt concentration for various periods of time. Ribosomal proteins were prepared and fractionated (40) using 4% polyacrylamide gels in the first dimension and 15% slab gels in the second dimension. Coomassie blue stained patterns of the labelled proteins in the region of the gels containing S6 are shown. The majority of the other ribosomal proteins are not included in the photographs. In panel A, the six forms of S6 containing increasing numbers of phosphate groups are labelled a-f respectively. A, S6 from control cells; B-E, S6 from cells subjected to hypertonic salt (120mM excess NaCl) for 5, 10, 20 and 45 min respectively; F-I, S6 from cells subjected to hypertonic salt for 20 min and then returned to isotonic conditions by dilution of the salt for 10, 20, 30 and 45 min respectively.

state (S6a) and in forms containing 1-5 phosphate groups (S6b-S6f) following treatment of exponentially growing cells with excess sodium chloride and the reversal of this treatment. Fig. 8 shows that addition of excess sodium chloride immediately inhibits protein synthesis and that dilution of the excess salt equally rapidly reverses the inhibition. Analysis of the S6 forms reveals, however, that dephosphorylation of S6 d-f (the most highly phosphorylated forms) occurs much more slowly and takes several minutes for half-maximal change to be effected. Similarly, upon reversal of hypertonic salt shock by dilution, conversion of the unphosphorylated S6a back to S6 d-f again takes 30-40 minutes. Levels of S6b and c do not change significantly during these treatments, as they represent intermediate states of phosphorylation. These results suggest that the changes in phosphorylation of S6 are far too slow to be the cause of the change in rate of protein synthesis, and this exception to the rule therefore raises serious doubts about the recent general conclusions (37-39) regarding the role of S6 in the control of initiation.

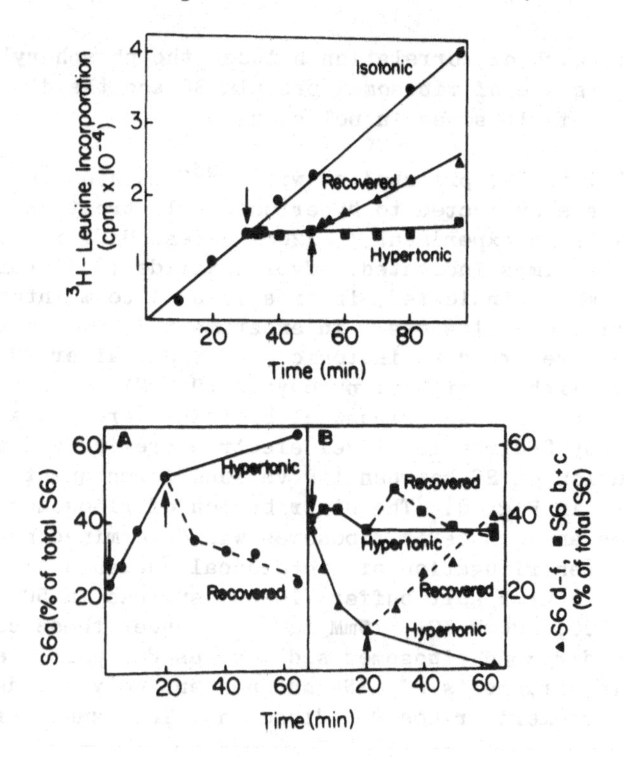

Fig. 8: Differential time-courses of changes in the rate of protein synthesis and in the phosphorylation state of ribosomal protein S6 in myeloma cells during hypertonic treatment and recovery

MPC11 cells were either incubated with [^{3}H]leucine (2μCi/ml) to monitor protein synthesis (acid-insoluble radioactivity - upper panel) or were labelled overnight with [^{35}S]methionine and the ribosomal proteins analysed on 2-dimensional gels as described in Fig. 7. The distribution of protein S6 between the unphosphorylated form S6a and the phosphorylated forms S6b-f was quantitated by excising each spot from the gels, oxidising the sample with H_2O_2 and determining the ^{35}S radioactivity. The results are expressed as % of total S6 in S6a (●), S6b+c (■) and S6d+e+f (▲) (lower panels). Cells were either kept under isotonic conditions or subjected to hypertonic salt (120mM excess NaCl, indicated by ↓) for the times shown. Portions of the cells were also allowed to recover by dilution of the salt (indicated by ↑) for the times shown.

Table 1: Lack of correlation between the phosphorylation state of ribosomal protein S6 and the distribution of ribosomes in polysomes

MPC11 cells, prelabelled with ^{35}S methionine as in Fig. 7, were subjected to hypertonic salt treatment (75mM excess NaCl in experiment 1; 120mM excess NaCl in experiment 2) for the times indicated. Cycloheximide (0.1 g/ml) was present, where indicated, in experiment 1 to maintain aggregation of polysomes. In experiment 2, portions of the cells were returned to isotonic conditions after 20 min and incubated with or without puromycin (0.2mM) for a further 40 min. In all cases ribosomal proteins were extracted and analysed by 2-dimensional gel electrophoresis, and the distribution of S6 between its various forms quantitated as described in Fig. 8. The distribution of ribosomes between polysomes and monomeric ribosomes was determined by sucrose gradient centrifugation of micrococcal nuclease treated cell extracts in high salt buffer (10-30% sucrose in 50mM Tris-HCl, pH 7.6, 800mM KCl, 15mM $MgCl_2$). Under these conditions polysome-derived ribosomes sediment as 80S particles containing fragments of mRNA and nascent polypeptide chains whereas monomeric ribosomes dissociate into subunits (41).

Incubation conditions	Time (min)	% ribosomes in polysomes	% of total S6 in various forms	
			S6a (unphosphorylated)	S6d-f (highly phosphorylated)
Expt. 1				
Isotonic	60	70	8	58
Hypertonic	60	42	22	30
*Isotonic+CHX	60	87	6	77
Hypertonic+ CHX	60	76	11	49
Expt. 2				
Hypertonic	20	15	48	0
Hypertonic	60	13	51	0
Hypertonic/ isotonic	20/40	68	22	30
Hypertonic/ isotonic+ puromycin	20/40	10	30	20

*CHX = cycloheximide

The possibility remained that S6 phosphorylation changes not as a cause but as a consequence of the change in the rate of protein synthesis or distribution of ribosomes between inactive monomers and polysomes. In order to investigate this we therefore used either a low concentration of cycloheximide (to make chain elongation rate limiting and accumulate ribosomes in polysomes) or puromycin (to disaggregate polysomes independently of the salt conditions to which the cells were exposed). Table 1 shows that hypertonic treatment of cells causes net dephosphorylation of S6 even in the presence of cycloheximide, thus suggesting that the loss of polysomes is not responsible for this effect. Conversely, when cells are allowed to recover from hypertonic salt in the presence of puromycin S6 becomes rephosphorylated even though the rate of protein synthesis remains low and very few polysomes are formed. These experiments therefore reveal conditions under which the state of phosphorylation of S6 appears to be unrelated to either the rate of protein synthesis or the distribution of ribosomes in polysomes and raise considerable doubts as to the mechanistic relationship between S6 phosphorylation and translation activity. It remains to be established exactly what physiological role can be assigned to changes in S6 phosphorylation. The fact that cells respond to such a variety of environmental changes by altering their S6 phosphorylation state does suggest that some important physiological mechanism may be involved. It simply remains to discover what this mechanism happens to be!

CONCLUSIONS

This survey has presented data concerning the role of phosphorylation of initiation factor eIF-2 and ribosomal protein S6 in the control of eukaryotic cell translation. A number of important questions about the relationships between protein phosphorylation and protein synthesis remain to be answered and additional areas of research may provide new leads in this connection. Perhaps the most interesting new development concerns the evidence for a role for amino-acyl-tRNA synthetases in the regulation of protein synthesis (33). The recent work of Damuni et al. (42) showing that these enzymes may themselves be subject to phosphorylation/dephosphorylation mechanisms raises some interesting questions for the future. It seems that at last the field is moving away from the reticulocyte, which has made immensely valuable contributions to our knowledge of translation control but may have its limitations. The use of cultured mammalian tumour cells, or of normal cells derived from tissues and maintained in primary culture, will hopefully provide equally important insights into translational control in the future.

ACKNOWLEDGEMENTS

The work described in this paper has been supported by The Cancer Research Campaign (M.J.C. and S.A.A.), the Deutsche Forschungsgemeinschaft (J.K.) and the Medical Research Council (C.G.P. and V.M.P.). M.J.C. holds a Career Development Award from The Cancer Research Campaign.

REFERENCES

1. Farrell, P.J., Balkow, K., Hunt, T., Jackson, R.J. & Trachsel, H. (1977) Cell 11, 187-200.
2. Clemens, M.J. In "The Biochemistry of Cellular Regulation" Vol. I (Clemens, M.J., ed.) CRC Press, Boca Raton, 1980, pp. 159-191.
3. Levin, D.H., Ernst, V., Leroux, A., Petryshyn, R., Fagard, R. & London, I.M. In "Protein Phosphorylation and Bio-regulation" (Thomas, G., Podesta, E.J. & Gordon, J., eds.) Karger, Basel, 1980, pp. 128-141.
4. Jackson, R.J. In "Protein Biosynthesis in Eukaryotes" (Perez-Bercoff, R., ed.) NATO Advanced Studies Institute Series, Vol. 41, 1982, pp. 363-418.
5. Pain, V.M. & Henshaw, E.C. (1975) Eur.J.Biochem. 57, 335-342.
6. Pain, V.M., Lewis, J.A., Huvos, P., Henshaw, E.C. & Clemens, M.J. (1980) J.Biol.Chem. 255, 1486-1491.
7. Storti, R.V., Scott, M.P., Rich, A. & Pardue, M.L. (1980) Cell 22, 825-834.
8. Nuss, D.L. & Koch, G. (1976) J.Mol.Biol. 102, 601-612.
9. Lodish, H.F. & Porter, M. (1980) J.Virol. 36, 719-733.
10. Wool, I.G. (1979) Ann.Rev.Biochem. 48, 719-754.
11. Traugh, J.A. In "Biochemical Actions of Hormones" Vol. 8 (Litwack, G., ed.) Academic Press, New York, 1981, pp. 167-208.
12. Trachsel, H. & Staehelin, T. (1978) Proc.Natl.Acad.Sci. U.S.A. 75, 204-208.
13. Safer, B., Peterson, D. & Merrick, W.C. In "Translation

of Natural and Synthetic Polynucleotides" (Legocki A.B., ed.) Poznan Agricultural University Press, Poznan, 1977, pp. 24-31.
14. Benne, R., Salimans, M., Goumans, H., Amesz, H. & Voorma, H.O. (1980) Eur.J.Biochem. 104, 501-509.
15. Clemens, M.J., Pain, V.M., Wong, S-T. & Henshaw, E.C. (1982) Nature (London) 296, 93-95.
16. Proud, C.G., Clemens,M.J. & Pain, V.M. (1982) FEBS Lett. 148, 214-220.
17. Pain, V.M. & Clemens, M.J. (1983) Biochemistry 22, 726-733.
18. Siekierka, J., Mauser, L. & Ochoa, S. (1982) Proc.Natl. Acad.Sci. U.S.A. 79, 2537-2540.
19. Konieczny, A. & Safer, B. (1983) J.Biol.Chem. 258, 3402-3408.
20. Siekierka, J., Manne, V., Mauser, L. & Ochoa, S. (1983) Proc.Natl.Acad.Sci. U.S.A. 80, 1232-1235.
21. Matts, R.L., Levin, D.H. & London, I.M. (1983) Proc.Natl. Acad.Sci. U.S.A. 80, 2559-2563.
22. Peterson, D.T., Safer, B. & Merrick, W.C. (1979) J.Biol. Chem. 254, 7730-7735.
23. Merrick, W.C. (1979) J.Biol.Chem. 254, 3708-3711.
24. Walton, G.M. & Gill, G.N. (1975) Biochim.Biophys.Acta 390, 231-245.
25. Siekierka, J., Mitsui, K-I. & Ochoa, S. (1981) Proc.Natl. Acad.Sci. U.S.A. 78, 220-223.
26. Panniers, L.R.V., Wong, S-T. & Henshaw, E.C. (1983) J. Biol.Chem. 258, (in press).
27. Farrell, P.J., Hunt, T. & Jackson, R.J. (1978) Eur.J. Biochem. 89, 517-521.
28. Leroux, A. & London, I.M. (1982) Proc.Natl.Acad.Sci. U.S.A. 79, 2147-2151.
29. Ralston, R.O., Das, A., Grace, M., Das, H. & Gupta, N.K. (1979) Proc.Natl.Acad.Sci. U.S.A. 76, 5490-5494.
30. Amesz, H., Goumans, H., Haubrich-Morree, T., Voorma, H.O. & Benne, R. (1979) Eur.J.Biochem. 98, 513-520.
31. Austin, S.A. & Kay, J.E. In "Essays in Biochemistry" Vol. 18 (Campbell, P.N. & Marshall, R.D., eds.) The Biochemical Society, London, 1982, pp. 79-120.
32. Austin, S.A., Pain, V.M., Lewis, J.A. & Clemens, M.J. (1982) Eur.J.Biochem. 122, 519-526.
33. Clemens, M.J. (1983) Nature (News and Views) 302, 110.
34. Wong, S-T., Mastropaolo, W. & Henshaw, E.C. (1982) J.Biol.Chem. 257, 5231-5238.
35. Austin, S.A. & Clemens, M.J. (1981) Eur.J.Biochem. 117, 601-607.

36. Thomas, G., Martin-Perez, J., Siegmann, M. & Otto, A.M. (1982) Cell 30, 235-242.
37. Nielsen, P.J., Duncan, R. & McConkey, E.H. (1981) Eur.J. Biochem. 120, 523-527.
38. Duncan, R. & McConkey, E.H. (1982) Eur.J.Biochem. 123, 535-538.
39. Duncan, R. & McConkey, E.H. (1982) Eur.J.Biochem. 123, 539-544.
40. Lastick, S.M. & McConkey, E.H. (1976) J.Biol.Chem. 251, 2867-2875.
41. Martin, T.E. (1973) Exp.Cell Res. 80, 496-498.
42. Damuni, Z., Caudwell, F.B. & Cohen, P. (1982) Eur.J. Biochem. 129, 57-65.

POST-TRANSLATIONAL MODIFICATION OF RNA POLYMERASE I BY PROTEIN KINASE NII AND SOME NOVEL IMMUNOLOGICAL ASPECTS OF THE TWO ENZYMES.

Samson T. Jacob, Dean A. Stetler and Kathleen M. Rose[+]
Department of Pharmacology
The Milton S. Hershey Medical Center
The Pennsylvania State University
Hershey, PA 17033, U.S.A.

[+] Present address : see below

Protein kinase NII from a rat tumor, Morris hepatoma 3924A, was purified essentially to homogeneity. It had a molecular weight of 140,000 and consisted of two subunits of molecular weights 42,000 and 25,000. A protein kinase with characteristics similar to protein kinase NII was associated with RNA polymerase I purified to homogeneity from the hepatoma. Based on several physiochemical criteria, the 42 and 25 kilodalton polypeptides of RNA polymerase I appear to correspond to the protein kinase NII. These two subunits were present in stoichiometric amounts in RNA polymerase I from the tumor and its proportion in the enzyme molecule was related to the RNA polymerase I activity. The protein kinase autophosphorylated its 25 kilodalton polypeptide as well as the 120, 65, 25 and 19.5 kilodalton polypeptides of RNA polymerase I. Polyamines, at physiological concentrations, augmented the phosphorylation reaction several fold, which resulted from phosphorylation at additional sites in the enzyme molecule. Phosphorylation of RNA polymerase I resulted in its activation and prevented premature termination of RNA chains. Antibodies raised in rabbits against purified RNA polymerase I inhibited RNA polymerase I activity as well as the activity of the protein kinase NII, but not that of RNA polymerase II. Sera from patients with rheumatic autoimmune diseases such as systemic lupus erythematosus (SLE), mixed connective tissue disease (MCTD) and rheumatoid arthritis (RA) contained antibodies to RNA polymerase I. Patients with specific autoimmune diseases exhibited antibodies either to a single polypeptide or to a distinct spectrum of polypeptides of RNA polymerase I. Some patients also produced antibodies to protein kinase NII. A clear correlation existed between the production of antibodies to protein kinase NII and of antibodies to the 42 and 25 kilodalton subunits of RNA polymerase I.

INTRODUCTION

There are at least two distinct protein kinases, designated NI and NII, in the cell nucleus, (Desjardins *et al.*, 1972). These protein kinases are cyclic nucleotide-independent, and utilize casein, but not histone, as a substrate. Since both kinases are largely confined to the nucleus (Thornburg and Lindell, 1977; Rose and Jacob, 1983) studies on the phosphorylation of nuclear proteins by these enzymes and possible functional modification of the phosphorylated proteins have received considerable attention (Stahl and Knippers, 1980; Hasuma *et al.*, 1980; Inoue *et al.*, 1980; Harrison and Jungmann, 1982). By virtue of their nuclear localization it is logical to conceive that DNA-dependent RNA polymerases might be the natural substrates for the protein kinases NI and NII. Previous studies have shown that RNA polymerases I and II can be phosphorylated *in vivo* and *in vitro* (see Rose and Jacob, 1983). For several years, our laboratory has been involved in studies on the structure, function and regulation of mammalian DNA-dependent RNA polymerases, particularly RNA polymerase I, the enzyme responsible for ribosomal RNA synthesis (see Rose *et al.*, 1983b). The present studies were undertaken to determine whether (a) RNA polymerase I can be phosphorylated by a specific protein kinase, (b) phosphorylation results in altered enzyme activity and (c) phosphorylation of RNA polymerase I plays a key role in the regulation of ribosomal gene transcription.

RESULTS

Identification of the protein kinase associated with highly purified RNA polymerase I.

RNA polymerase I, purified essentially to homogeneity

from Morris hepatoma 3924A, consists of two subclasses of this enzyme (IA and IB) which could be separated by gel electrophoresis under nondenaturing conditions. It is comprised of 8 subunits; S1 (M_r 190,000), S2 (M_r 120,000), S3 (M_r 65,000), S4 (M_r 42,000), S5 (M_r 25,000), S6 (M_r 21,000), S7 (M_r 19,500), and S8 (M_r 17,500) (Rose et al., 1981c). The purified enzyme exhibited protein kinase activity; following incubation with [γ-^{32}P]ATP and Mg^{2+}, both IA and IB were phosphorylated (Fig. 1). No detectable radioactivity due to phosphorylation of any minor contaminants by polymerase-associated kinase was evident, which further indicates the purity of RNA polymerase I. No other protein kinase activity was present in regions of the gel that did not exhibit RNA polymerase I activity (data not shown). RNA polymerase IB (slower electrophoretic mobility) incorporated considerably more phosphate in vitro than RNA polymerase IA. Unlike RNA polymerase I, purified RNA polymerase II from the same rat tumor did not contain a protein kinase (Stetler and Rose, 1983).

Similarities between RNA polymerase I-associated protein kinase and protein kinase NII.

The protein kinase associated with RNA polymerase I resembled protein kinase NII (M_r 140,000 with two subunits of 42 and 25 kilodaltons) purified to homogeneity from Morris hepatoma 3924A (Rose et al., 1981c) in several respects. Thus, both kinases utilized ATP and GTP as substrates, exhibited Km ATP of 10-12 µM and were cyclic nucleotide-independent. The two kinases were sensitive to heparin, whereas another nuclear protein kinase, NI, purified from the rat hepatoma was not inhibited by this polyanion. Protein kinase NII activity (30 pmol) was completely inhibited by 0.10 µg/ml of heparin.

The two protein kinases also resembled each other with respect to the subunits of RNA polymerase I used as substrates. Thus, four polypeptides of molecular weights 120,000 (S2), 65,000 (S3), 25,000 (S5) and 19,500 (S6) were the major substrates for the kinase (Fig. 2). The kinase could be autophosphorylated under these conditions, the 25 kilodalton subunit of the kinase being the major acceptor protein.

RNA polymerase I-associated protein kinase and protein kinase NII were both inhibited by antibodies raised in rabbits against purified RNA polymerase I whereas purified RNA polymerase II and bovine heart cyclic AMP-dependent (cytoplasmic) protein kinase were relatively unaffected. Concentrations of IgG which inhibited the first two kinases by 90% inhibited the bovine cytoplasmic kinase only by about 20%; at these concentrations, immune IgG did not inhibit purified RNA polymerase II (Table 1). Anti-RNA polymerase I antibodies interacted with six (S1 through S6) out of eight polypeptides of RNA polymerase I and both subunits of the purified protein kinase NII (Fig. 2). None of the RNA polymerase II subunits interacted with these

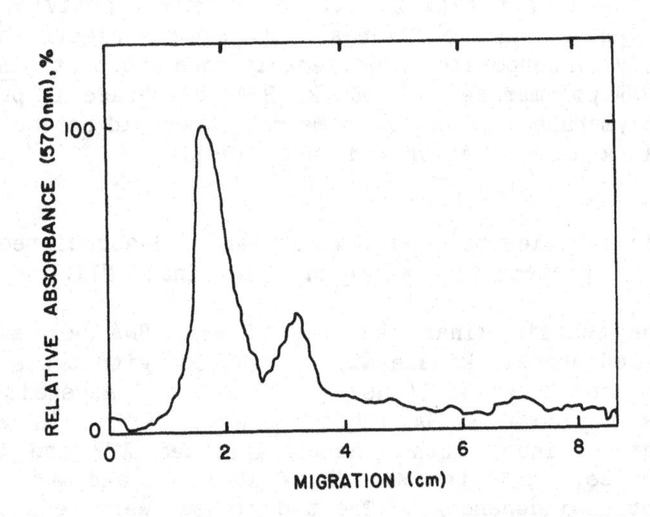

Fig. 1. Phosphorylation of purified RNA polymerase I by endogenous protein kinase NII. Purified RNA polymerase I containing both IA and IB was incubated with [γ^{32}P]ATP under conditions described previously (Rose et al., 1981c). Autoradiograms were scanned at 570 nm with a Transidyne 2955 densitometer. The peak on the left corresponds to RNA polymerase IB and the peak on the right corresponds to IA.

TABLE 1 EFFECT OF ANTI-RNA POLYMERASE I ANTIBODIES ON RNA POLYMERASE AND PROTEIN KINASE ACTIVITIES

Amount of IgG (μg)	RNA Polymerase activity (% inhibition)			Protein Kinase activity (% inhibition)		
	I	II	E.coli	NII	Polymerase Associated	Cytoplasmic Kinase
8	20	0	0	19	0	0
70	48	0	0	50	20	10
100	57	0	0	78	45	15
500	68	0	0	88	85	25
750	79	0	0	-	-	-

RNA polymerase I and protein kinase NII were purified from Morris hepatoma 3924A as described previously (Rose *et al.*, 1981a,c). Anti-RNA polymerase I antibodies were produced in rabbits following injection at biweekly intervals with 50-100 μg of purified RNA polymerase I emulsified in Freund's complete adjuvant (Rose *et al.*, 1981a,b). An immunoglobulin (IgG) fraction was obtained from the sera by $(NH_4)_2SO_4$ precipitation (50% saturation at 4°C), followed by chromatography on Sephacryl S-200 and protein A-Sepharose. Nonimmune IgG samples were prepared in the same manner from the sera of animals prior to injection with the antigen. Samples were preincubated for 10 min. at 30°C with a constant amount of total IgG, containing increasing quantities (as indicated) of the immune IgG. RNA polymerase II used in these studies was purified from the same tumor using identical fractionation techniques as those used for RNA polymerase I, except that the 0.4 M $(NH_4)_2SO_4$ fraction from the DEAE-sephadex were used as starting material. Reactions contained one of the following enzymes: 130 units of RNA polymerase I, 126 units of RNA polymerase II, 207 units of E.coli RNA polymerase, RNA polymerase I equivalent to 12 protein kinase NII units, 8 units of protein kinase NII and 7 units of bovine cytoplasmic cAMP-dependent protein kinase (Sigma). RNA polymerase and protein kinase reactions were performed as described previously (Rose *et al.*, 1981c). Results are the average of triplicate samples and are expressed as the percent inhibition relative to control reactions containing preimmune IgG.

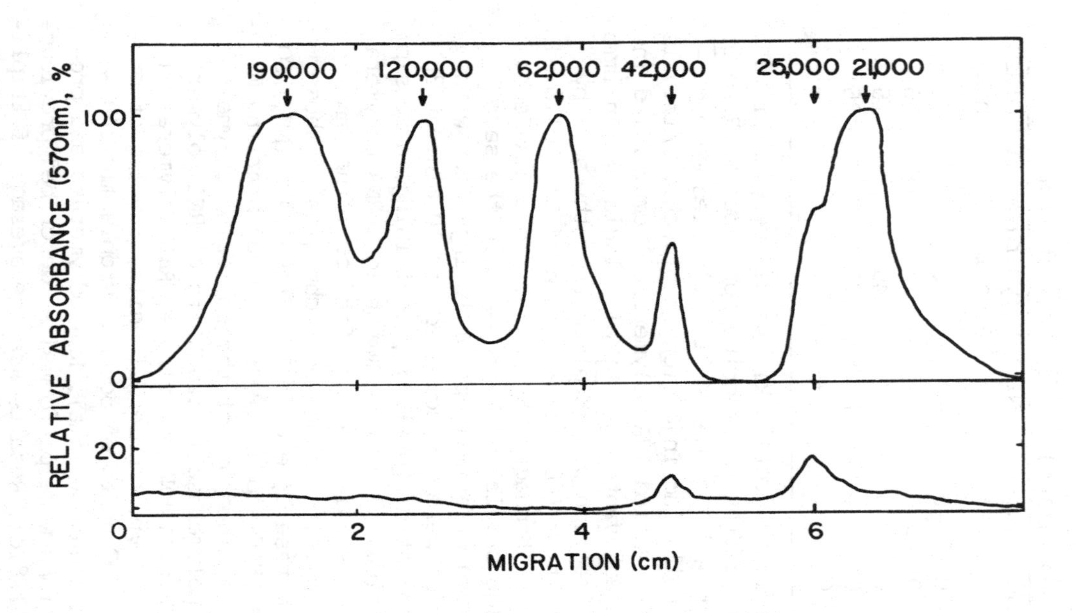

Fig. 2. Interaction of anti-RNA polymerase I antibodies with subunits of RNA polymerase I and protein kinase NII. Subunits of RNA polymerase I and protein kinase NII immobilized in DBM filters were incubated with sera from rabbits that had been immunized with purified RNA polymerase I and immune complexes were detected using [^{125}I]-Protein A (Rose *et al*., 1981c). Autoradiograms were scanned (570nm) with a Transidyne 2955 densitometer. Numbers above the peaks correspond to the molecular weights of the subunits. Markers were run on parallel gel tracks and stained with Coomassie blue.

TABLE 2 EFFECT OF SPERMINE ON PHOSPHORYLATION OF INDIVIDUAL RNA POLYMERASE I POLYPEPTIDES

Polypeptide	Extent of Phosphorylation (area in Cm^2 under the peak generated by densitometric scanning of autoradiograms) -Spermine	+Spermine	+Spermine/-Spermine
120,000	0.34	17.08	50
65,000	0.11	10.74	98
25,000	0.02	12.51	626
19,500	0.56	15.85	28

Purified RNA polymerase I was incubated with [$\gamma^{32}P$] ATP in the presence and absence of 2.5 mM spermine and subjected to polyacrylamide gel electrophoresis under denaturing conditions as described in the legend to Figure 3. The extent of phosphorylation of each polymerase polypeptide was quantitated by scanning autoradiograms of the gels at 570 mm with a densitometer equipped with an integrator. Results are the areas (cm^2) under the densitometer peaks normalized for 20 h film exposure.

antibodies, which further attests to the specificity of the anti-RNA polymerase I antibodies. Such specificity has been substantiated by the recent studies demonstrating the selective inhibition of ribosomal RNA synthesis following microinjection of immune IgG into viable cells (Mercer, Baserga, Rose and Jacob, unpublished data).

Another striking similarity between the two kinases was in regard to their stimulation by polyamines (Jacob et al., 1981, 1982). Phosphorylation of casein by protein kinase NII was stimulated as much as 20-fold by 2.5 mM spermine. Physiological concentrations of spermine and spermidine (1-2 mM) could stimulate the phosphorylation of casein by protein kinase several-fold. Putrescine had no significant stimulatory effect; even at a final concentration of 10 mM, this phosphorylation was enhanced only about 50% over the control samples. Spermine could also stimulate phosphorylation of RNA polymerase I by both endogenous (Fig. 3; Table 2) or exogenous protein kinase NII (Jacob et al., 1981).

In earlier studies (Rose et al., 1981c) phosphorylation of the 19.5 kilodalton subunit was not fully appreciated due to lack of complete separation of 25,000 and 19,5000 dalton subunits. Using longer gels for electrophoresis, we have now achieved much better separation of these two subunits. In the absence of polyamines, maximal phosphorylation by the polymerase I-associated protein kinase was in the 19,500 dalton subunit. Spermine stimulated the endogenous kinase-catalyzed phosphorylation

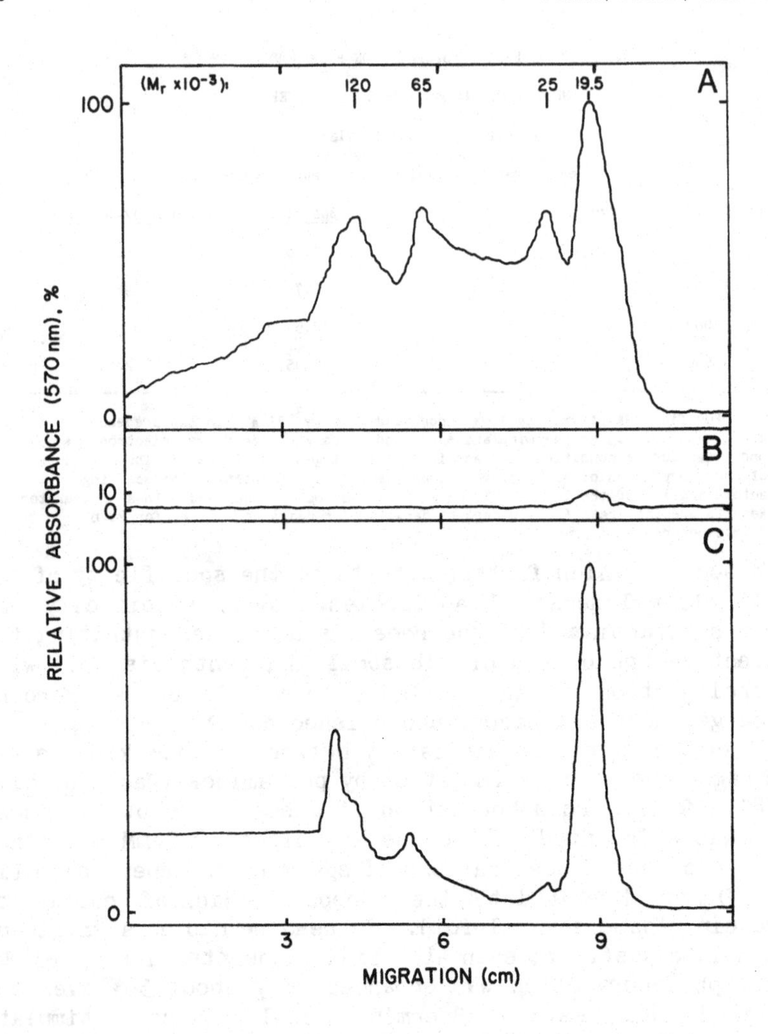

Fig. 3. Phosphorylation of RNA polymerase I subunits by the endogenous protein kinase NII in the presence or absence of spermine.
Panel A : phosphorylation in the presence of 2.5mM spermine, 20 h film exposure.
Panels B and C : phosphorylation in the absence of spermine, 20 h and 168 h film exposures, respectively.

TABLE 3. EFFECT OF ATP ON ANTIBODY BINDING TO RNA POLYMERASE I AND PROTEIN KINASE NII SUBUNITS

Enzyme	Subunits (Mr)	Antibody binding % control
Protein kinase NII	42,000	27
	25,000	23
RNA polymerase I	190,000	108
	120,000	104
	65,000	93
	42,000	47
	25,000	35
	21,000	99

DBM-papers containing the subunits of RNA polymerase or protein kinase NII were incubated with anti-RNA polymerase I antibodies in the absence or presence of ATP (1mM) and the immune complexes were detected with [^{125}I]-labeled protein A and quantitated as described previously (Rose et al., 1981c). Results are expressed as the area under each peak in the presence of ATP x 100 divided by that in the absence of ATP and are the average of duplicate experiments (taken from Rose et al., 1981c).

of 120, 65, 25 and 19.5 kilodalton polypeptides 50-fold, 98-fold, 626-fold and 28-fold respectively. It should, however, be emphasized that the higher degree of phosphorylation in the 19,500 dalton polypeptide might be due to the fact that this subunit has been dephosphorylated extensively during storage of the enzyme. Further studies are required to distinguish between the different possibilities.

Finally, the two subunits of the kinase (M_r, 42,000 and 25,000) were shown to bind ATP in a manner similar to the polymerase I polypeptides of the same molecular weight. In these studies, RNA polymerase I and protein kinase NII polypeptides were transferred to diazobenzyloxymethyl (DBM) paper and incubated with anti-RNA polymerase I antibodies in the presence or absence of ATP; binding of antibodies to 42 and 25 kilodalton polypeptides, but not to other subunits of RNA polymerase I, was inhibited in the presence of ATP (Table 3). Likewise, ATP reduced binding of the same antibodies to the two kinase subunits.

Activation of RNA polymerase I by phosphorylation.

As much as a 3-4 fold stimulation of RNA polymerase I activity was observed within 30 minutes of incubation with

TABLE 4 EFFECT OF PROTEIN KINASES NI AND NII ON RNA POLYMERASES I AND II

	Activity (pmol/30 min)	
	Mg^{2+}	Mn^{2+}
RNA polymerase I	1.7 ± 0.7	4.2 ± 1.8
Plus NII	6.4 ± 1.1	6.4 ± 2.3
Plus NI	2.9 ± 1.0	4.2 ± 0.4
RNA polymerase II	32 ± 2.1	88 ± 3.9
Plus NII	56 ± 3.2	72 ± 0.8
Plus NI	32 ± 1.7	72 ± 0.8

Purified RNA polymerases I, II and protein kinase NII were prepared as described previously (Rose *et al.*, 1981c). Protein kinase NI was purified from Morris hepatoma 3924A (Rose and Jacob, 1979). Results are the mean of triplicates (adapted from Rose *et al.*, 1983a).

excess protein kinase NII in the presence of Mg^{2+}. The activation of RNA polymerase I was evident only in the presence of Mg^{2+} which has been shown to be indispensable for the function of protein kinase NII (Table 4). When Mg^{2+} was replaced by Mn^{2+}, the stimulatory effect of the kinase was minimal, which suggests that the activation of RNA polymerase I by protein kinase NII is largely due to phosphorylation. Further, activation of the polymerase by exogenous kinase was not observed when ATP and GTP were replaced by 5'-adenylyl imidodiphosphate and 5'-guanylyl imidodiphosphate, respectively, the substrates containing nonhydrolyzable terminal phosphate (Rose *et al.*, 1983a) It should be noted that protein kinase NI which is known to activate another nuclear enzyme, poly(A) polymerase (Rose and Jacob, 1979, 1980) had no marked effect on RNA polymerase I in the presence of Mg^{2+} or Mn^{2+} both of which were utilized by this kinase (Table 4).

The mechanism by which phosphorylation augments RNA polymerase I activity was investigated by determining the average chain length and the number of 3' termini of the product. In the absence of nucleotidase or nucleosidase activities, measurement of radioactivity in the nucleoside derived from RNA synthesized *in vitro* is considered a fairly accurate estimate of the number of RNA chains initiated (Cox, 1976; Duceman and Jacob, 1980). Since purified RNA polymerase I and protein kinase NII were devoid of any detectable hydrolytic activities, it was possible to adapt this technique for elucidating the mechanism of activation of RNA polymerase I by phosphorylation. Phosphorylation of RNA polymerase I resulted in a marked increase in the average chain length whereas no significant change in the number of 3' termini was detected (Duceman *et al.*, 1981).

Anti-RNA polymerase I and anti-protein kinase NII antibodies in the sera of patients with rheumatic autoimmune diseases.

A hallmark of rheumatic autoimmune diseases is the presence of serum autoantibodies to nuclear proteins (antinuclear antibodies or ANA) (Tan, 1982). Anti-ribonucleoprotein (RNP) antibodies occur in sera of most patients with systemic lupus erythematosus (SLE) and mixed connective tissue disease (MCTD) (Sharp *et al.*, 1972; Notman *et al.*, 1975; Busch *et al.*, 1982). Sera from some patients have been shown to precipitate RNPs containing small nuclear RNAs (Lerner and Steitz, 1979). Since antibodies are also known to be directed against nucleolar antigens (Pinnas *et al.*, 1973; Miyawaki and Ritchie, 1973; Bernstein *et al.*, 1982; Reddy *et al.*, 1983), it was of considerable interest to determine whether one of the nucleolar antigens is RNA polymerase I. To test this possibility, sera from patients with SLE, MCTD and rheumatoid arthritis (RA) were screened for anti-RNA polymerase I antibodies by using a solid-phase radioimmunoassay. Significant levels of antibodies were detected in all patients with SLE and MCTD and in most patients with RA (Table 5). None of the normal individuals produced these antibodies. However, a distinct spectrum of antibodies was elicted in specific autoimmune diseases. Thus, sera from SLE patients produced immune complexes with S3, (M_r 65,000) and S5 (M_r 25,000), or S2 and S3 or with S2, S3 and S5, whereas sera from MCTD patients contained immunoglobulins directed against S4 (M_r, 42,000) or S3 and S4, or against S3, S4 and S5. RA patients produced antibodies against only the 65,000 dalton polypeptide of RNA polymerase I.

TABLE 6 INTERACTION OF SERA WITH RNA POLYMERASE I AND PROTEIN KINASE NII

Patient	Diagnosis	Anti-RNA Polymerase I Antibodies	Antibodies to Individual Polypeptides of RNA Polymerase I (cpm x 10^{-3})				Anti-protein Kinase NII Antibodies (cpm x 10^{-2})
			S2	S3	S4	S5	
S.S.	SLE	5.3 ± 0.5	3.2	2.3	0	4.1	8.3 ± 0.04
V.D.	SLE	3.4 ± 0.2	3.0	2.0	0	2.2	3.3 ± 0.05
H.S.	SLE	1.9 ± 0.4	3.6	5.4	0	0.4	2.4 ± 0.04
A.F.	SLE	10.2 ± 0.8	0	11.9	0	1.8	3.9 ± 0.08
R.S.	SLE	23.7 ± 1.7	0	20.0	0	3.2	7.2 ± 0.02
A.S.	SLE	3.2 ± 0.6	2.2	1.6	0	0	0 ± 0.50
D.G.	SLE	3.4 ± 0.1	2.6	1.6	0	0	0 ± 0.04
D.L.	SLE	2.6 ± 0.3	2.8	3.1	0	0	0 ± 0.12
C.M.	SLE	22.5 ± 1.2	17.4	4.0	0	0	0 ± 0.25
J.E.	RA	2.6 ± 0.1	0	0.5	0	0	0 ± 0.01
E.K.	RA	1.7 ± 0.1	0	0.6	0	0	0 ± 0.03
V.S.	RA	0.1 ± 0.04	0	1.2	0	0	0 ± 0.03
S.S.	RA	0.6 ± 0.1	0	3.7	0	0	0 ± 0.15
E.D.	RA	3.5 ± 0.5	0	5.4	0	0	0 ± 0.04
S.B.	MCTD	5.4 ± 0.5	0	8.0	4.2	20.5	8.0 ± 0.06
M.B.	MCTD	8.9 ± 0.6	0	7.1	2.3	22.3	8.6 ± 0.07
I.S.	MCTD	6.0 ± 0.2	0	15.8	2.7	0	8.0 ± 0.04
C.C.	MCTD	9.5 ± 0.3	0	0	4.7	0	8.4 ± 0.02
A.B.	normal	0 ± 0.1	0	0	0	0	0 ± 0.15
D.B.	normal	0 ± 0.1	0	0	0	0	0 ± 0.07
C.K.	normal	0 ± 0.1	0	0	0	0	0 ± 0.05

The quantity of anti-RNA polymerase I antibodies present in the sera was determined by using a solid phase radioimmunoassay as described previously (Stetler et al., 1982). A similar method was used to detect anti-protein kinase NII antibodies, except that the kinase was radioactively labelled by autophosphorylation and consequently [^{125}I]-labelled Staph protein A was eliminated from the assay. RNA polymerase I and protein kinase NII used as the antigens were purified from a rat hepatoma (Rose et al., 1981a,c).

To determine which subunits of RNA polymerase I formed immunocomplexes with the serum samples, the individual polypeptides of the enzyme were separated by polyacrylamide gel electrophoresis under denaturing conditions and then transferred to diazobenzyloxymethyl-(DBM) paper. Immune complexes were detected with [^{125}I]-Protein A and quantitated as described previously (Stetler et al., 1982).

None of the autoimmune sera tested so far contained antibodies to terminal deoxynucleotidyl transferase, reverse transcriptase or E. coli RNA polymerase (Table 6). On the contrary, 4 out of 7 SLE patients and 1 out of 2 RA patients had in their sera antibodies to RNA polymerase III which were not found in normal subjects. In view of the immunological cross-reactivity of RNA polymerases I and III (Hossenlopp et al., 1975), it is conceivable that at least in some patients autoantibodies are produced against the antigenic determinants common to these two enzymes. On the other hand, autoantibodies against RNA polymerase III might be directed against antigenic determinants unique to this enzyme. Further studies are needed to validate this issue.

The close association of RNA polymerase I and protein kinase NII, the striking similarities between the 42 and 25

TABLE 6 AUTOANTIBODIES IN HUMAN SERA AGAINST PURIFIED ENZYMES

	Antigen				
Diagnosis	RNA Polymerase I	RNA Polymerase III	E. Coli RNA Polymerase	Terminal Deoxynucleotidyl Transferase	Reverse Transcriptase
SLE	18/18	4/7	0/7	0/6	0/6
MCTD	4/4	0/2	0/2	0/2	0/2
RA	7/9	1/2	0/2	0/2	0/2
Normal	0/15	0/5	0/5	0/5	1/5

Sera from patients and normal individuals were tested for the presence of antibodies against various enzymes by a solid phase radioimmunoassay (Stetler et al., 1982). RNA polymerases I and III were purified to homogeneity from isolated nuclei of Morris hepatoma 3924A. E. coli RNA polymerase was obtained from Miles Laboratories (Elkhart, IN). Terminal deoxynucleotidyl transferase and reverse transcriptase were gifts from Dr. Sue Coleman and Dr. James Beard, respectively. Values given are the number of individuals whose sera contained antibodies against the particular enzyme over the number of individuals tested.

kilodalton polypeptides of RNA polymerase I and the protein kinase, and the presence of antibodies to the above polypeptides in rheumatic autoimmune diseases raised the possibility that patients producing antibodies to 42 and 25 kilodalton polymerase subunits may also produce anti-protein kinase NII antibodies. Since it was technically difficult to obtain adequate amounts of highly purified protein kinase NII for screening several serum samples, it was necessary to radioactively label this protein prior to solid phase radioimmunoassay (RIA). This was achieved by autophosphorylating the 25 kilodalton subunit of the purified kinase with [γ-^{32}P]ATP. The solid-phase RIA was performed by incubating the [^{32}P]-labelled protein kinase with patients' IgG adsorbed to microtiter wells. Sera from SLE patients which contained antibodies against the 25 kilodalton subunit always reacted with purified protein kinase NII. More important, the amount of kinase bound to serum IgG was proportional to the amount of antibodies produced against the 25 kilodalton subunit of RNA polymerase I (data not shown). There was no apparent relationship between the amount of anti-S2 or anti-S3 antibodies and anti-protein kinase NII antibodies. A correlation also existed between anti-S4 (M_r 42,000) antibodies and anti-kinase antibodies, as evident by the presence of these antibodies in sera of the MCTD patients which contained antibodies against S4 subunit of RNA polymerase I (Table 5). The close association of both 42 and 25 kilodalton polypeptides in the protein kinase molecule could explain the formation of radioactively labelled immune complex between sera containing anti-42K (polymerase) antibodies and [^{32}P]-labelled 25 kilodalton subunit of the kinase. Sera from SLE and RA patients which do not contain antibodies against RNA polymerase I subunits other than S4 or S5 did not form immune complex with protein kinase. Thus, there is a direct correlation between the production of antibodies that react with S4 and S5 subunits of RNA polymerase I and those that complex with protein kinase NII.

The specificities of anti RNA polymerase I and anti-protein kinase NII antibodies were further confirmed by determining the effect of IgG derived from the patients' sera on the activities of these two enzymes. Sera from SLE, MCTD and RA patients could inhibit RNA polymerase I activity whereas only sera containing antibodies to either S4 or S5 polypeptide of RNA polymerase I inhibited the

TABLE 7 INHIBITION OF RNA POLYMERASE I AND PROTEIN KINASE NII ACTIVITIES BY IgG FROM HUMAN SERA

Patient	Diagnosis	RNA Polymerase I, % inhibition (subunits)	Protein Kinase NII, % inhibition	Cytoplasmic Kinase, % inhibition
S.S.	SLE	32 (S_2,S_3,S_5)	33	6
A.F.	SLE	49 (S_3,S_5)	38	0
C.M.	SLE	ND (S_2,S_3)	0	0
S.B.	MCTD	47 (S_3,S_4,S_5)	40	0
M.B.	MCTD	58 (S_3,S_4,S_5)	54	7
J.E.	RA	33 (S_3)	0	0

RNA polymerase I was purified and assayed as described previously (Rose et al., 1981c) in the presence of 1.0 mg IgG from the patient indicated. Results are the mean of triplicate samples and are expressed as the percent inhibition of enzyme activity; 100% RNA polymerase I activity (% inhibition) obtained in the presence of 1.0 mg IgG from a normal individual (which had no significant effect on the enzymes) was equivalent to 3.4 ± 0.2 (mean ± SEM) pmol of UMP incorporated into RNA in 30 min. at 30°C and 100% protein kinase NII and cytoplasmic bovine heart cytoplasmic protein kinase activities were equivalent to 18 ± 0.6 and 29 ± 2.7 pmol (mean ± SEM) of terminal phosphate transferred from ATP to casein or histone/30 min. at 30°C respectively. S_2-S_5 (in parentheses) represent the subunit(s) of RNA polymerase I with which specific sera reacted.

ND, not determined.

activity of the kinase; under these conditions, an unrelated protein kinase, bovine heart cytoplasmic cAMP-dependent kinase, was unaffected (Table 7).

DISCUSSION

The present studies have demonstrated that RNA polymerase I can be phosphorylated in vitro by an endogenous protein kinase which is functionally related to protein kinase NII. Phosphorylation of RNA polymerase I resulted in its activation and activation augmented elongation of RNA chains transcribed from calf thymus DNA. It is possible that phosphorylation of RNA polymerase might also have an effect on chain initiation in a cell-free system capable of accurately transcribing cloned rDNA segment. Studies to test this possibility are currently underway in our laboratory.

The effect of polyamines on phosphorylation of RNA polymerase I deserves some comment. First, physiological concentrations of polyamines can stimulate phosphorylation dramatically. Second, polyamines preferentially increase phosphorylation of specific polypeptides. Such a selective increase in phosphorylation of specific regions of the enzyme molecule might be related to a shift in the phosphorylation profile (increase in the ratio of threonine-P to serine-P) observed in the presence of polyamines (Jacob et al., 1981, 1982).

It seems highly unlikely that the protein kinase associated with tumor RNA polymerase I is a minor contaminant of the polymerase preparation. First, only two visible protein bands corresponding to subclasses IA and IB are seen after electrophoresis of purified RNA polymerase I in nondenaturing gel and the only proteins phosphorylated in vitro by the endogenous protein kinase corresponded to IA and IB (Rose et al., 1981c). Second, the 42 and 25 kilodalton polypeptides of purified RNA polymerase I, which correspond in size to protein kinase NII subunits, are present in stoichiometric amounts. These two polypeptides are present in RNA polymerase I preparations purified from several sources (Roeder, 1976; Rose et al., 1981b). Third, rat liver enzyme which usually contains less than molar quantities of the 42 and 25 kilodalton subunits has correspondingly less kinase activity (Rose et al., 1981c, Duceman et al., 1981). One such preparation containing only traces of protein kinase had an extremely low specific activity (10 units/mg protein vs 400-500 units/mg of protein observed with active enzyme preparations). This enzyme could be stimulated more than 13-fold by exogenous protein kinase NII (Duceman, Rose and Jacob, unpublished data). Finally, there is a close relationship between the antibodies produced against S_4 and S_5 polypeptides of RNA polymerase I and against protein kinase NII in the rheumatic autoimmune diseases, systemic lupus erythematosus (SLE) and mixed connective tissue disease (MCTD). Sera containing autoantibodies against other subunits of RNA polymerase I did not contain antibodies against protein kinase NII. Only sera containing autoantibodies against S_4 or S_5 could inhibit protein kinase NII activity. These sera were also capable of inhibiting RNA polymerase I activity. These data collectively suggest a regulatory role for protein kinase NII in RNA polymerase I-directed transcriptional process. Whether such a role persists in vivo is a question which remains to be answered. It is plausible that the protein kinase may represent only a fraction of the 42 and 25 kilodalton proteins. Only a structural comparison by techniques such as peptide mapping between the polypeptides of RNA polymerase I and protein kinase NII will resolve this issue. The limited availability of adequate amounts of highly purified protein kinase NII has prevented us from vigorously pursuing these studies; only 2-4 μg of the pure nuclear kinase is obtained from 200-300 g of Morris hepatoma 3924A (Rose _et al_., 1981a).

Recently, RNA polymerase I from calf thymus has been shown to be structurally unrelated to casein kinase II (Dahmus, 1981b). A structural study with this casein kinase was made possible due to the availability of fairly large quantities of highly purified enzyme from calf thymus; in fact, casein kinase II is present in relative abundance (compared to protein kinase NII) in most tissues; approximately 1 mg of pure casein kinase II can be obtained from 500 g of calf thymus (Dahmus, 1981a). It is unlikely that casein kinase II used by Dahmus (1981) is identical to protein kinase NII for the following reasons: (a) the casein kinase was obtained from whole cells (designated KII); (b) Unlike the nuclear kinase, the kinase from whole calf thymus is a serine-specific kinase; (c) casein kinase (KII) does not phosphorylate the 120 kilodalton polypeptide of RNA polymerase I whereas protein kinase NII can phosphorylate this polypeptide; (d) casein kinase KII does not activate RNA polymerase I whereas exogenous addition of protein kinase NII phosphorylates and activates the polymerase; (e) even partial removal of protein kinase NII from RNA polymerase I results in a dramatic decrease in the activity of the latter enzyme, whereas complete dissociation of casein kinase KII from calf thymus RNA polymerase I has no effect on the activity of RNA polymerase I. It is feasible that protein kinase NII usually becomes associated with "core" RNA polymerase I only in rapidly proliferating cells (such as Morris hepatoma 3924A) or in response to physiological stimuli. It is possible that phosphorylation of the 120 kilodalton subunit might be a pre-requisite for activation of RNA polymerase I. This subunit of the calf thymus enzyme might already be phosphorylated _in vivo_, which prevents further phosphorylation _in vitro_ and consequent activation in response to exogenous kinase. Thus, although protein kinases KII and NII appear to exhibit certain common characteristics, they are probably not identical proteins.

Finally, association of protein kinase NII and RNA polymerase I should not be viewed as an isolated instance of an enzyme complex. In fact, purified RNA polymerase I from yeast contains RNase H activity (Huet _et al_., 1975).

Similarly, Sigma factor co-purifies with Escherichia coli RNA polymerase (Burgess, 1976). As in the case of protein kinase NII from rat liver, sigma factor can be dissociated from the core polymerase during extensive purification with concomitant loss of transcriptional activity on specific templates (Burgess, 1976). Other unrelated enzymes are also known to be either complexed with or an integral part of nucleic acid-synthesizing enzymes. Thus, purified poly(A) polymerase from liver contains a 3' exonuclease activity capable of degrading poly(A) (Abraham and Jacob, 1978). A DNA 3'→5' exonuclease activity is associated with prokaryotic (Gefter, 1975), eukaryotic (Byrnes et al., 1976) and viral (Knopf, 1976) DNA polymerase preparations. Similarly, RNAse H activity is present in purified reverse transcriptase preparations (Molling et al., 1971; Verma, 1975)

These studies have clearly demonstrated that protein kinase NII is inseparable from homogeneous preparation of RNA polymerase I from Morris hepatoma 3924A. Further studies are required to determine whether (a) the kinase becomes associated in response to growth-promoting stimuli, which makes it a regulatory factor for "core" RNA polymerase I (b) phosphorylation plays a key role in regulation of ribosomal gene expression in vivo and (c) phosphorylation exerts a positive control in the initiation of ribosomal RNA synthesis in vivo in addition to its effect on RNA chain elongation observed in vitro.

The authors thank Mrs. Carla Romanoski for her assistance in the preparation of this manuscript. This work was supported in part by the USPHS grants CA 25078 (STJ), CA 31894 (STJ) and GM 26740 (KMR). The authors wish to thank Dr. Wayne Criss for the generous supply of animals bearing Morris hepatoma 3924A, and Yohan Park for expert technical assistance.

+Present address : Department of Pharmacology, The University of Texas, Medical School at Houston, Houston, Texas 77025, U.S.A.

REFERENCES

Abraham, K.A., and Jacob, S.T. (1978). Proc. Natl. Acad. Sci. U.S., 75, 2085-2087.

Bernstein, R.M., Steigerwald, J.C., and Tan, E.M. (1982). Clin. Exp. Immunol., 48, 43-51.

Burgess, R.R. (1976), in RNA polymerase (Losick, R., and Chamberlin, M., eds) Cold Spring Harbor Laboratory, Cold Spring Harbor, New York, pp. 69-100.

Busch, H., Reddy, R., Rothblum, L. and Chor, Y.C. (1982). Ann. Rev. Biochem., 51, 617-?

Byrnes, J.J., Downey, K.M., Black, V.L., and So, A.G. (1976). Biochemistry, 15, 2817-2823.

Cox, R.F. (1976). Cell, 7, 455-465.

Dahmus, M.E. (1981a). J. Biol. Chem., 256, 3319-3325.

Dahmus, M.E. (1981b). J. Biol. Chem., 256, 11239-11243.

Desjardins, P.R., Lue, P.F., Liew, C.C., and Gornall, A.G. (1972). Can. J. Biochem. 50, 1249-1259.

Duceman, B.W., and Jacob, S.T. (1980). Biochem. J. 190, 781-789.

Duceman, B.W., Rose, K.M., and Jacob, S.T. (1981). J. Biol. Chem. 256, 10755-10758.

Gefter, M. (1975). Ann. Rev. Biochem., 44, 45-78.

Harrison, J.J., and Jungmann, R.A. (1982). Biochem. Biophys. Res. Commun., 108, 1204-1209.

Hasuma, T., Yukioka, M., Nakajima, S. Morisawa, S., and Inoue, A. (1980). Eur. J. Biochem., 109, 349-357.

Hossenlopp, P., Wells, D., and Chambon, P. (1975). Eur. J. Biochem., 58, 237-251.

Huet, J., Buhler, J.M., Sentenae, A., and Fromageot, P. (1977). J. Biol. Chem. 252, 8848-8855.

Inoue, A., Tei, Y., Hasuma, T., Yukioka, M., and Morisawa, S. (1980). Eur. J. Biochem., 109, 349-357.

Jacob, S.T., Duceman, B.W., and Rose, K.M. (1981). Med. Biol., 59, 381-388.

Jacob, S.T., Rose, K.M., and Canellakis, Z-N. (1982). Adv. Polyamine Res., 4, 631-646.

Knopf, K.W. (1979). Eur. J. Biochem., 98, 231-234.

Lerner, M.R., and Steitz, J.A. (1979). Proc. Natl. Acad. Sci. U.S.A., 76, 5495-5499.

Miyawaki, S., and Ritchie, R.-F. (1973). Arthritis Rheum. 16, 726-736.

Molling, K., Bolognesi, D.P., Bauer, W., Busen, W., Plassmann, H.W., and Hausen, P. (1971). Nature, New Biol., 234, 240-243.

Notman, D.D., Kurata, N. and Tan, E.M. (1975). Ann. Intern. Med., 83, 464-469.

Pinnas, J.L., Northway, J.D., and Tan, E.M. (1973). J. Immunol., 111, 996-1004.

Reddy, R., Tan, E.M., Henning, D., Nohga, K., and Busch, H. (1983). J. Biol. Chem., 258, 1383-1386.

Roeder, R.G. (1976), in RNA Polymerase (Losick, R., and Chamberlin, M., eds.) Cold Spring Harbor Laboratory, Cold Spring Harbor, New York, pp. 285-329.

Rose, K.M., Bell, L.E., Siefken, D.A., and Jacob, S.T. (1981a), J. Biol. Chem., 256, 7468-7477.

Rose, K.M., Duceman, B.W., and Jacob, S.T. (1981b). Isozymes: Current Topics in Biological and Medical Research, 5, 115-141.

Rose, K.M., Stetler, D.A., and Jacob, S.T. (1981c). Proc. Natl. Acad. Sci. U.S.A., 78, 2833-2837.

Rose, K.M., and Jacob, S.T. (1979). J. Biol. Chem., 254, 10256-10261.

Rose, K.M., and Jacob, S.T. (1980). Biochemistry, 19, 1472-1476.

Rose, K.M., and Jacob, S.T. (1983). Molecular aspects of cellular regulation, 3, in press.

Rose, K.M., Duceman, B.W., and Jacob, S.T. (1983a). Adv. Enzym. Regul., 21, 307-319.

Rose, K.M., Stetler, D.A., and Jacob, S.T. (1983b), in Enzymes of Nucleic Acid Synthesis and Modification (S.T. Jacob, Ed.), CRC Press, Boca Raton, Florida, pp. 135-157.

Sharp, G.C., Irwin, W.S., Tan, E.M., Gould, R.C., and Holman, H.R. (1972). Am. J. Med., 52, 148-159.

Stahl, H., and Knippers, R. (1980). Biochim. Biophys. Acta., 614, 71-80.

Stetler, D.A., and Rose, K.M. (1983). Biochem. Biophys. Acta., 739, 105-113.

Stetler, D.A., Rose, K.M., Wenger, M.E., Berlin, C.M., and Jacob, S.T. (1982). Proc. Natl. Acad. Sci. U.S.A., 79, 7499-7503.

Tan, E.M. (1982). Adv. Immunol., 33, 167-240.

Thornburg, W., and Lindell, T.J. (1977). J. Biol. Chem., 252, 6660-6665.

Verma, I.M. (1975). J. Virol., 15, 121-126.

CONTROL OF TRANSLATION BY REGULATION OF PROTEIN PHOSPHORYLATION AND DEPHOSPHORYLATION

B.Hardesty, G.Kramer, E.Wollny, S.Fullilove,
G.Zardeneta, S.-C.Chen, and J.Tipper

Clayton Foundation Biochemical Institute and
the Department of Chemistry
The University of Texas at Austin
Austin, Texas 78712

INTRODUCTION

The activity of a seemingly exponentially increasing number of well characterized, specific enzymes and enzyme systems has been shown to be regulated by phosphorylation and dephosphorylation of their constituent peptides. Depending on the specific enzyme system, increased phosphorylation may cause either a decrease or an increase in enzymatic activity. Cognately with this change in enzymatic activity, a large number of phosphorylated peptides can be detected in extract from intact cells or cell lysates. Subtle to dramatic changes in the phosphorylation state of a single given target protein may be brought about by a variety of components or conditions, such as hormones and growth factors, cAMP, temperature and nutrients or virus infection that affect the biochemistry and physiology of intact cells. These types of observations have led to the hypothesis that protein phosphorylation and dephosphorylation reactions constitute a primary regulatory mechanism in eukaryotic cells.

The extent of phosphorylation of most proteins is dependent on a dynamic equilibrium between the phosphorylation reaction, catalyzed by a protein kinase from a nucleotide triphosphate, usually ATP, and a dephosphorylation reaction promoted by a phosphoprotein phosphatase:

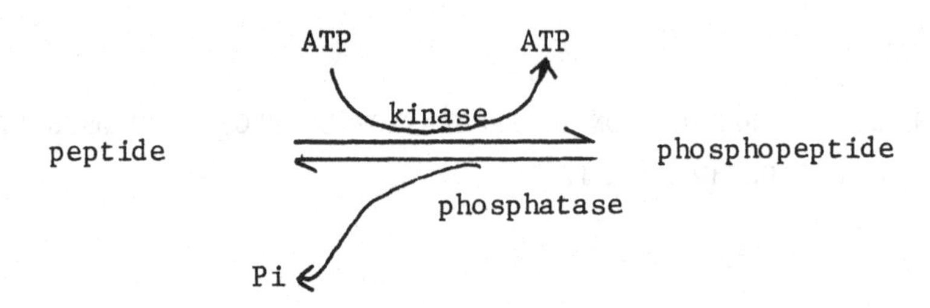

The overall equilibrium between the phosphorylated and dephosphorylated state of an enzyme subject to such a regulatory system can be shifted by activation or inhibition of the enzymatic activity of either the kinase or phosphatase.

Insulin provides an extensively investigated example of a peptide hormone that causes both increased and decreased phosphorylation of specific proteins. Glycogen synthase, glycogen phosphorylase, pyruvate dehydrogenase and hydroxymethylglutaryl-CoA reductase are enzymes that show decreased phosphorylation following insulin treatment of target cells, while increased phosphorylation is observed in acetyl-CoA carboxylase, ATP-citrate lyase and ribosomal protein S6. Following insulin treatment, increase in the activity of both protein kinases and phosphatases in target cells have been reported (Seals and Czech, 1982; Jarett _et al_., 1982). Although there is considerable insight into the mechanism by which the cyclic nucleotide-dependent protein kinases are regulated, little is known about the specific reactions involved in the regulation of the protein phosphatases and the cyclic nucleotide-independent protein kinases.

Many components of the translational apparatus have been demonstrated to undergo phosphorylation under _in vivo_ or _in vitro_ conditions. A partial list includes eukaryotic peptide initiation factors 2, 3, 4B and 5, a number of ribosomal proteins including most notably S6, and aminoacyl-tRNA synthetases (as reviewed by Traugh, 1981). Of these components, the effect of phosphorylation on the function of S6 and eIF-2 have received the most attention. Although dramatic changes in S6 phosphorylation are closely correlated with the physiological state of intact cells, phosphorylation of S6 has not been clearly demonstrated to affect ribosome function _in vitro_ in protein synthesis or any

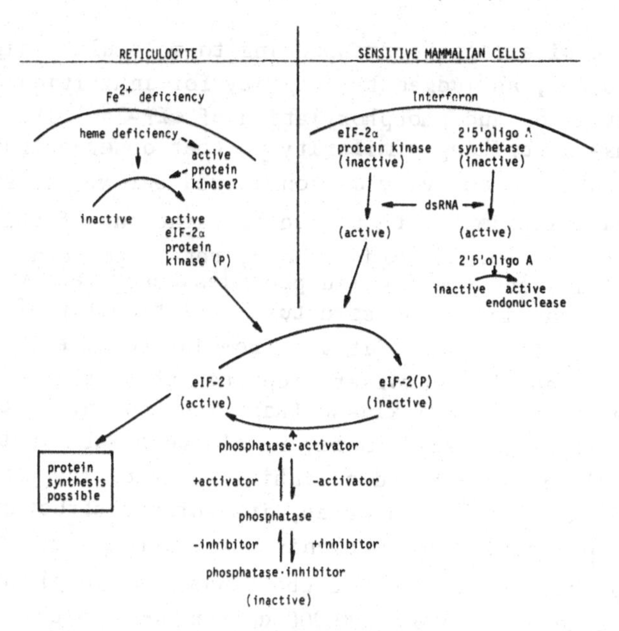

Figure 1: Regulation of peptide initiation through regulation of eIF-2α phosphorylation.

other specific biochemical reaction.

Peptide initiation factor 2, eIF-2, can be phosphorylated by specific kinases in either its α or β subunits. Only the former has been shown to affect its activity for peptide initiation dramatically (as reviewed recently, (Ochoa et al., 1981; Hardesty et al., 1981; London et al., 1981; Jackson et al., 1983). The recognized enzyme systems that affect the phosphorylation state of eIF-2α are depicted in Figure 1. These enzyme systems and their effect on the activity of eIF-2 are considered below.

eIF-2α KINASES

Two distinct protein kinases are recognized that phosphorylate the smallest, 38,000 dalton, α subunit of eIF-2. One of the enzymes is activated under conditions of heme deficiency and is associated with what has been called the heme-controlled repressor HCR. This enzyme is present in the postribosomal supernatant of reticulocytes. During gel filtration chromatography of this fraction, it is distributed over a wide range corresponding to molecular weights of up to 500,000, as judged by activity for inhibition of peptide initiation and phosphorylation of eIF-2α. The distribution has a striking similarity to that observed for protein phosphatase activity as considered below. It seems more than coincidental that some preparations of this kinase, purified 10,000 fold or more, retain protein phosphatase activity and the regulin peptides described below. However, the functional and structural relationship of these peptides to the kinase, if any, remains to be established.

London and his coworkers reported the isolation to homogeneity of the HCR kinase (Ranu and London, 1976; Trachsel et al., 1978). However, although purifications of 30,000 fold or more based on inhibitory activity have been obtained in other laboratories, its subunit structure, physical properties and mechanism by which it is activated are in doubt. Activation of the kinase is closely coupled with phosphorylation of a 100,000 dalton peptide that is thought to be a constituent of the holoenzyme (Gross and Mendelewski, 1978). Evidence has been presented which was interpreted to indicate that the kinase carries out autophosphorylation of the 100,000 dalton peptide during activation and that heme interacts directly with the kinase to prevent this reaction (Trachsel et al., 1978). However, Wallis and coworkers (Wallis et al., 1980) demonstrated that at least one additional component containing a peptide of 90,000 dalton participates in the activation process. The role of this 90,000 dalton activator peptide is not firmly established, but results are consistent with its being a kinase for the 100,000 dalton peptide, a kinase-kinase that might form part of a reaction cascade. It is very difficult to demonstrate conclusively that trace amounts of such a kinase are not present as an undetected contaminant in apparently homogeneous enzyme preparations.

The other eIF-2α kinase is associated with ribosomes from which it may be removed with solutions containing 0.5 M KCl. This kinase is activated by double stranded RNA, dsRNA. Its activation is closely associated with phosphorylation of a peptide of 67,000 daltons (Grosfeld and Ochoa, 1980; Levin et al., 1981) which may be part of the eIF-2α kinase itself. It has been suggested that the enzyme carries out autophosphorylation in the presence of dsRNA but this remains to be established. A second enzyme, the (2'-5')oligo A synthetase, also is activated by dsRNA. This enzyme carries out the synthesis of an oligoadenylate with an unusual 2' to 5' phosphodiester bond. This oligonucleotide activates what appears to be a unique endoribonuclease that can degrade mRNA and tRNA.

Both of the dsRNA-activated enzymes, the eIF-2α kinase and the (2'-5')oligo A synthetase, are present in latent form in reticulocytes. Their physiological significance in these cells is unclear. However, in many other types of mammalian cells these enzymes are induced by interaction of an interferon with a specific cell surface receptor. Interferons induce an antiviral state in many types of target cells. Although the physiological effects of interferons are complex, it is assumed that their antiviral activity is a result of the dsRNA-dependent eIF-2α kinase and the (2'-5')oligo A-dependent endonuclease functioning in concert to block protein synthesis and thus virus replication.

Both the heme-controlled and dsRNA-activated protein kinases appear to phosphorylate the same serine residues in eIF-2α (Farrell _et al_., 1977, Samuel, 1979). We have shown that by limited proteolysis a 4000 dalton terminal fragment can be cleaved from one end of the α subunit. All phosphorylation site(s) for the heme-controlled eIF-2α kinase are located on this fragment (Zardeneta _et al_., 1982). Limited proteolysis of eIF-2 eliminates first the β subunit (Mitsui _et al_., 1981) with concomitant loss of activity of eIF-2 to bind Met-$tRNA_f$ to 40S ribosomal subunits (Zardeneta _et al_., 1982). The ability of eIF-2 to form a ternary complex with Met-$tRNA_f$ and GTP is not significantly impaired when the β subunit is hydrolyzed but is lost with the cleavage of the α subunit into a 34,000 dalton peptide and the terminal 4000 dalton fragment (Zardeneta _et al_., 1982). These data are summarized in Figure 2. Siekierka _et al_. (1982) and Clemens _et al_. (1982) have presented evidence indicating that the immediate effect of eIF-2α phosphorylation is to inhibit the interaction of one or more factors with an eIF-2·GDP complex thus blocking the exchange of GTP for GDP and hence leading to the observed inhibition. A close analogy is drawn with the role of EF-Ts in displacing GDP from EF-Tu in the prokaryotic peptide elongation process.

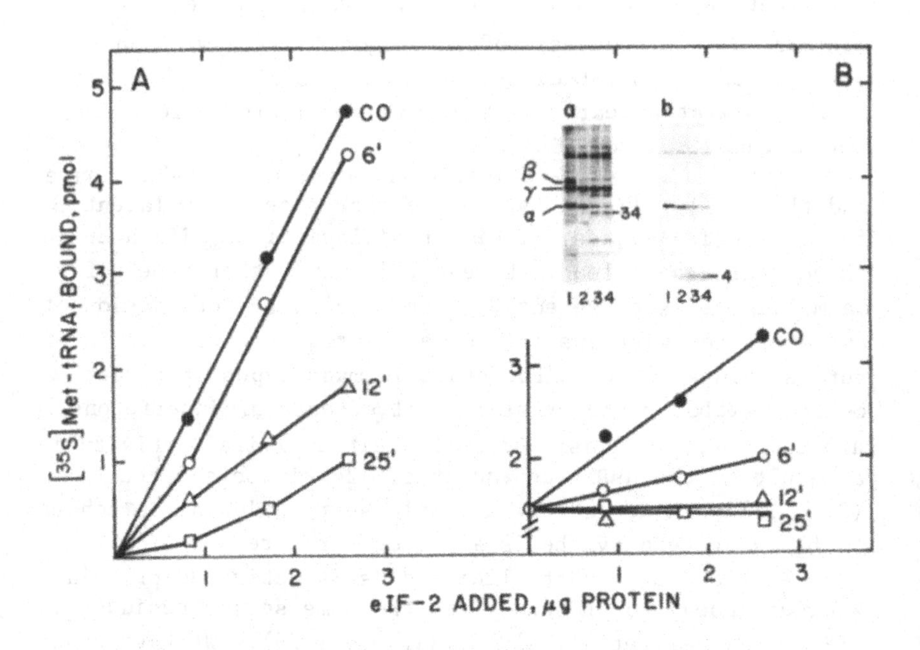

Figure 2: Biological activity of trypsin-treated eIF-2. A. Ternary complex formation with GTP and [^{35}S]Met-tRNA$_f$. B. Binding of [^{35}S]Met-tRNA$_f$ to reticulocyte 40S ribosomal subunits. These assays following trypsin treatment of eIF-2 were carried out as described by Zardeneta et al. (1982). The amount of [^{35}S]Met-tRNA bound to Millipore filters is dependent on eIF-2 added as shown. These amounts refer to intact eIF-2 present at the beginning of the trypsin digestion. CO, 6', 12', 25' indicate treatment with trypsin for 0, 6, 12, and 25 min, respectively. Inset. The time course of trypsin digestion is analyzed by gel electrophoresis in SDS followed by Coomassie Blue staining (a) and autoradiography (b). Tracks 1-4 contain samples from CO, 6', 12', and 25', respectively. α, β, and γ refer to the 3 subunits of eIF-2; 34 and 4 indicate the positions of the fragments obtained from the α subunit.

PROTEIN PHOSPHATASES

Phosphoprotein phosphatases have been studied most intensively in relation to glycogen metabolism using enzymes isolated from muscle or liver (cf. recent reviews by Krebs and Beavo, 1979; Lee et al., 1980). A striking characteristic of protein phosphatase activity in crude extracts from these tissues is that it is found in large complexes of M_r 250,000 or more from which smaller forms may be derived. Enzyme species of M_r 70,000, 85,000, 50,000 and 35,000 have been observed frequently. Conversion to smaller forms, generally with concomitant increase in enzymatic activity, can be accomplished by treatment with ethanol, urea, or proteases. Protein phosphatases have proven to be unusually difficult to isolate in homogeneous form. The basis for this difficulty is not fully understood; however, their ability to interact with other proteins appears to be a factor. Fischer and his coworkers have postulated that the lower molecular weight species are derived by Mn^{2+}-dependent proteolysis from larger precursors (Brautigan et al., 1982). Most of the protein phosphatases except the 35,000 dalton species exhibit an unusual Mn^{2+} requirement for maximum activity.

Isolation and Characterization

Protein phosphatase activity from reticulocytes also appears to occur in high molecular weight complexes, as judged by gel filtration chromatography of crude extracts. We obtained the partial purification of a protein phosphatase with relatively high activity for phosphorylated eIF-2α using ethanol precipitation of the reticulocyte postribosomal supernatant fraction followed by a series of steps involving conventional techniques of protein fractionation. The partially purified phosphatase chromatographed by gel filtration with an apparent molecular weight of 76,000 and was strongly dependent on Mn^{2+} (Grankowski et al., 1980a). The substrate specificity of the enzyme with a number of phosphoproteins was relatively low. It should be noted, however, that considerable variation in K_m and V_{max} was noted with different substrates, and that the latter parameter can be influenced by what appears to be regulatory peptides (Grankowski _et al_., 1980b). This enzyme was shown to reverse the inhibition of eIF-2 activity caused by phosphorylation with the HCR eIF-2α kinase (Grankowski _et al_., (1980a). These data were important in establishing conclusively that eIF-2α phosphorylation is a causal event in inhibition of peptide initiation by HCR.

Recently, we have purified to apparent homogeneity a 56,000 dalton peptide with endogenous protein phosphatase activity. The peptide was isolated from the postribosomal supernatant fraction from reticulocytes (Wollny _et al_., (1983).Successful fractionation of the peptide is dependent upon chromatography through Sephacryl S300 in a solution containing 6 M urea. This procedure appears to dissociate and separate components that otherwise exist as relatively stable complexes. Isolation of the peptide in homogeneous form is dependent upon its unusual retention on a size exclusion column during high performance liquid chromatography (HPLC). The column used gives separation of most globular proteins as an approximately linear function of their molecular weight similar to the preparation that might be obtained with Sephadex G100 in conventional gel filtration chromatography. However, the 56,000 dalton protein phosphatase is retained on the column beyond the elution time of small molecules such as β-mercaptoethanol. These results are presented in Figure 3B. The physiochemical basis for retention of the 56,000 dalton peptide is not known; however, it provides a novel and most useful procedure for purification of this protein phosphatase to apparent homogeneity. The peptide was determined to have a molecular weight of 56,000 by equilibrium centrifugation and about 55,000 by SDS gel electrophoresis (Figure 3A).

Activation and Regulation

Endogenous protein phosphatase activity in partially purified fractions is low but is markedly increased following exposure to denaturants such as urea or ethanol, mild proteolysis, or incubation with ATP and a phosphatase activator protein called F_A (Yang _et al_., 1980; Vandenheede _et al_. 1980). Recently, F_A has been shown to be a protein kinase with activity for phosphorylation of glycogen synthase inhibitor 2 (Hemmings _et al_., 1982). We have purified F_A

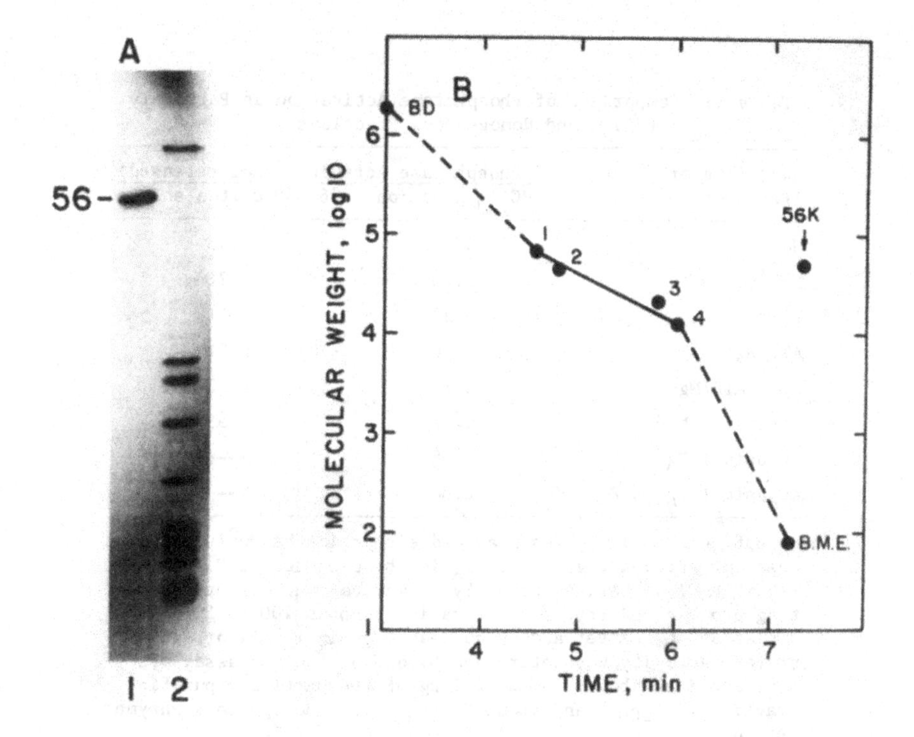

Figure 3: A. Polyacrylamide gel electrophoresis of the homogeneous phosphatase (track 1) and of proteins from the E. coli 30S ribosomal subunit as markers (track 2). The peptides were stained with Coomassie Blue. B. Size exclusion HPLC of molecular weight standards. Elution times of standard proteins against the logarithm of their molecular weight is plotted. Bovine serum albumin (1), ovalbumin (2), soybean trypsin inhibitor (3) and cytochrome c (4) were used. BD (Blue dextran) and BME (β-mercaptoethanol) are markers for V_o and V_t, respectively). 56K indicates the retention time for the M_r 56,000 phosphatase.

to about 90% purity from the reticulocyte postribosomal supernatant and shown it to be composed of a single peptide species of about 52,000 dalton as judged by SDS gel electrophoresis (unpublished results).

Typical results for activation of the partially purified reticulocyte phosphatase are shown in Table I. With the assay conditions used, the phosphate released is in-

Table I: Comparison of Phosphatase Activation in Partially Purified and Homogeneous Fractions

Additions or Treatment	Phosphatase Activity (pmol released)	
	PC_{100} fraction	56,000 dalton enzyme
None	0.42	0.38
Mn^{2+}, 4 mM	1.70	3.21
F_A	0.38	0.41
ATP/Mg^{2+}	0.40	0.21
F_A + ATP/Mg^{2+}	2.89	0.22
trypsin + Mn^{2+}	4.12	2.99
trypsin + F_A	0.68	----
trypsin + F_A, ATP/Mg^{2+}	0.68	----

Phosphatase activity was measured either in the Mn^{2+}-dependent assay or after activation by F_A in the presence of 0.1 mM ATP, 0.5 mM $MgCl_2$. Limited proteolysis was carried out by incubating protein and trypsin in a ratio of about 400 to 1 for 3 min at 35° C. After adding soybean trypsin inhibitor, phosphatase activity was determined by one of the two assay systems outlined above. About 2.2 μg of the partially purified fraction ("PC_{100}") and about 0.4 μg of the homogeneous enzyme were used.

increased from 0.42 pmol to 1.70 pmol in the presence of 4 mM Mn^{2+} or to 2.89 pmol following preincubation of the enzyme fraction with F_A + ATP/Mg^{2+}. The latter activity is not increased further by the addition of Mn^{2+} to the assay reaction mixture. Limited proteolysis of the enzyme fraction by trypsin followed by assay in the presence of Mn^{2+} gives 4.12 pmol of phosphate released. The enzymatic activity of the trypsin-activated enzyme is strongly dependent upon Mn^{2+}, and it does not respond to activation by F_A + ATP/Mg^{2+}. The corresponding results for experiments in which the homogeneous M_r 56,000 protein phosphatase was substituted for the partially purified enzyme also are shown in Table I. The principal difference is that the homogeneous enzyme is not activated by either F_A + ATP/Mg^{2+} or trypsin, even though it is dependent upon Mn^{2+}.

During the course of purification of the eIF-2α kinase of the HCR system described above, monoclonal hybridomas

Table II: Effect of Monoclonal Antibodies on the Enzymatic Activity of the Partially Purified and Homogeneous Protein Phosphatase

Conditions		Phosphate released[a] (pmol)
A. Partially purified enzyme		
1. Mn^{2+}-dependent assay		
PC_{100} only		2.18
+ Ab[b]		0.91
+ trypsin[c]		5.83
+ trypsin + Ab		5.47
2. F_A assay system	ATP/Mg^{2+}	
PC_{100} only	-	0.39
	+	1.89
PC_{100} + Ab	-	0.50
	+	2.36
B. M_r 56,000 enzyme (4 mM Mn^{2+})		
enzyme only		0.61
+ Ab		0.63

[a]Phosphatase activity was determined as described in the text. For part A and B, 2.5 μg and 0.3 μg protein, respectively, were used.

[b]Ab = monoclonal antibodies.

[c]Trypsin to protein ratio was 1:400.

were produced from spleen cells of mice that had been immunized with an enzyme fraction enriched more than 10,000 fold in kinase activity. To our surprise, the monoclonal antibodies secreted by these hybridomas caused an apparent increase rather than inhibition of kinase activity. Subsequent work demonstrated that they inhibited a protein phosphatase, apparently present as a contaminant in the kinase preparation. The effects of the monoclonal antibodies on the phosphatase activity of a partially purified phosphatase preparation from which the M_r 56,000 enzyme was isolated are shown in Table II. The antibodies do not cause inhibition if the partially purified phosphatase has been activated previously with trypsin or F_A nor do they inhibit the activity of the homogeneous enzyme. They recognize a series of peptides in the partially purified enzyme preparations that range in size from about 230,000 dalton down to 35,000 dalton or less (Figure 4B, track 1). A number of

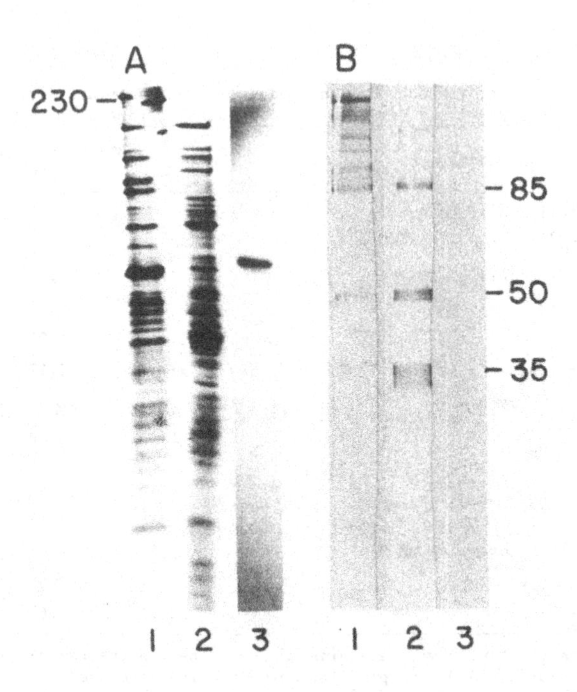

Figure 4: Monoclonal antibodies recognize peptides of the partially purified phosphatase fraction but not the 56,000 dalton enzyme. Proteins of the partially purified fraction were separated on a 15% polyacrylamide gel in SDS after incubation without (tracks 1) or with trypsin in a ratio of 400:1 (tracks 2). Tracks 3 show the homogeneous 56,000 dalton phosphatase. A, after electrophoresis peptides were stained with Coomassie Blue. B, proteins from the other tracks were electrophoretically transferred to a nitrocellulose sheet. Antigen peptides were detected by an ELISA and visualized with dianisidine following the procedure of Towbin et al., (1979) as described in detail by Fullilove et al., (1983).

monoclonal hybridoma lines producing antibodies that recognize the 230,000 dalton protein have been isolated using spleen cells from mice that had been immunized with the partially purified phosphatase. Monoclonal antibodies from most of these lines interact with the same peptides. The frequency with which these lines are obtained leads to the

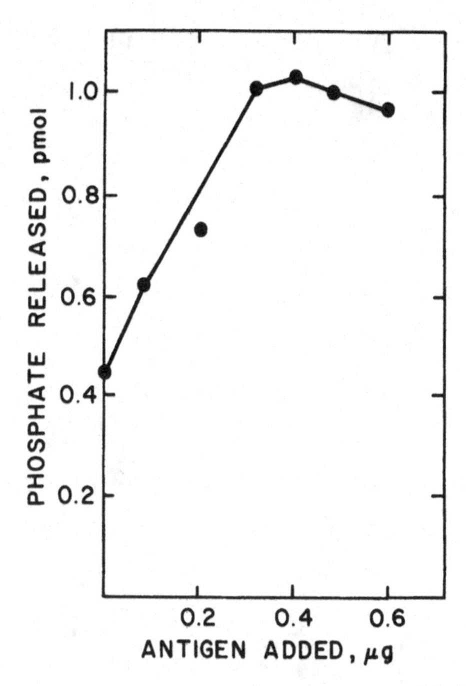

Figure 5: Stimulation of the homogeneous 56,000 dalton phosphase by the antigen fraction. Antigen peptides were isolated by antibody-affinity chromatography. The 56,000 dalton phosphatase was preincubated with the indicated amounts of the antigen fraction in the presence of Mn^{2+}. Then dephosphorylation of the substrate was determined.

conclusion that peptides of the series are unusually antigenic.

Limited proteolysis of the partially purified phosphatase comparable to that used for activation of phosphatase activity, converts larger members of the series of antigen peptides to smaller forms with prominent species at about M_r 85,000, 50,000 and 35,000 (compare tracks 1 and 2, Figure 4B). These molecular weights are provocatively similar to those of protein phosphatases from many sources; however, the M_r 56,000 enzyme is not recognized by the antibodies (track 3, Figure 4B). The antigen peptides were isolated from the partially purified phosphatase preparation by affinity chromatography using the monoclonal antibodies. The

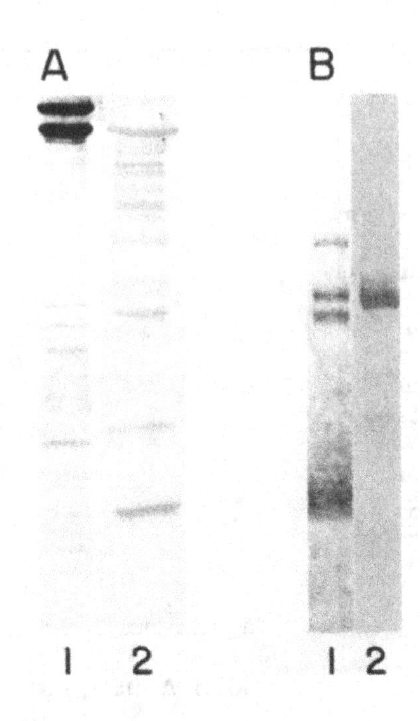

Figure 6: Presence of the antigen peptides in spectrin preparations from rabbit reticulocytes. Spectrin was prepared by the method of Litman et al. (1980). Aliquots were analyzed on 15% polyacrylamide gels (A) or on 5% gels (B), both in the presence of SDS. Tracks 1, peptides were visualized by staining with Coomassie Blue. Tracks 2, identical aliquots as shown in tracks 2 were subjected to an ELISA with monoclonal antibodies following gel electrophoresis and nitrocellulose blotting.

isolated antigen peptides cause a 3-4 fold increase in the Mn^{2+}-dependent enzymatic activity (Figure 5). The antigen fraction does not stimulate phosphatase activity in the absence of Mn^{2+} (data not shown).

A surprising observation was that the antigen peptides were relatively abundant in preparations of spectrin isolated from erythrocyte ghosts or the membrane/cytoskeleton fraction from reticulocytes by extraction at high pH in a solution of low ionic strength (Litman et al., 1980). The subunits of spectrin and the size distribution of the anti-

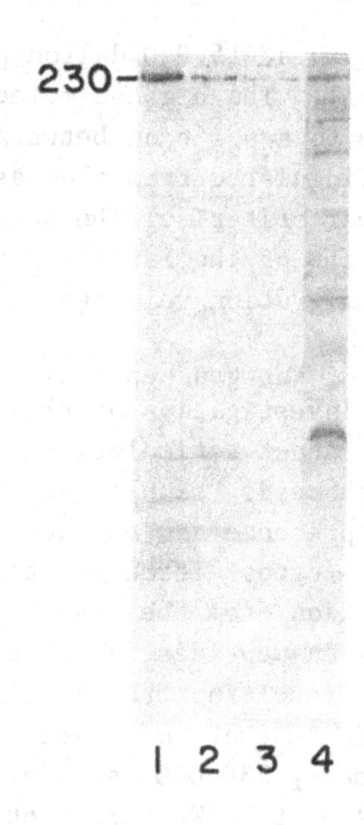

Figure 7: The 230,000 dalton antigen peptide, regulin, is found predominantly in the membrane fraction of reticulocytes. The cells were lysed either by resuspending them directly into electrophoresis "sample buffer" (track 1) or by suspension in a hypotonic solution in the presence of protease inhibitors (PMSF, EGTA, NEM). The membrane and soluble fractions were separated by low speed centrifugation. An aliquot of the membrane fraction was resuspended in "sample buffer" (track 2). An equivalent aliquot of the soluble fraction after ammonium sulfate precipitation and resuspension is analyzed in track 3. Track 4 shows an aliquot from the membrane fraction of a parallel preparation in which the protease inhibitors were omitted.

gen peptides in a preparation of reticulocyte spectrin are shown in Figure 6. Spectrin is a major cytoskeletal protein in cells of the erythroid series. It is a highly elongated protein with an (α_2, β_2) structure in which the α and β subunits have molecular weights of 240,000 and 220,000, respectively (Shotton et al., 1981). Spectrin is attached to the cell membrane by a 215,000 dalton protein, ankyrin (Branton et al., 1981). The highest molecular weight antigen peptide migrates to a position between the spectrin subunits and is distinguished from them as well as from ankyrin by several other criteria. The assignment of 230,000 as the molecular weight of the largest peptide is based on its electrophoretic migration relative to spectrin subunit peptides.

The occurrence of antigen peptides in spectrin preparations prompted an investigation of their subcellular and size distribution. Intact reticulocytes were taken directly into hot electrophoresis "sample buffer" containing 2% SDS or lysed in the presence and absence of protease inhibitors. The membrane/cytoskeleton fraction was isolated by low speed centrifugation from the lysed cells. The size distribution of antigen peptides in quantitatively equivalent aliquots of the resulting fractions is shown in Figure 7. The results indicate that in intact cells most of the antigen occurs as the M_r 230,000 peptide. Based on the intensity of the ELISA of the "Western Blot," we estimate that about 70% of the total antigen in intact reticulocytes is present in the membrane fraction, primarily as the M_r 230,000 species. It is degraded to smaller peptides in the membrane fraction if proteolysis is not prevented. A serine protease is implicated.

CONCLUSION AND DISCUSSION

The results described above indicate that regulin, a 230,000 membrane or cytoskeletal protein, and peptides derived from it, are involved in the regulation of a protein phosphatase and possibly the eIF-2α kinase of the HCR system. Proteolysis of regulin occurs when the cell membrane is disrupted. Activation of a membrane-associated serine protease appears to be involved. Activation of the protease in intact cells and the physiological significance of the regulin remain to be demonstrated.

ACKNOWLEDGEMENTS

This work was supported in part by Grant No. CA-16608

(to B.H.) from the National Cancer Institute, Department of Health, Education and Welfare. E.W. is the recipient of a fellowship from the Deutche Forschungsgemeinschaft. G.Z. and J.T. are supported by Training Grant No. CA-09182 from the National Cancer Institute.

REFERENCES

Branton,D., Cohen,C.M., and Tyler,J.(1981)Cell 24, 24-32.

Brautigen,D.L., Ballou,L.M., and Fischer,E.H.(1982) Biochemistry 21, 1977-1982.

Clemens,M.J., Pain,V.M., Wong,S-T., and Henshaw,E.C. (1982) Nature 296, 93-95.

Farrell,P.J., Balkow,K., Hunt,T., Jackson,R., and Trachsel,H.(1977) Cell 11, 187-200.

Fullilove,S., Wollny,E., Stearns,G., Chen,S.-C., Kramer, G., and Hardesty,B.(1983) manuscript submitted.

Grankowski,N., Lehmusvirta,D., Kramer,G., and Hardesty, B.(1980a) J.Biol.Chem. 255, 310-316.

Grankowski,N., Lehusvirta,D., Stearns,G., Kramer,G., and Hardesty,B.(1980b) J.Biol.Chem. 255, 5755-5762.

Grosfeld,H., and Ochoa,S.(1980) Proc.Natl.Acad.Sci.USA 77, 6526-6530.

Gross,M., and Mendelewski,J.(1978) Biochim.Biophys. Acta 520, 650-663.

Hardesty,B., and Kramer,G.(1981) in Protein Phosphorylation: Cold Spring Harbor Conferences on Cell Proliferation Vol.8 (Rosen,O.M., and Krebs,E.G., eds). Cold Spring Harbor, pp.959-977.

Hemmings,E.A., Yellowlees,D., Kernohan,J.C., and Cohen, P.(1981) Eur.J.Biochem. 119, 443-451.

Hemmings,B.A., Resink,T.J., and Cohen,P.(1982) FEBS Lett. 150, 319-324.

Jackson,R.J., Herbert,P., Campbell,E.A., and Hunt,T. (1983) Eur.J.Bopchem. 131, 313-324.

Jarett,L., Kiechle,F.L., and Parker,J.C.(1982) Fed.Proc. 41, 2736-2741.

Krebs,E.G., and Beavo,J.A.(1979) Ann.Rev.Biochem. 48, 923-959.

Lee,E.Y.C., Silberman,S.R., Granapathi,M.K., Petrovic, S., and Paris,H.(1980) in Advances in Cyclic Nucleotide Research Vol.13(Greengard,P., and Robison,G.A.,eds) Raven Press, New York, pp. 95-131.

Levin,D.H., Petryshyn,R., and London,I.M.(1981)J.Biol. Chem. 256, 7638-7641.

Litman,D.,Hsu,C.J., and Marchesi,V.T.(1980) J.Cell. Sci. 42, 1-22.

London,L.M., Ernst,V., Fagard,R., Lerourx,A., Levin,H. and Petryshyn,R.(1981) in Protein Phosphorylation: Cold Spring Harbor Conferences on Cell Proliferation, Vol.8 (Rosen,O.M., and Krebs,E.G., eds.) Cold Spring Harbor, pp.941-958.

Mitsui,K.-I., Datta,A., and Ochoa,S.(1981) Proc.Natl. Acad.Sci.USA 78, 4128-4132.

Ochoa,S., Siekierka,J., Mitsui,K., deHaro,C., and Grosfeld,H(1981) in Protein Phosphorylation: Cold Spring Harbor Conferences on Cell Proliferation, Vol.8 (Rosen,O.M., and Krebs,E.G., eds.) Cold Spring Harbor, pp. 931-940.

Ranu,R.S., and London,I.M.(1976) Proc.Natl.Acad.Sci.USA 73, 4349-4353.

Samuel,C.E.(1979) Proc.Natl.Acad.Sci.USA 76, 600-604.

Seals, J.R., and Czech,M.P.(1982) Fed.Proc. 41, 2730-2735.

Shotton,D.M., Burke,B.E., and Branton,D.(1979) J.Mol.Biol. 131, 303-329.

Sierkierka,J., Mause,L., and Ochoa,S.(1982) Proc.Natl.Acad. Sci.USA 79, 2537-2540.

Towbin,H., Staehelin,T., and Gordon,J.(1979) Proc.Natl. Acad.Sci.USA 76, 4350-4354.

Trachel,H., Ranu,R.S., and London,I.M.(1978) Proc.Natl. Acad.Sci.USA 75, 3654-3658.

Traugh,J.A.(1981) in Biochemical Actions of Hormones,Vol. VIII (Litwack,G. ed.) Academic Press, New York, pp. 167-208.

Vandenheede,J.R., Yang,S.-D., Goris,J., and Merlevede,W. (1980) J.Biol.Chem. 255, 11768-11774.

Wallis,M.H., Kramer,G., and Hardesty,B.(1980) Biochemistry 19, 798-804.

Wollny,E., Watkins,K., Kramer,G., and Hardesty,B.(1983) Manuscript submitted.

Yang,S.-D., Vandenheede,J.R., Goris,J., and Merlevede,W. (1980) J.Biol.Chem. 255, 11759-11767.

Yang,S.-D., Vandenheede,J.R., and Merlevede,W.(1981) J. Biol.Chem. 256, 10231-10234.

Zardeneta,G., Kramer,G., and Hardesty,B.(1982) Proc.Natl. Acad.Sci.USA 79, 3158-3161.

HORMONAL CONTROL OF ADIPOSE TISSUE LIPOLYSIS BY PHOSPHORYLATION/DEPHOSPHORYLATION OF HORMONE-SENSITIVE LIPASE

Håkan Olsson, Gudrun Fredrikson,
Peter Strålfors and Per Belfrage

Dept. of Physiological Chemistry, University of Lund, P.O.B. 750, S-220 07 Lund, Sweden

HORMONE-SENSITIVE LIPASE OF ADIPOSE TISSUE: FUNCTION AND PROPERTIES

The quantitatively most important energy substrate in mammals are free fatty acids (FFA), stored as triacylglycerol in the adipocytes. Hormones and the sympathetic nervous system regulate the mobilization of FFA from the adipose tissue, and the rate of this process mainly reflects the hydrolysis of the intracellular triacylglycerols, in the process of adipose tissue lipolysis. Fast-acting lipolytic hormones increase, and insulin inhibits this lipolysis by controlling the activity of the hormone-sensitive lipase (HSL), the rate-limiting enzyme. The lipase catalyzes hydrolysis of the first ester bond of the triacylglycerol substrate, which is the rate-limiting step of fat breakdown, and also the degradation of the 1,2(2,3)-diacylglycerols to 2-monoacylglycerols. The final hydrolysis of the monoacylglycerols to FFA and glycerol is catalyzed by a separate monoacylglycerol lipase (15).

HSL from rat adipose tissue has been detergent-solubilized, extensively purified and partially characterized (7), and its properties and regulation have recently been reviewed (1,4,14). Microgram amounts of approximately 50% pure enzyme can be obtained from several hundred rats. The enzyme has a minimum M_r of 84000 (sodium dodecylsulphate polyacrylamide gel electrophoresis (SDS-PAGE)), an apparent molecular size (gel chromatography) of about 150000, which may indicate that it forms a dimer, and a pI of 6.8 (4°C). The enzyme has a marked preference for the 1(3)-ester bonds of the acylglycerol substrates, and hydrolyzes tri-: 1,2(2,3)-di-:2-monoacylglycerol:cholesterol ester at the maximal relative rates of 1:10:1:1.5 (G. Fredrikson and P. Belfrage, unpublished). Like other lipases, e.g. pancreatic lipase and lipoprotein lipase, it has a high specific activity, 400 µmoles of fatty acids released per min per mg enzyme with 1,2-dioleoylglycerol as substrate, accounting for a high level of activity in adipose tissue in spite of a low tissue concentration. The enzyme is inhibited by micromolar diisopropylfluorophosphate and cysteine-directed reagents (Hg^{2+}, N-ethylmaleimide) indicating a reactive serine group in the catalytic site, and one or several functional sulfhydryl groups.

It was recently demonstrated that the neutral, cytosolic cholesterol ester hydrolase in bovine adrenal cortex (5) and corpus luteum (6) is identical, or closely similar, to the hormone-sensitive lipase. The enzyme may thus have other important functions besides the control of adipose tissue lipolysis. In fact, it seems likely that hormone-sensitive lipase will turn out to be a hormone-regulated, multi-functional tissue lipase (for discussion, see ref. 14).

REGULATION OF ISOLATED HORMONE-SENSITIVE LIPASE BY REVERSIBLE PHOSPHORYLATION

HSL is rapidly phosphorylated and activated by cyclic AMP-dependent protein kinase (cAMP-PrK) (Fig. 1) at a rate which is comparable to that found in intact adipocytes (see below). Addition of the specific inhibitor protein of cAMP-PrK arrests the phosphorylation immediately, demonstrating that the kinase acts directly on HSL without any intervening lipase kinase (7). The incorporated phosphate is exclusively recovered as phosphoserine, and after proteolytic digestion of ^{32}P-HSL with Staph. aur. V8 protease and trypsin, the phosphoserine is found in a small peptide of approximately 10 amino acids (P. Strålfors and P. Belfrage, unpublished). From these findings, and since the maximal incorporation is one mol of phosphate per mol of enzyme subunit (13), it can be concluded that the enzyme is phosphorylated on a single serine residue by cAMP-PrK.

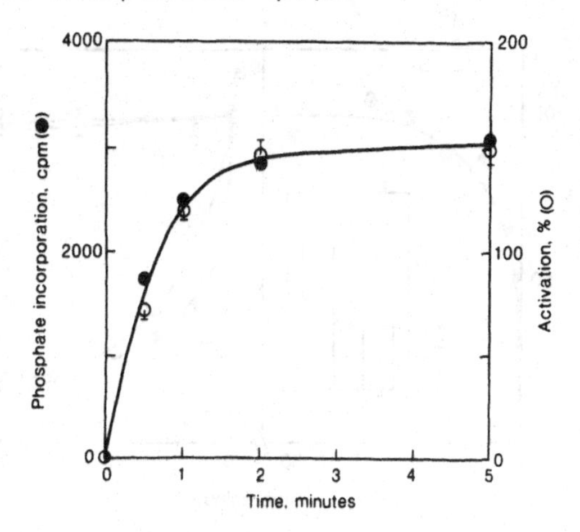

Fig. 1. Time-course of hormone-sensitive lipase phosphorylation and activation. The enzyme (approx. 30 nM) was incubated for 5 min with [γ-^{32}P]ATP-Mg^{2+} and the catalytic subunit of cAMP-PrK (0.3 μM). At the indicated time-points aliquots were withdrawn, and [^{32}P]phosphate incorporation into the enzyme protein and activation against an emulsified trioleoylglycerol substrate determined. Enzyme activation has been calculated as percent increase over control. ●, phosphate incorporation, mean of duplicates; O, activation, mean of four determinations; vertical bars indicate S.E.

Partially purified protein phosphatase from rat adipose tissue dephosphorylates and deactivates HSL, phosphorylated by cAMP-PrK (Fig. 2) (H. Olsson, P. Strålfors and P. Belfrage, unpublished). The enzyme preparation used contained protein phosphatase classified as type 2A (80%) and type 1 (see ref. 8). Rabbit skeletal muscle protein phosphatases 1, 2A and 2C, but not 2B, have been found to be able to dephosphorylate HSL. However, the relative importance of these protein phosphatases in the dephosphorylation of the enzyme in the adipocyte is not yet known.

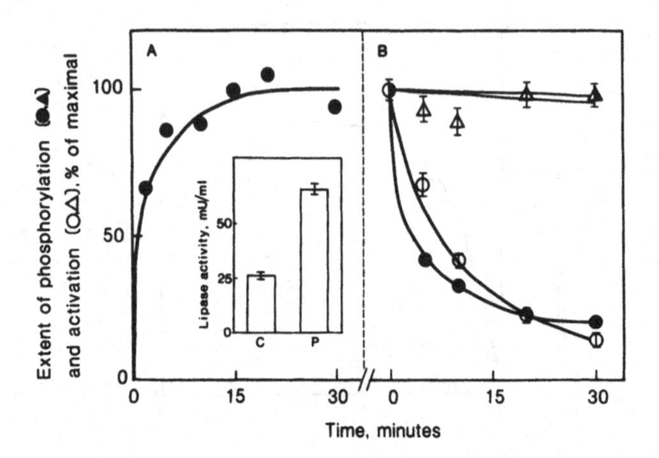

Fig. 2. Dephosphorylation and deactivation of hormone-sensitive lipase phosphorylated by cAMP-PrK. A. Phosphorylation of HSL, essentially as in Fig. 1, but at a lower concentration of the catalytic subunit of cAMP-PrK. At 45 min the [γ-^{32}P]ATP was removed by desalting and consumption by hexokinase and glucose. Inset: Hormone-sensitive lipase activity in control (C) and after maximal phosphorylation (P); one mU represents the release of one nanomol of fatty acid per min. B. Time-course of dephosphorylation and deactivation of HSL by partially purified protein phosphatase. At indicated time points aliquots were withdrawn for determination of extent of HSL phosphorylation and activation as in Fig. 1. Controls (triangles) without protein phosphatase were run in parallel. Each point is the mean of three (phosphorylation) or five (activation) determinations, ± S.E. (vertical bars, in some cases covered by the points).

PHOSPHORYLATION OF HORMONE-SENSITIVE LIPASE IN INTACT FAT CELLS

HSL from fat cells, preincubated with [^{32}P]ortophosphate to label intracellular ATP, has been extensively purified (2), and it has been demonstrated that HSL is phosphorylated in intact rat adipocytes. Recent experiments have demonstrated that this phosphorylation occurs on two phosphorylatable serine residues (13; P. Strålfors and P. Belfrage, unpublished). Peptide mapping of the phosphopeptides obtained from ^{32}P-HSL, isolated from the fat cells and proteolytically degraded with V8 protease and trypsin, showed that one of these sites is identical to the site phosphorylated by cAMP-PrK in isolated HSL. Since this site is only phosphorylated when the fat cells have been exposed to lipolytic hormones (e.g. catecholamines, ACTH, glucagon), and is directly involved in the control of HSL activity, it will be referred to as the 'regulatory' site. The other site, referred to as the 'basal' site, is phosphorylated regardless of hormonal stimulation of the fat cells. As with the 'regulatory' site, the phosphate is incorporated into a serine residue, but after peptide mapping the 'basal' site is found in a 15-amino acid phosphopeptide, well separated from the peptide containing the 'regulatory' site.

EFFECT OF FAST-ACTING LIPOLYTIC HORMONES AND INSULIN ON THE HORMONE-SENSITIVE LIPASE PHOSPHORYLATION AND ACTIVITY IN INTACT ADIPOCYTES

The biological activity of HSL in intact adipocytes can be measured continuously in a fat cell suspension as the release of FFA by using a pH-stat titration technique (Fig. 3) (11). Measurement of the glycerol release has shown that the reesterification of the FFA is negligible under the conditions used. The extent of HSL phosphorylation in the intact fat cells can be determined by measuring

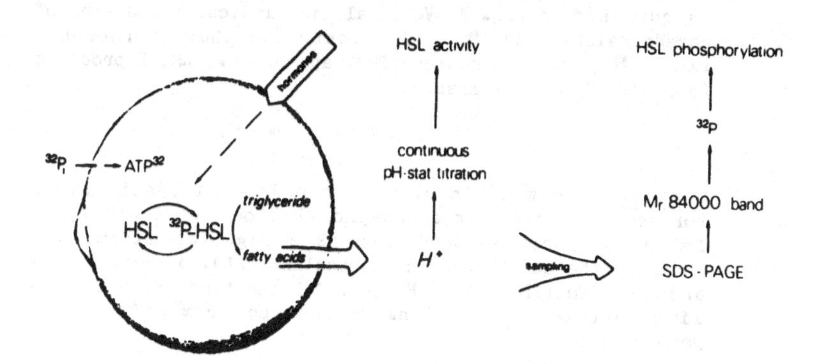

Fig. 3. Outline of methods for determination of the activity and the extent of phosphorylation in intact rat adipocytes. HSL, hormone-sensitive lipase; SDS-PAGE, sodium dodecylsulphate polyacrylamide gel electrophoresis.

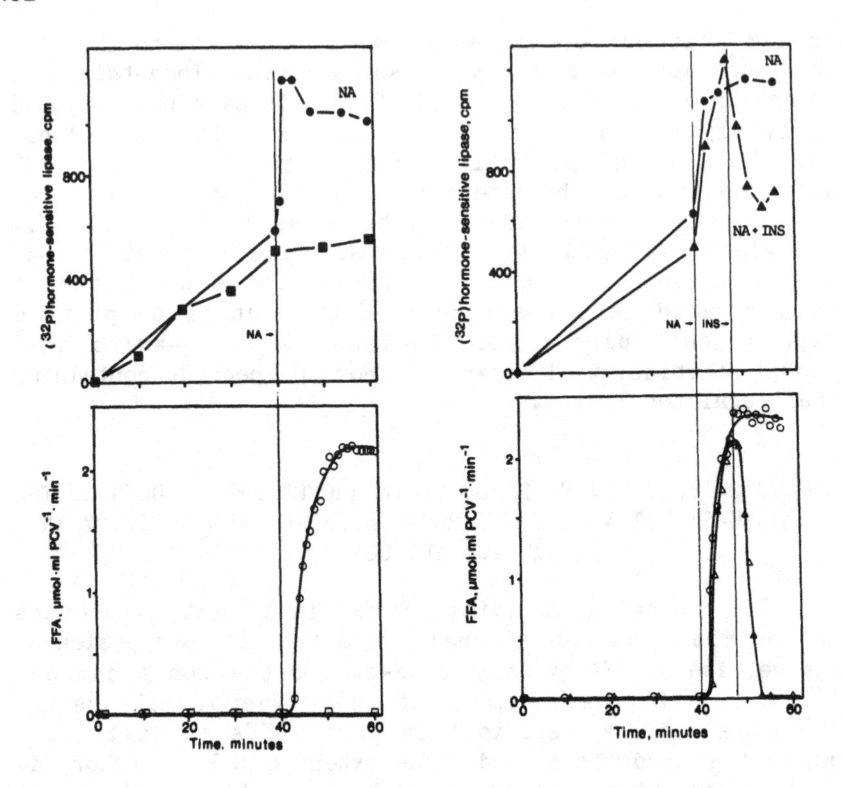

Fig. 4. Effect of noradrenaline on hormone-sensitive lipase phosphorylation and activity in intact adipocytes. Isolated rat fat cells were incubated with [^{32}P]ortophosphate and the extent of HSL phosphorylation and activity determined as outlined in Fig. 3. Vertical line indicates addition of noradrenaline (NA); ■ , □ , control without noradrenaline; ● , ○ , noradrenaline, approx. 0.3 μM. Reproduced from ref. 10, by permission.

Fig. 5. Effect of insulin on noradrenaline-stimulated hormone-sensitive lipase phosphorylation and activity in intact adipocytes. Conditions as in Fig. 4. Vertical bars indicate the addition of noradrenaline (NA), approx. 0.3 μM and insulin (INS), 700 pM. ▲ , △ , addition of noradrenaline followed by insulin. Reproduced from ref. 10, by permission.

the ^{32}P-radioactivity of the M_r=84000 [^{32}P] phosphopeptide band, obtained by SDS-PAGE of the total proteins of fat cells preincubated with [^{32}P]ortophosphate (Fig. 3) (9, 10, 14).

Incubation of fat cells with [^{32}P]ortophosphate leads to incorporation of ^{32}P into HSL to a steady state level after about 40 min (Fig. 4), with no detectable effect on the HSL activity measured as FFA release. Recent analysis of the phosphorylatable sites has demonstrated that this HSL phosphorylation is due to incorporation of phosphate in the 'basal' site (13; P. Strålfors and P. Belfrage, unpublished). The phosphorylation of this site is not affected by exposure of the cells to adenosine, or to insulin. Exposure of the cells to a fast-acting lipolytic hormone, e.g. noradrenaline (Fig. 4), rapidly increases HSL phosphorylation and, after a short time lag, the HSL activity (10). Half-maximal effect is found at 20-30 nM noradrenaline (9). The rapid increase of the extent of HSL phosphorylation and activity can be readily reversed by the β-adrenergic antagonist propranolol. Analysis of the phosphorylated sites has shown that this reversible phosphorylation occurres at the 'regulatory' site, i.e. the site phosphorylated by cAMP-PrK in the isolated HSL. Approximately equal amounts of ^{32}P are incorporated into the 'basal' and 'regulatory' sites (see also Fig. 4), indicating that a single serine residue is phosphorylated in both sites.

Exposure of maximally noradrenaline-stimulated fat cells to 700 pM (100 μU/ml) insulin rapidly decreases the extent of HSL phosphorylation, accompanied by decreased HSL activity (Fig. 5) (9, 10). When added shortly before, or together with noradrenaline, insulin (700 pM) completely prevents the anticipated increase of HSL phosphorylation and activity (data not illustrated). Half-maximal inhibition is found at 20-30 pM insulin. Analysis of the phosphorylated sites showed that the insulin-induced decrease of HSL phosphorylation is due to net dephosphorylation of the 'regulatory' site (13; P. Strålfors and P. Belfrage, unpublished).

MODES OF ACTION OF THE FAST-ACTING LIPOLYTIC HORMONES AND INSULIN

Exposure of the fat cells to millimolar concentration of dibutyryl cyclic AMP enhances the HSL phosphorylation and activity to the same extent as maximal noradrenaline stimulation (3). Noradrenaline enhances cellular cyclic AMP, HSL phosphorylation, and activity in a dose-dependent manner (N.Ö. Nilsson, P. Strålfors, P. Björgell, J.N. Fain and P. Belfrage, unpublished), and the cAMP-PrK activity

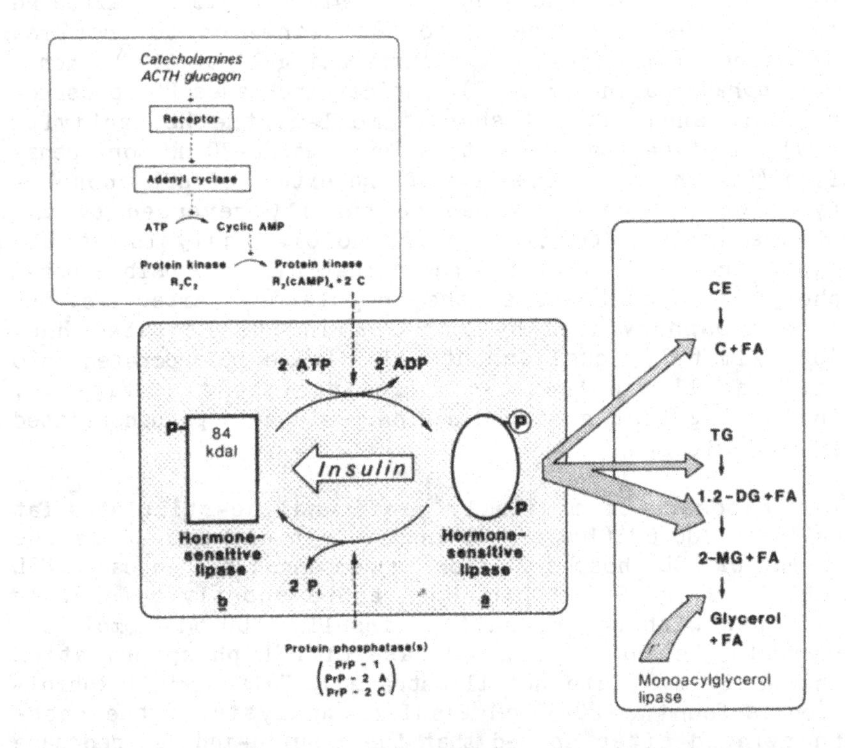

Fig. 6. Mechanisms for the hormonal control of adipose tissue lipolysis through reversible phosphorylation of hormone-sensitive lipase. Arrow, with insulin, indicates the net dephosphorylation of the 'regulatory' site, which accounts for insulin's anti-lipolytic effect. Encircled-P, phosphorylation of 'regulatory' site. CE, cholesterol ester; TG, DG, MG, tri-, di- and monoacylglycerol; FA, fatty acid.

ratio (+cAMP/-cAMP) in the fat cells increases from about 0.4 to 0.8 after maximal noradrenaline stimulation, with a time-course parallelling that of the phosphorylation of the HSL 'regulatory' site (P. Björgell and P. Belfrage, unpublished). Taken together these data establish that noradrenaline enhances HSL activity, and thus lipolysis, through cyclic AMP and cAMP-PrK mediated phosphorylation of the 'regulatory' site of HSL (Fig. 6), essentially as proposed in the 'lipolytic activation cascade' hypothesis a long time ago (12).

In contrast, the mechanism(s) through which insulin causes net dephosphorylation of the 'regulatory' site of HSL (Fig. 6) can not be established by the presented results. This insulin effect can be mediated through a decrease of cAMP-PrK activity, an increase of the activity of one or several of the protein phosphatases involved in the dephosphorylation of HSL (see above), or both. It has been found that prior exposure of fat cells to insulin substantially reduces the cellular cyclic AMP induced by subsequent maximal catecholamine stimulation (N.Ö. Nilsson, P. Strålfors, P. Björgell, J.N. Fain and P. Belfrage, unpublished). Thus cyclic AMP reduction, decreased cAMP-PrK activity, and a decreased rate of phosphorylation of HSL are certainly involved in the anti-lipolytic effect of insulin, but to an unknown extent.

ACKNOWLEDGEMENT

The contributions of Drs. Nils Östen Nilsson, Per Björgell and Staffan Nilsson to the work underlying the present review is gratefully acknowledged, as is the skilful technical assistance by Ingrid Nordh, Birgitta Danielsson, Aniela Szulezynski-Klein and Stina Fors. Financial support has been obtained from the following sources: A. Påhlsson, Malmö; T. and E. Segerfalk, Helsingborg; A.O. Swärd, Stockholm; Syskonen Svensson, Malmö; the Swedish Diabetes, Stockholm; Nordic Insulin, Copenhagen; P. Håkansson, Eslöv; The Medical Faculty, University of Lund and the Swedish Medical Research Council (project No. 3362).

REFERENCES

1. Belfrage, P. (1983): Hormonal control of lipid degradation. In: New Perspectives in Adipose Tissue Structure, Function and Development, edited by A. Cryer and R.L.R. Van. Butterworths, London. In press.
2. Belfrage, P., Fredrikson, G., Nilsson, N.Ö. and Strålfors, P. (1980): Phosporylation of hormone-sensitive lipase in intact rat adipocytes. FEBS Lett., 111:120-124.
3. Belfrage, P., Fredrikson, G., Nilsson, N.Ö. and Strålfors, P. (1981): Regulation of adipose tissue lipolysis by phosphorylation of hormone-sensitive lipase. Int. J. Obesity, 5:635-641.
4. Belfrage, P., Fredrikson, G., Strålfors, P. and Tornqvist, H. (1983): Adipose tissue lipases. In: Lipases, edited by B. Borgström and H. Brockman. Elsevier/North Holland, Amsterdam. In press.
5. Cook, K.G. and Yeaman, S.J.; Strålfors, P., Fredrikson, G. and Belfrage, P. (1982): Direct evidence that adrenal cortex cholesterol ester hydrolase is the same enzyme as adipose tissue hormone-sensitive lipase. Eur. J. Biochem., 125:245-249.
6. Cook, K.G., Colbran, R.J., Snee, J. and Yeaman, S.J. (1983): Cytosolic cholesterol ester hydrolase from bovine corpus luteum: its purification, identification and relationship to hormone-sensitive lipase. Biochim. Biophys. Acta. In press.
7. Fredrikson, G., Strålfors, P., Nilsson, N.Ö. and Belfrage, P. (1981): Hormone-sensitive lipase of rat adipose tissue: purification and some properties. J. Biol. Chem., 256:6311-6320.
8. Ingebritsen, T.S. and Cohen, P. (1983): The protein phosphatases involved in cellular regulation. 1. Classification and substrate specificities. Eur. J. Biochem., 132:255-261.
9. Nilsson, N.Ö. (1981): Studies on the short-term regulation of lipolysis in rat adipocytes with special regard to the anti-lipolytic effect of insulin. Thesis, University of Lund, Sweden. ISBN 91-7222-402-9.
10. Nilsson, N.Ö., Strålfors, P., Fredrikson, G. and Belfrage, P. (1980): Effects of noradrenaline and insulin on phosphorylation of hormone-sensitive lipase and on lipolysis in intact rat adipcytes. FEBS Lett., 111:125-130.

11. Nilsson, N.Ö. and Belfrage, P. (1981): Continuous measurement of free fatty acid release from intact adipocytes by pH-stat titration. In: Meth. Enzymol., vol. 72, edited by J.M. Lowenstein, pp. 319-325. Acad. Press, New York.
12. Steinberg, D. and Huttunen, J.K. (1972): The role of cyclic AMP in activation of hormone-sensitive lipase of adipose tissue. Adv. Cyclic Nucl. Res., 1:47-62.
13. Strålfors, P. (1983): Regulation of hormone-sensitive lipase through reversible phosphorylation. Role of cyclic AMP-dependent protein kinase. Thesis. University of Lund, Sweden. ISBN 91-722-627-7.
14. Strålfors, P. and Belfrage, P. (1983): Reversible phosphorylation of hormone-sensitive lipase/cholesterol ester hydrolase in the hormonal control of adipose tissue lipolysis and of adrenal steroidogenesis. In: Molecular Aspects of Cellular Regulation. vol. 3. Recently Discovered systems of Enzyme Regulation by Reversible Phosphorylation. Part 2., edited by P. Cohen. Elsevier/North Holland, Amsterdam. In press.
15. Tornqvist, H. and Belfrage, P. (1976): Purification and some properties of a monoacylglycerol-hydrolyzing enzyme of rat adipose tissue. J. Biol. Chem., 251:813-819.

Protein Phosphorylation as a Regulatory Mechanism

Michael J. Clemens

CRC Group, Department of Biochemistry,
St George's Hospital Medical School, London
SW17 ORE, UK.

The topics covered so far have included detailed considerations of the mechanism of protein synthesis in eukaryotes and of the numerous post-translational events which can affect the fate of newly synthesized polypeptide products. Post-translational phosphorylation of proteins is a widely occurring phenomenon which, since it is often readily reversible, can have important regulatory significance for numerous cellular activities. Included among these activities is the process of protein synthesis itself, so that in a sense we can reckon to have turned full circle in considering protein phosphorylation in relation to translation. Boyd Hardesty's paper, in the translational initiation section, and my own contribution here both deal with aspects of protein phosphorylation as a translational control mechanism. It is clear that several proteins involved in the machinery of protein synthesis, most notably initiation factor eIF-2 and ribosomal protein S6, undergo a dynamic process of phosphorylation and that the extent to which a given site on a peptide chain is modified depends on the balance between specific kinase and phosphatase activities. The protein kinases are complex enough, but the protein phosphatases are even less well defined yet in terms of their structures, regulation and substrate specificities.

Intriguing as the information derived from reticulocytes is, an important concern now is to establish whether phosphorylation (especially of eIF-2) plays any role in controlling protein synthesis in non-erythroid cell types.

There is some indirect evidence in support of this possibility but no direct demonstration that eIF-2 changes its phosphorylation state in response to changes in cellular environment. However, absence of proof is not proof of absence and we may have to face the possibility that very small, subtle changes in the extent of phosphorylation of eIF-2, S6 or other proteins may be sufficient to bring about large changes in protein synthesis. This could occur, for instance, if a phosphorylated protein stoichiometrically bound (or released) a factor which was present in low concentration and was rate-limiting for overall translation.

The kind of influences which affect protein synthesis and which could act through phosphorylation mechanisms include essential nutrients, heat shock, hypertonic conditions, virus infection, polypeptide growth factors and hormones. In the case of the hormones we know of plenty of examples where particular rate-limiting enzymes in metabolic pathways are controlled by reversible phosphorylation. Hakan Olsson's paper on the hormone-sensitive lipase of adipose tissue describes a good example. Will future research reveal specific eIF-2 or S6 kinases and phosphatases which are switched on or off in response to insulin, glucagon, trophic hormones, etc.?

We can extend the spectrum of processes potentially regulated by phosphorylation back from the level of specific enzymes or the translational apparatus to the level of transcription of specific DNA sequences. Samson Jacob's paper presents the intriguing possibility that the activity of RNA polymerase I (and thus the synthesis of new ribosomes) may be controlled by phosphorylation. Certainly rRNA synthesis is a highly and rapidly regulated event in eukaryotic cells. Phosphorylation would be a sensitive means by which this regulation could be achieved.

In conclusion, I would like to summarize the characteristics that need to be determined for a complete understanding of any protein phosphorylation system in living cells. The things we need to know are :

1. The identity of the protein which is phosphorylated.

2. The nature of the phosphorylation site(s). Which amino acids are modified; what are the peptide sequences in which they occur and what positions do they occupy in

the polypeptide chain; how many phosphate groups are incorporated per molecule?

3. The identity of the protein kinase(s) and phosphatase(s) which phosphorylate and dephosphorylate the substrate; are they related to other enzymes known to act on other proteins? Do the kinases prefer ATP or GTP as phosphate donor and what are the kinetic parameters of the phosphorylation reaction?

4. How does the activity of the kinase balance against that of the phosphatase and what factors regulate these enzymes?

5. What are the functional consequences of the protein phosphorylation or dephosphorylation? Do they have regulatory significance to the metabolic coordination of the cell or tissue?

Not surprisingly, in relatively few cases have all these questions been answered and it is safe to say that much work remains to be done in this important field of eukaryotic cell biochemistry.

Index